Performance-Related Material Properties of Asphalt Mixture Components (Second Edition)

Performance-Related Material Properties of Asphalt Mixture Components (Second Edition)

Editors

Yao Zhang
Haibo Ding
Yu Chen
Meng Ling

Basel • Beijing • Wuhan • Barcelona • Belgrade • Novi Sad • Cluj • Manchester

Editors

Yao Zhang
College of Architectural
Science and Engineering
Yangzhou University
Yangzhou
China

Haibo Ding
School of Civil Engineering
Southwest Jiaotong University
Chengdu
China

Yu Chen
School of Transportation and
Logistics Engineering
Wuhan University
of Technology
Wuhan
China

Meng Ling
School of Civil and
Transportation Engineering
Beijing University of Civil
Engineering and Architecture
Beijing
China

Editorial Office
MDPI AG
Grosspeteranlage 5
4052 Basel, Switzerland

This is a reprint of articles from the Special Issue published online in the open access journal *Materials* (ISSN 1996-1944) (available at: www.mdpi.com/journal/materials/special_issues/KSG93331RF).

For citation purposes, cite each article independently as indicated on the article page online and as indicated below:

Lastname, A.A.; Lastname, B.B. Article Title. *Journal Name* **Year**, *Volume Number*, Page Range.

ISBN 978-3-7258-1992-8 (Hbk)
ISBN 978-3-7258-1991-1 (PDF)
doi.org/10.3390/books978-3-7258-1991-1

© 2024 by the authors. Articles in this book are Open Access and distributed under the Creative Commons Attribution (CC BY) license. The book as a whole is distributed by MDPI under the terms and conditions of the Creative Commons Attribution-NonCommercial-NoDerivs (CC BY-NC-ND) license.

Contents

Preface vii

Junyao Tang, Siyu Chen, Tao Ma, Binshuang Zheng and Xiaoming Huang
Calculation of and Key Influencing Factors Analysis on Equivalent Resilient Modulus of a Submerged Subgrade
Reprinted from: *Materials* 2024, 17, 949, doi:10.3390/ma17040949 1

Min Xiao, Yu Chen, Haohao Feng, Tingting Huang, Kai Xiong and Yaoting Zhu
Evaluation of Fatigue Behavior of Asphalt Field Cores Using Discrete Element Modeling
Reprinted from: *Materials* 2024, 17, 3108, doi:10.3390/ma17133108 18

Bo Li, Minghao Liu, Aihong Kang, Yao Zhang and Zhetao Zheng
Effect of Basalt Fiber Diameter on the Properties of Asphalt Mastic and Asphalt Mixture
Reprinted from: *Materials* 2023, 16, 6711, doi:10.3390/ma16206711 33

Carina Emminger, Umut D. Cakmak and Zoltan Major
Multi-Step Relaxation Characterization and Viscoelastic Modeling to Predict the Long-Term Behavior of Bitumen-Free Road Pavements Based on Polymeric Resin and Thixotropic Filler
Reprinted from: *Materials* 2024, 17, 3511, doi:10.3390/ma17143511 56

Yujuan Zhang, Pei Qian, Peng Xiao, Aihong Kang, Chenguang Jiang, Changjiang Kou, et al.
Probing the Effect of Linear and Crosslinked POE-*g*-GMA on the Properties of Asphalt
Reprinted from: *Materials* 2023, 16, 6564, doi:10.3390/ma16196564 73

Lei Jiang, Junan Shen and Wei Wang
Performance of High-Dose Reclaimed Asphalt Mixtures (RAPs) in Hot In-Place Recycling Based on Balanced Design
Reprinted from: *Materials* 2024, 17, 2096, doi:10.3390/ma17092096 90

Lieguang Wang, Lei Wang, Junxian Huang, Mingfei Wu, Kezhen Yan and Zirui Zhang
Investigation of Phenolic Resin-Modified Asphalt and Its Mixtures
Reprinted from: *Materials* 2024, 17, 436, doi:10.3390/ma17020436 109

Xiao Meng, Yunhe Liu, Zhiyuan Ning, Jing Dong and Gang Liang
Experimental Investigation of the Size Effect on Roller-Compacted Hydraulic Asphalt Concrete under Different Strain Rates of Loading
Reprinted from: *Materials* 2024, 17, 353, doi:10.3390/ma17020353 126

Zhiyuan Ji, Xing Wu, Yao Zhang and Gabriele Milani
Aging Behavior and Mechanism Evolution of Nano-Al_2O_3/Styrene-Butadiene-Styrene-Modified Asphalt under Thermal-Oxidative Aging
Reprinted from: *Materials* 2023, 16, 5866, doi:10.3390/ma16175866 142

Grzegorz Mazurek, Przemysław Buczyński, Marek Iwański, Marcin Podsiadło, Przemysław Pypeć and Artur Kowalczyk
Effects of the Mixing Process on the Rheological Properties of Waste PET-Modified Bitumen
Reprinted from: *Materials* 2023, 16, 7271, doi:10.3390/ma16237271 162

Jia Zhao, Weigang Zhao, Kaize Xie and Yong Yang
A Fractional Creep Constitutive Model Considering the Viscoelastic–Viscoplastic Coexistence Mechanism
Reprinted from: *Materials* 2023, 16, 6131, doi:10.3390/ma16186131 183

Min Li, Jian Wang, Zibao Guo, Jingchun Chen, Zedong Zhao and Jiaolong Ren
Evaluation of the Adhesion between Aggregate and Asphalt Binder Based on Image Processing Techniques Considering Aggregate Characteristics
Reprinted from: *Materials* **2023**, *16*, 5097, doi:10.3390/ma16145097 . . . . **201**

Yuanshuai Dong, Zihao Wang, Wanyan Ren, Tianhao Jiang, Yun Hou and Yanhong Zhang
Influence of Morphological Characteristics of Coarse Aggregates on Skid Resistance of Asphalt Pavement
Reprinted from: *Materials* **2023**, *16*, 4926, doi:10.3390/ma16144926 . . . . **217**

Jingchun Chen, Jian Wang, Min Li, Zedong Zhao and Jiaolong Ren
Mesoscopic Mechanical Properties of Aggregate Structure in Asphalt Mixtures and Gradation Optimization
Reprinted from: *Materials* **2023**, *16*, 4709, doi:10.3390/ma16134709 . . . . **232**

Preface

The Reprint of "Performance-Related Material Properties of Asphalt Mixture Components (Second Edition)" brings together a collection of 14 scientific papers that represent the most recent advances in the field of pavement engineering. It is dedicated to investigating the material properties that directly influence the performance of asphalt mixtures and pavements. The scope of the papers spans a wide range of topics, including the analysis of resilient modulus, fatigue behavior, the effects of fibers and polymers on asphalt properties, aging mechanisms, viscoelastic modeling, and the impact of recycling processes. The aim of this Reprint is to provide researchers, engineers, and industry professionals with a valuable resource that captures the most recent research and developments in the field, offering practical solutions and innovative approaches to improving pavement materials and performance.

The authors would like to express our sincere appreciation to the publisher and editors for their invaluable support in the production of this Reprint. Their assistance has been instrumental in ensuring the quality and accessibility of the content presented here. We hope that this Reprint will serve as a useful and insightful resource for all those involved in the study and practice of pavement engineering and that it will contribute to the ongoing efforts to improve pavement materials, performance, and evaluation for the benefit of all.

Yao Zhang, Haibo Ding, Yu Chen, and Meng Ling
Editors

Article

Calculation of and Key Influencing Factors Analysis on Equivalent Resilient Modulus of a Submerged Subgrade

Junyao Tang [1], Siyu Chen [1], Tao Ma [1,*], Binshuang Zheng [2] and Xiaoming Huang [1,3,*]

1. School of Transportation, Southeast University, Nanjing 211189, China; tjyseu@163.com (J.T.); chsy@seu.edu.cn (S.C.)
2. School of Modern Posts, Nanjing University of Posts and Telecommunications, Nanjing 210023, China; zhengbs@njupt.edu.cn
3. National Demonstration Center for Experimental Education of Road and Traffic Engineering, Southeast University, Nanjing 211189, China
* Correspondence: matao@seu.edu.cn (T.M.); huangxm@seu.edu.cn (X.H.); Tel.: +86-15805160021 (T.M.); +86-13905174081 (X.H.)

Abstract: To calculate and analyze the equivalent resilient modulus of a submerged subgrade, a constitutive model considering the effect of saturation and matrix suction was introduced using ABAQUS's user-defined material (UMAT)subroutine. The pavement response under falling weight deflectometer (FWD) load was simulated at various water levels based on the derived distribution of the resilient modulus within the subgrade. The equivalent resilient modulus of the subgrade was then calculated using the equivalent iteration and weighted average methods. Based on this, the influence of the material and structural parameters of the subgrade was analyzed. The results indicate that the effect of water level rise on the tensile strain at the bottom of the asphalt layer and the compressive strain at the top of the subgrade is obvious, and its trend is similar to an exponential change. The equivalent resilient modulus of the subgrade basically decreases linearly with the rise in the water level, and there is high consistency between the equivalent iteration and weighted average methods. The saturated permeability coefficient and subgrade height have the most significant effect on the resilient modulus of the subgrade, which should be emphasized in the design of submerged subgrades, and the suggested values of the resilient modulus of the subgrade should be proposed according to the relevant construction conditions.

Keywords: submerged subgrade; equivalent resilient modulus; constitutive model; finite element method; fluid–solid coupling

1. Introduction

Abundant rainfall in tropical regions like Africa and Southeast Asia causes the seasonal submersion of the subgrade. Variations in moisture greatly affect the subgrade's resilient modulus, which, in turn, reduces soil stiffness and increases structural reactivity under pavement loads [1–3]. A large number of studies have been carried out to discuss the effects of moisture content [4,5], matrix suction [6,7], loading frequency [8,9], confining stress [10], and soil type [11–13] on the resilient modulus. In practice, it is difficult to take into account the influence of so many factors at the same time. The most important ones are stress state and moisture state [14]. Through a large number of laboratory dynamic resilient modulus tests, the resilient modulus prediction equation was established to quantitatively describe the relationship between the resilient modulus and physical property parameters, state variables and environmental variables [15–17]. The mechanistic-empirical pavement design guide (MEPDG) design method incorporates the resilient modulus prediction equation as an intrinsic model of the subgrade and the granular layer to carry out the structural nonlinear analysis of pavement [18–21], as shown in Equation (1):

$$M_r = k_1 p_a \left(\frac{\theta}{p_a}\right)^{k_2} \left(\frac{\tau_{oct}}{p_a} + 1\right)^{k_3} \quad (1)$$

where M_r is the resilient modulus, MPa; θ is the bulk stress, kPa; τ_{oct} is the octahedral shear stress, kPa; p_a is the atmospheric pressure, usually taken as 100 kPa; and k_1, k_2, and k_3 are material parameters.

On this basis, state variables characterizing moisture are added to the equation, and equations that consider both the stress and moisture states of the material were developed [22]. Liang et al. [23] proposed a moisture-dependent prediction equation based on the Bishop effective stress. Similarly, Qian et al. [24] replaced the parameter reflecting the contribution of matrix suction with a saturation-related parameter. Gu et al. [25], on the other hand, introduced volumetric moisture content to capture the effect of humidity, while adding an adjustable saturation factor for error reduction. Furthermore, Zhang et al. [26] proposed a new resilient modulus model to incorporate relative compaction in addition to matrix suction and the stress state. Liu et al. [27] collected and analyzed the existing resilient modulus models of subgrade unsaturated soils, and summarized the characteristics and scope of application of each model.

Due to the stress and moisture dependence of the resilience modulus, its spatial distribution is inhomogeneous [28,29]. Determining the equivalent resilient modulus entails replacing a series of resilient modulus values with a representative value based on the equivalence principle. In the design of pavement structures, the commonly chosen equivalence metric is the mechanical response or the service life of the structure in question [30]. Among the available mechanical response metrics, the most widely used one is deflection [31]. The equivalent resilient modulus is determined using the iterative inverse calculation method according to the principle of deflection equivalence [32,33]. In AASHTO, the current service index of pavement is used as an index to determine the empirical equation for the relative loss coefficient for different subgrade moduli, and the effective resilient modulus of the subgrade is back-calculated based on the average relative loss coefficient [34]. By referring to this idea, researchers predicted the attenuation of the resilient modulus of a subgrade using easily measured indicators, such as the compressive strain at the top of the subgrade, the number of load repetitions, etc., and established corresponding empirical models [35,36]. These empirical methods facilitate the determination of the modulus, but their accuracy depends on a large number of field tests and is highly dependent on the region in question.

In summary, although many moisture-dependent soil resilient modulus constitutive models have been proposed and verified using triaxial tests, most of them stay in the theoretical research stage, and there are few studies incorporating the finite element method to study the distribution of the modulus field and the structural mechanical response of the subgrade structure. Meanwhile, the methods for calculating the equivalent resilient modulus of subgrade are mostly empirical, and the process is simple and lacks systematic theoretical guidance; thus, whether it can be directly used for the resilient modulus calculation of a submerged subgrade needs to be further discussed. Consequently, analyses on the equivalent resilient modulus of a subgrade under different submerged conditions and subgrade parameters are less frequent, making it difficult to provide effective guidance for submerged subgrade design. Therefore, it is necessary to propose a procedural approach to simplify the calculation and analysis of the equivalent resilient modulus of a subgrade.

Accordingly, in order to efficiently calculate the mechanical response and equivalent resilient modulus of a submerged subgrade and analyze the key influencing factors of the resilient modulus, a suction-dependent resilient modulus constitutive model was applied to reflect the effect of the moisture field on the mechanical properties of a subgrade based on ABAQUS's user-defined material (UMAT) subroutine. Subsequently, the pavement response was analyzed, and the equivalent resilient modulus was calculated through the equivalent iteration and weighted average methods via ABAQUS python script. Furthermore, the influence of key factors on the equivalent resilient modulus was analyzed, including material parameters and the structural parameters of the subgrade. This study provides a complete framework for the rapid calculation of the subgrade equivalent resilient modulus, and the analyzed results can be utilized as a guide for subgrade design and water damage mitigation.

2. Materials and Methods

2.1. Typical Submerged Subgrade Model

The typical highway structure applied in this study is shown in Figure 1, together with information on highway construction and hydrogeology in Africa. The pavement is composed of 4 cm AC13 + 6 cm AC20 + 20 cm graded aggregate with a width of 16 m. The slope of the subgrade was set to 1:1.5, with a height of 6 m. The foundation was assumed to be 12 m deep, with the initial water level being 6 m underground.

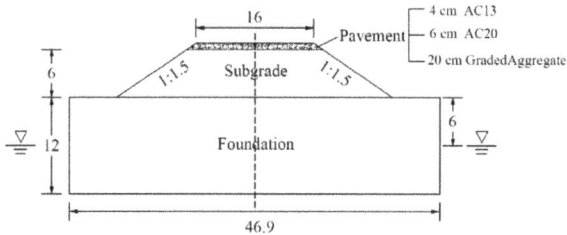

Figure 1. Typical highway structure (unit: m).

The parameters of the pavement material are shown in Table 1. We used dynamic moduli for all the elastic moduli E in order to replicate the structural performance under the field conditions.

Table 1. Pavement material parameters.

Layer	Material	Thickness (cm)	Elastic Modulus (MPa)	Poisson's Ratio	Dry Density (g·cm^{-3})
Top layer	AC13	4	8200	0.25	2.40
Bottom layer	AC20	6	8000	0.25	2.40
Base course	Graded Aggregate	20	300	0.35	2.32

The parameters of the subgrade and the foundation were set as follows: elastic modulus E = 35 MPa, Poisson's ratio ν = 0.3, cohesion c = 15 kPa, and friction angle φ = 30°. The relationship between permeability coefficient K_w and saturation S_r in the unsaturated zone is shown in Equation (2):

$$K_w = S_r^3 K_{ws} \tag{2}$$

where K_{ws} is saturated permeability coefficient, taken as 5×10^{-7} m·s^{-1}.

To account for the seepage problems in unsaturated soil, the SWCC (Soil–Water Characteristic Curve) was defined using the Fredlund and Xing model [37]:

$$\theta_w = C(\psi) \frac{\theta_s}{\left\{ ln\left[e + \left(\frac{\psi}{a}\right)^n\right]\right\}^m} \tag{3}$$

$$C(\psi) = 1 - \frac{ln\left(1 + \frac{\psi}{\psi_r}\right)}{ln\left(1 + \frac{10^6}{\psi_r}\right)} \tag{4}$$

where θ_w is volumetric moisture content; θ_s is saturated volumetric water content; $C(\psi)$ is correction function; ψ is matrix suction, given in kPa; ψ_r is residual matrix suction, given in kPa; and a, n, and m are fitting parameters, taken as 2813, 0.4836, and 3.7106, respectively.

The corresponding hydraulic parameter curves developed according to Equations (2)–(4) are shown in Figure 2.

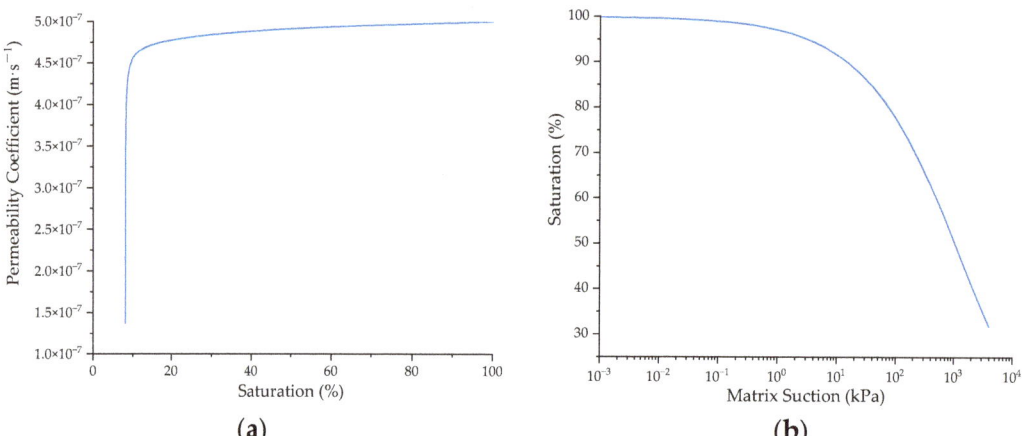

Figure 2. Hydraulic parameter curves of soil material: (**a**) relationship between permeability coefficient and saturation; (**b**) SWCC (Soil–Water Characteristic Curve).

The sides of the slope and foundation were set as the submerged boundary, and the boundary conditions related to pore water pressure were incorporated. In order to effectively carry out the fluid–solid coupling analysis, the plane pore water pressure units, CPE8P, were adopted in the subgrade and foundation. Since falling weight deflectometer (FWD) loads need to be applied subsequently to observe the mechanical response of the subgrade and pavement and back-calculate the resilient modulus of the subgrade accordingly, in order to improve the accuracy of the results, the mesh was divided into a local refinement set in addition to the global distribution. The middle part of the mesh was gradually encrypted from the outside to the inside, and the meshes of the surface layer and the base layer were encrypted vertically. The corresponding results are shown in Figure 3.

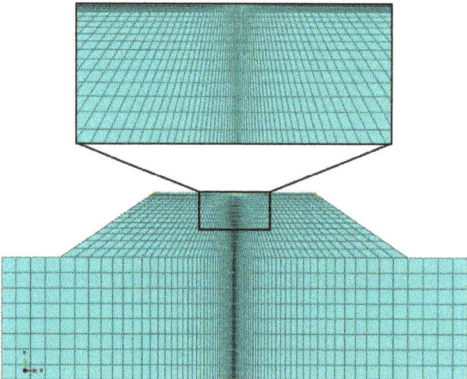

Figure 3. Mesh partition and local encryption.

2.2. Constitutive Model of Resilient Modulus

Since a constitutive model of the unsaturated soil resilient modulus is not provided in ABAQUS, it was necessary to complete the definition with the assistance of the UMAT subroutine. Given the stress dependence and moisture dependence of the resilient modulus in unsaturated soil, a constitutive model considering the effect of saturation and matrix suction was applied based on Equation (1), as shown in Equation (5):

$$M_r = k_1 p_a \left(\frac{\theta + 3 S_r \psi}{p_a} \right)^{k_2} \left(\frac{\tau_{oct}}{p_a} + 1 \right)^{k_3} \qquad (5)$$

where the fitting parameters k_1, k_2, and k_3 were determined using nonlinear regression analysis, and the source data came from the predicted resilient modulus of unsaturated clay in Nantong, Jiangsu Province, from Qian's research [24]. As shown in Figure 4, the horizontal and vertical coordinates of the dots represent the results of the resilient modulus predicted using the two different models respectively. The fitting parameters were obtained as k_1 = 0.4351, k_2 = 0.9698, and k_3 = −1.6522, and the goodness-of-fit value of the two models was 0.9542, with a significant correlation.

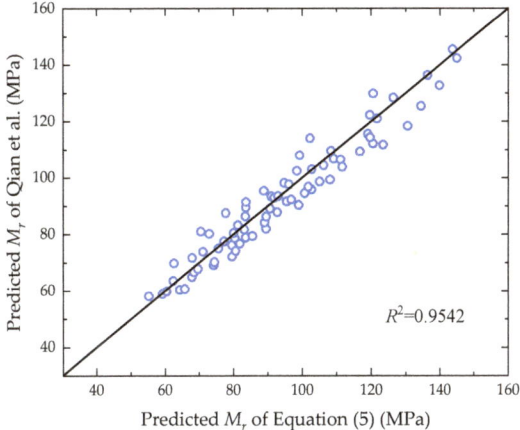

Figure 4. Correlation analysis of two prediction models [24].

Under the submerged condition, the stress state at each node of the subgrade was extracted by the UMAT subroutine through the interface with ABAQUS, and the corresponding resilient modulus was calculated based on the constitutive model. Subsequently, the stress state was updated according to the new distribution of the resilient modulus. When the water level rises, the states of stress and moisture change correspondingly, leading to differences in the resilient modulus and the pavement response.

2.3. Calculation of Equivalent Resilient Modulus

Due to the stress and moisture dependence of the resilient modulus in unsaturated soil, the stress and moisture states in the subgrade vary from position to position, resulting in an inhomogeneous spatial distribution. The equivalent resilient modulus is a representative value of the spatial distribution according to a given equivalence principle used to reflect the overall bearing capacity of the subgrade. For this purpose, the equivalent resilient modulus of the subgrade was calculated using the equivalent iteration and weighted average methods.

2.3.1. Equivalent Iteration

For the design of flexible base asphalt pavement, the tensile strain at the bottom of the asphalt layer, ε_t, and the compressive strain at the top of the subgrade, ε_c, are regarded as key indicators in structural calculations, controlling fatigue damage and permanent deformation, respectively, and both can be adopted as equivalent indicators for iteration [38,39].

The iteration process based on ε_t is shown in Figure 5. The iteration algorithm was implemented via a python script. The steps are as follows:

1. Set the reference value. The value ε_t was calculated using the nonlinear model under the peak load.
2. Calculate the linear elastic response. A linear elastic model was established with the initial modulus E_i between the maximum and minimum values in the modulus distribution of the nonlinear model, and the corresponding response ε_t' under the same load was calculated.
3. Iterative convergence algorithm. The convergence criterion is that the error between the linear elastic response and the reference value must be less than the permitted value (0.5%), and the bisection method was applied to ensure convergence of the iterations.

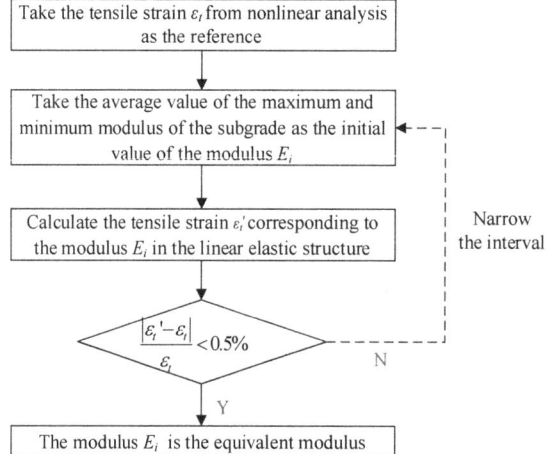

Figure 5. Equivalent iteration process of subgrade equivalent resilient modulus.

2.3.2. Weighted Average

The resilient modulus distribution of the subgrade was calculated according to Equation (5). On this basis, determining the weighted average involved finding a suitable weight function to calculate the equivalent resilient modulus of the subgrade directly.

Generally speaking, the closer the soil is to the center of the load, the higher the load it is subjected to, and, accordingly, the greater its contribution to the equivalent resilient modulus. Thus, the weight function should reflect this characteristic. To this end, a falling weight deflectometer load was applied to the model to analyze the dynamic response. The FWD load was simplified as a circular homogeneous load with a half-period sinusoidal function with a peak load of 0.714 MPa, an action time of 30 ms, and a loading radius of 0.15 m, as shown in Figure 6. Considering the two-dimensional plane model used in this study, the circular homogeneous load was converted to a line load with a peak load of 168.23 kPa.

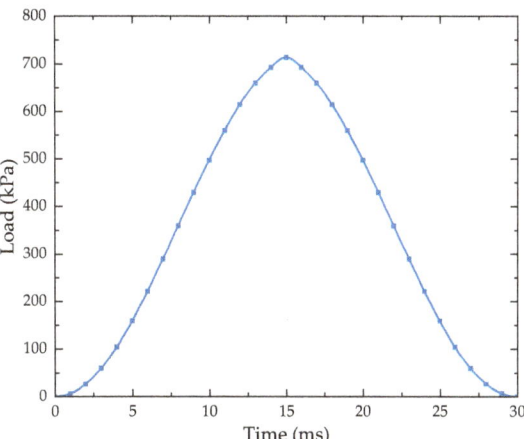

Figure 6. FWD load time-history curve.

The result of the top compressive strain of the subgrade ε_c at the peak load is shown in Figure 7, from which it can be seen that the spatial distribution of the compressive strain meets the requirements for the weight function, which can be used to calculate the weighted average of the equivalent resilient modulus according to Equation (6):

$$E = \frac{\sum E_i \cdot \varepsilon_{ci}}{\sum \varepsilon_{ci}} \tag{6}$$

where E is the equivalent resilient modulus; E_i and ε_{ci} are the resilient modulus and maximum compressive strain at the center of the elements within the subgrade.

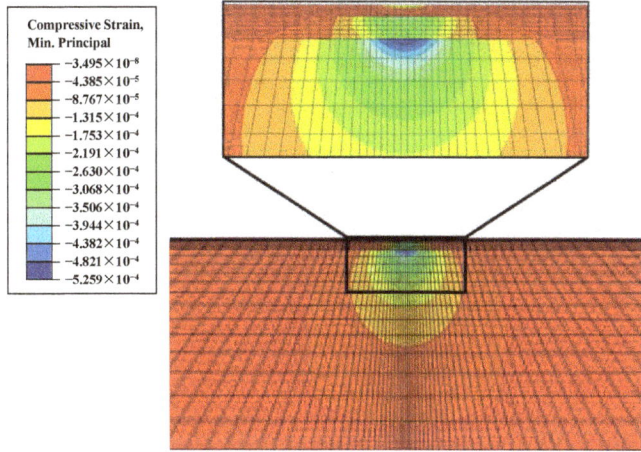

Figure 7. Distribution of maximum compressive strain at FWD peak load. (Unit: $\mu\varepsilon$).

3. Results

3.1. Validation of Constitutive Model

In order to verify the validity of the proposed constitutive model, as shown in Equation (5), the results of laboratory tests reported in the study by Qian et al. [24] and Liang et al. [23], which include clay samples from different regions of China and the USA, were selected for analysis, and the results of the resilient modulus of the soil samples under different matrix suction conditions were obtained through triaxial tests.

The model of Equation (5) was used to fit the above test data nonlinearly, and the obtained fitting parameters are summarized in Table 2. The fitting results are shown in Figure 8, wherein the goodness of fit is above 0.94. It can be seen that the results predicted by the model have a high degree of fitting with respect to the measured results of the laboratory tests, making this model highly applicable to different soil samples of subgrades in different regions.

Table 2. Fitting parameters of different soil samples.

Soil Samples	Fitting Parameters			
	k_1	k_2	k_3	R^2
Qian et al. Shanghai [24]	0.3712	0.9822	−1.6248	0.9627
Qian et al. Shandong [24]	0.3942	0.9973	−1.5841	0.9415
Liang et al. A-4 [23]	0.4544	0.8842	−1.7612	0.9638
Liang et al. A-6 [23]	0.4832	0.9161	−1.8033	0.9725

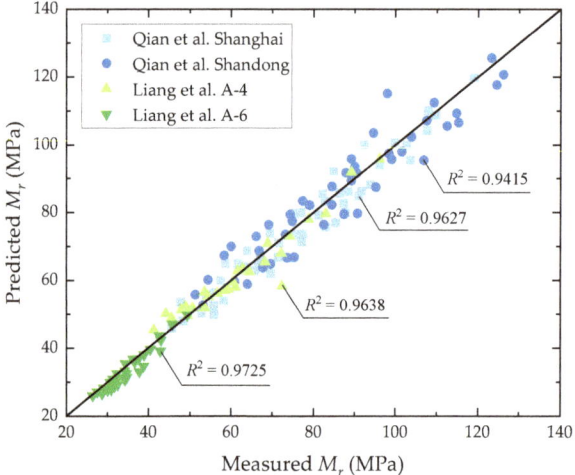

Figure 8. Comparison of predicted and measured resilient moduli for different soil samples in Qian et al. [24] and Liang et al. [23].

3.2. Distribution of Resilient Modulus

In order to illustrate the effect of water level variations on the distribution of field variable outputs, the final states with the initial water level and the water level raised to 2 m were chosen for analysis.

The vertical effective stress distribution of the initial state obtained from the analysis is shown in Figure 9a. Due to the existence of side slopes, the vertical effective stress shows a gradual increase from the surface of the side slopes inwards. As the water level rises and reaches stability, the vertical effective stresses in all parts of the structure are reduced significantly, as shown in Figure 9b, which is consistent with the results of the theoretical analysis.

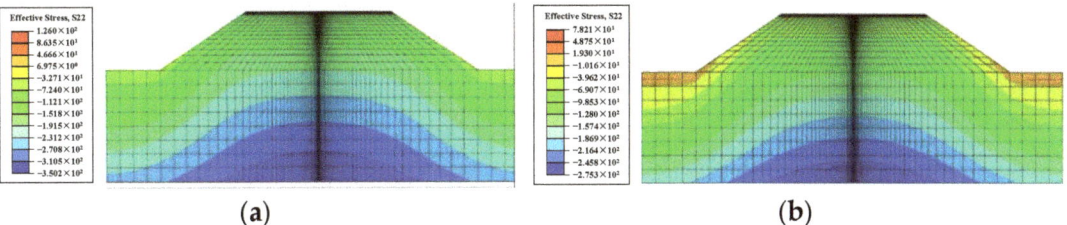

Figure 9. Distribution of vertical effective stress (**a**) at a water level of 6 m underground and (**b**) at a water level of 2 m (Unit: kPa).

The distribution of the resilient modulus at a raised water level of 2 m is shown in Figure 10. In general, the resilient modulus gradually increases from the top of the subgrade and the slope to the depth of the foundation, which is consistent with the stress field distribution in Figure 7. From Equation (5), it can be gleaned that the contributions of bulk stress and shear stress to the resilient modulus are opposite, which means that the value of bulk stress is relatively large compared with that of shear stress and plays a dominant role in the magnitude of the resilient modulus. Meanwhile, the effect of matrix suction is reflected in the stress results through Bishop's effective stress principle, and the foot of the slope has the smallest resilient modulus due to the small bulk stress and the presence of pore water pressure. At the top of the subgrade, the resilient modulus is larger than that at the foot of the slope; this is because the presence of matrix suction makes the effective stress here significantly higher.

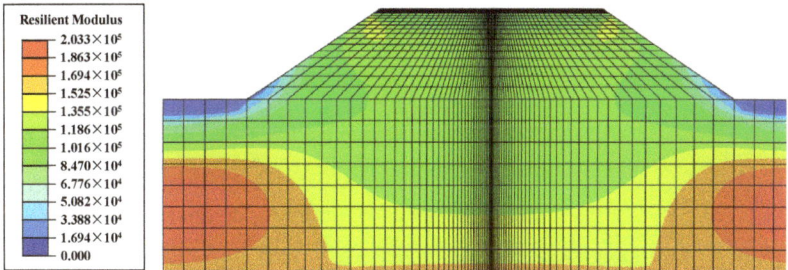

Figure 10. Distribution of resilient moduli at a water level of 2 m (Unit: kPa).

3.3. Dynamic Response

In order to analyze the dynamic response characteristics of the submerged subgrade, the variations in the tensile strain at the bottom of the asphalt layer ε_t and the compressive strain at the top of the subgrade ε_c at different water levels were analyzed. The strain results discussed in the following are the maximum principal strains of the elements at the FWD peak load in Figure 6.

From Figure 11, it can be seen that the tensile strain is similar to the exponential change with the elevation of the water level. When the water level is located underground (i.e., the water level is 0 m or less), the amount of change is slight, the water level rises from −6 m to 0 m, and the tensile strain increases 11.6%; when the water level is located above the ground, a saturated zone appears in the part of the subgrade, the tensile strain change is obvious, the water level rises from 0 m to 5 m, and the tensile strain increases by 36.3%. In particular, when the water level rises above 5 m, the tensile strain increases rapidly, corresponding to 39.3% between 5 m and 6 m.

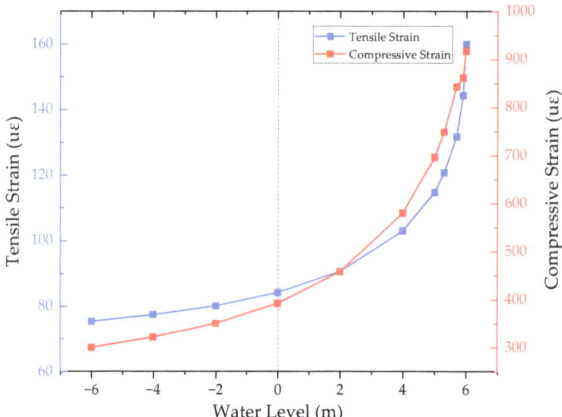

Figure 11. Curves of tensile and compressive strains versus water level.

The change rule of compressive strain was similar to that of tensile strain, increasing by 30.8%, 77.4%, and 31.6%, respectively, when the water level was at −6~0, 0~5, and 5~6 m. It can be seen that the variation in compressive strain is more obvious than that for tensile strain at a low water level, while the tensile strain better reflects the influence of the submerged water level at a high water level on the subgrade.

Specifically, the rapid increase in structural strain in the high-water-level case may be due to the influence of excess pore water pressure. The pore water pressure distributions corresponding to the peak loads at water levels of 0 m and 6 m, respectively, are given in Figure 12. Since the loading time was quite short, the FWD load was loaded to the peak value in only 0.015 s, and the permeability coefficient of the soil was quite small. Therefore, the excess pore water pressure accumulated within the subgrade could not be dissipated within such a short period of time, and the effective stresses were further narrowed down, leading to a decrease in the resilient modulus of the soil. However, when the water level of the subgrade was not high, the saturated zone of the soil was located at a deeper position, and only a small amount of excess pore water pressure was induced by the transfer of the pavement load to the saturated zone; additionally, the modulus of the soil at this depth contributes a small amount to the values of the strains, and thus the change in strain is not obvious.

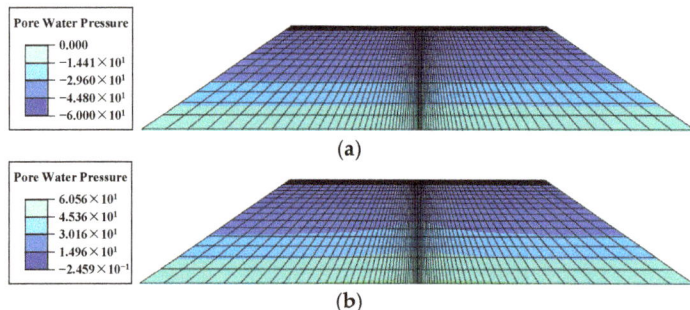

Figure 12. Pore water pressure distributions corresponding to peak load (**a**) at a water level of 0 m and (**b**) at a water level of 6 m.

3.4. Equivalent Resilient Modulus

The two methods described in Section 2.3 were employed to calculate the equivalent resilient modulus of the subgrade, and the results are shown in Figure 13. Comparing the results of equivalent iteration by the two indicators of tensile strain ε_t and compressive strain ε_c, it can be seen that the modulus values of the two curves are different at the same water level due to the different indicators chosen, but the two show similar trends. The moduli of the subgrade all decrease significantly with the increase in water level. The variation range is 20~190 MPa, and the linear correlation coefficients are 0.9867 and 0.9783, respectively, which can essentially be regarded as corresponding to a linear variation.

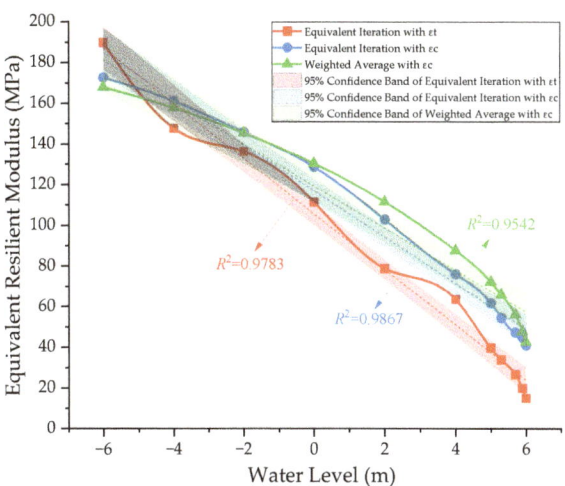

Figure 13. Equivalent resilient modulus results of subgrade obtained using different methods.

Based on the same indicator of compressive strain ε_c, the results of equivalent iteration and weighted average were compared. It was revealed that the weighted average results are in high agreement with the more accurate equivalent iteration results, so the method of determining the weighted average according to the distribution of the compressive strain can be considered feasible.

In conclusion, both methods for determining the equivalent resilient modulus value of a subgrade show a strong moisture correlation. This also implies that the effect of subgrade moisture should be emphasized in the design of actual pavement structures that and necessary anti-drainage measures should be taken. Between the two methods, the equivalent iteration method is more complicated, but its calculation results are more accurate with the aid of UMAT subroutine. The weighted average method is simpler, but the determination of the weight function is the main difficulty in this regard. The calculation method can be selected according to actual needs.

4. Discussion

In this section, the proposed equivalent modulus calculation method is used to analyze the influence of key influencing factors, including subgrade material and structural parameters, on the value of the equivalent resilient modulus so as to provide a reference for the design of submerged subgrades. Moreover, the suggested values for the resilient modulus of subgrade can also be presented in relation to local construction conditions.

In order to effectively control the variables, if not specified, the following simulation parameters were set so as to be consistent with those of the previous model, the water level was set to 2 m, and the compressive strain equivalent iteration method was applied to calculate the equivalent modulus. In particular, only the analysis results under several common working conditions are provided, and the reasonableness of the trend was analyzed with respect to the mechanism, while the specific values under other special working conditions need to be further explored.

4.1. Influence of Material Parameters

The SWCC of unsaturated soil determines the relationship between matrix suction and saturation, constituting an important parameter in the fluid–solid coupling analysis of unsaturated soil. Meanwhile, the saturated permeability coefficient of the subgrade directly affects the speed of moisture change inside the subgrade. Therefore, three SWCCs (Figure 14) and different saturated permeability coefficients of clay subgrade were selected to investigate the influence of the above two hydraulic parameters on the results of the equivalent modulus calculation.

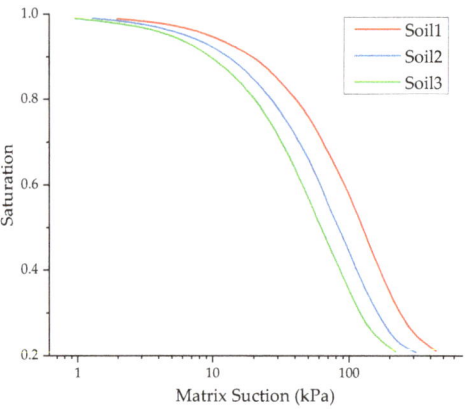

Figure 14. Three types of SWCC of clay.

As the SWCC gradually shifts towards the low-suction region, the saturation of the soil decreases under the same suction conditions, resulting in a decrease in the effective stress and modulus in the unsaturated zone, as shown in Figure 15. However, the magnitude of modulus reduction is not obvious, presumably due to the small permeability coefficient setting. With the continuous increase in the permeability coefficient, the modulus shows obvious attenuation. When the permeability coefficient is increased 10-fold, the modulus decreases by 9.9%, and when it is further increased 100-fold, the modulus only decreases by 3.2%, and there is a certain saturation effect, indicating that at this time, the permeability coefficient is already large enough, and the moisture at the side slope can enter the roadbed rapidly.

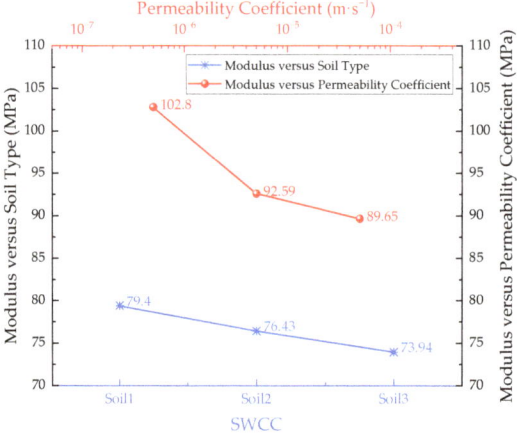

Figure 15. Equivalent modulus of subgrade with different hydraulic parameters.

4.2. Influence of Structural Parameters

In order to investigate the influence of subgrade structural parameters on the equivalent resilient modulus, the width, height, and slope of the subgrade were selected as the key influencing factors. For the width, three conditions, namely, 12 m, 16 m, and 26 m, were selected according to different highway grades. For the height, 6, 8, and 10 m were selected. For the slope, three conditions, namely, 1:1, 1:1.5, and 1:2, were selected for analysis.

As can be seen in Figure 16, the equivalent modulus of the subgrade increases gradually with the increase in subgrade width, but the overall change is not large, and the modulus only increases by 5.8% when the subgrade width increases from 12 m to 26 m. The increase in subgrade width prolongs the duration of water inflow from the slope to the inside of the subgrade when the water level rises; additionally, the hysteresis phenomenon of moisture inside the subgrade is obvious, and the equivalent modulus of the subgrade increases. Conversely, if the water level decreases, the moisture inside the subgrade is discharged slowly, resulting in a larger modulus in a wider subgrade. Therefore, in the determination of the correlation between subgrade width and equivalent modulus, one should also consider the water level change conditions.

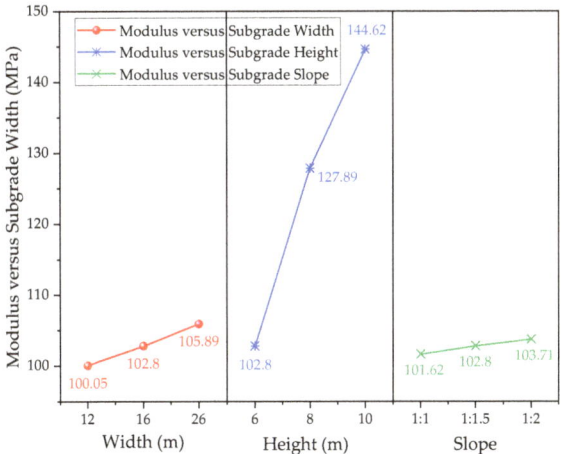

Figure 16. Equivalent modulus of subgrade with different structural parameters.

Compared with subgrade width, the effect of subgrade height on the equivalent modulus of the subgrade is fairly obvious, and the modulus increases by 40.7% when the subgrade height increases from 6 m to 8 m. With the increase in subgrade height, the range of the unsaturated zone of the subgrade enlarges, and the overall matrix suction in the unsaturated zone increases; meanwhile, the effective stress in the saturated zone of the subgrade obviously increases due to the increase in soil self-gravitational stress, which, in turn, leads to an increase in the overall modulus of the subgrade.

There are no significant differences in the results of the resilient modulus of the subgrade under different slopes. Theoretically, the effect produced by the increase in slope is similar to the increase in subgrade width; i.e., it can be regarded as a narrow subgrade when the slope rate is 1:1 and a wide subgrade when the slope rate is 1:2. However, the hysteresis effect of moisture when the slope decreases is not obvious, and therefore the effect of slope on the resilient modulus of the subgrade can nearly be neglected in the actual engineering process.

4.3. Design Framework for Submerged Subgrade

Based on the calculation method and analysis process proposed in this study, a preliminary framework for the design of a submerged subgrade was developed, as shown in

Figure 17. The framework allows designers to quickly calculate the equivalent resilient modulus of a submerged subgrade, assess whether the overall stiffness of the subgrade meets the design requirements, and guide the optimization of the initial design based on the results of the analysis of the key influencing factors. Unlike the current resilient modulus prediction model for soil samples under specified conditions, this method considers the spatial inhomogeneous distribution of the modulus of the subgrade and calculates the equivalent resilient modulus of the subgrade as a whole by means of the finite element method, which is of strong engineering practicability, and the framework can generally be divided into four parts:

1. Determine the model parameters, including boundary conditions, the constitutive model, structural parameters, and material parameters. The constitutive model of Equation (5) must be adopted, and other parameters need to be preliminarily determined in combination with local construction conditions and in consideration of engineering experience.
2. Calculate the mechanical response. Calculate the tensile strain at the bottom of the asphalt layer, ε_t, and the compressive strain at the top of the subgrade, ε_c, under FWD loading as representative dynamic response indicators. At the same time, these two indicators will also be used as the basis for calculating the equivalent resilient modulus of the subgrade.
3. Calculate the equivalent resilient modulus. The equivalent resilient modulus can be calculated via equivalent iteration with tensile strain and weighted average with compressive strain. Theoretically, the equivalent iteration method has higher accuracy, while the weighted average method has higher computational efficiency. Designers can choose a method according to the requirements of the task at hand.
4. Carry out decision making. Judge whether the results of the equivalent resilient modulus meet the requirements. If so, then complete this design. If not, the design parameters can be modified with reference to the results of the key influencing factors analysis given in this study until the design requirements are met.

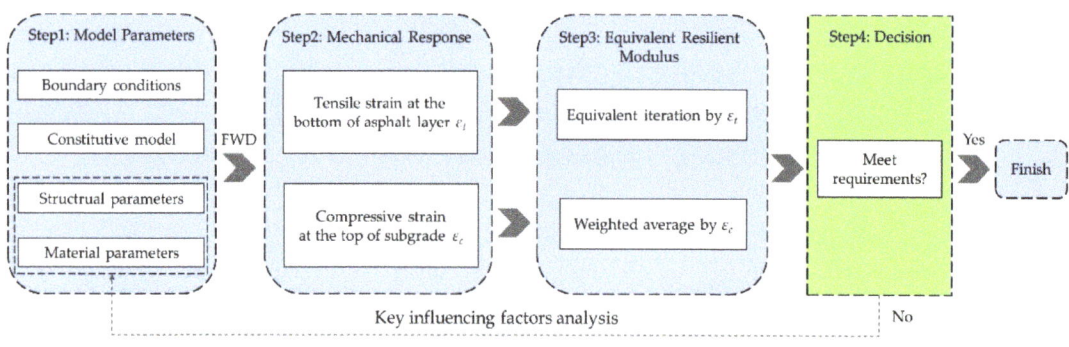

Figure 17. Flowchart for design of submerged subgrade.

5. Conclusions

In this study, a finite element model of a typical submerged subgrade was established, and a constitutive model considering the effect of saturation and matrix suction was introduced via a UMAT subroutine. The equivalent iteration and weighted average methods were used to calculate the equivalent resilient modulus. Based on this, the dynamic response and equivalent resilient modulus of the subgrade under different water levels were obtained. Finally, the influence of material parameters and the structural parameters of the subgrade were analyzed, and the results can provide a reference for the design and a suitable modulus value of the subgrade. The conclusions are as follows:

1. The effect of water level rise on the tensile strain at the bottom of the asphalt layer and the compressive strain at the top of the subgrade is obvious, and its trend is similar to an exponential change. At a low water level, the change in compressive strain is more obvious, while the change in tensile strain is more significant when the water level rises to a point where the subgrade is close to saturation. In fact, a situation in which near-saturation occurs is very rare, so the indicator of compressive strain is especially important in the design of a submerged subgrade.
2. The equivalent resilient modulus of the subgrade calculated using the equivalent iteration and weighted average methods has a strong correlation with the moisture content of the subgrade, and the modulus of the subgrade basically decreases linearly with the increase in the water level. The results of the weighted average based on the distribution of compressive strains at the top of the subgrade under FWD load are in high agreement with the results of the equivalent iteration, which is a more accurate method in theory. Therefore, it can be concluded that the method of determining the weighted average based on the distribution of compressive strain is feasible.
3. Among the subgrade materials and structural parameters considered in this study, the saturated permeability coefficient and subgrade height have the most significant effect on the resilient modulus of the subgrade, while SWCC and subgrade width have a slight effect on the modulus, and the effect of slope can be approximately ignored. Therefore, during the process of designing a submerged subgrade, the influence of the above parameters on the dynamic response of the structure should be emphasized, and the corresponding suggested values of the resilient modulus of the subgrade should be proposed according to the actual construction conditions.

This study provides an effective means for calculating and analyzing the equivalent resilient modulus of a submerged subgrade, but there is still room for improvement in this methodology. Only the results obtained under individual working conditions were considered. In fact, subgrades experience seasonal wet and dry cyclic effects, and the applicability of the constitutive model proposed in this study under such effects needs to be further discussed.

In addition, a preliminary framework for the design of a submerged subgrade was proposed in this study, but the overall process is still slightly cumbersome, and the computational efficiency is low in the case of a large quantity of data, making it to apply this framework on a wide scale in engineering practice. Therefore, it is possible to try to program the framework and develop a corresponding graphical user interface (GUI) so as to achieve rapid modelling and automatic computation, constituting the focus of our future research.

Author Contributions: Conceptualization, J.T. and S.C.; methodology, J.T.; software, J.T.; validation, J.T.; formal analysis, S.C.; investigation, S.C.; resources, T.M.; writing—original draft preparation, J.T.; writing—review and editing, B.Z.; visualization, B.Z.; supervision, X.H.; project administration, J.T. and X.H. All authors have read and agreed to the published version of the manuscript.

Funding: This research was financially supported by Postgraduate Research & Practice Innovation Program of Jiangsu Province (Grant No: KYCX23_0301) and National Key R&D Project of China (Grant Number: 2021YFB2600601, 2021YFB2600600).

Institutional Review Board Statement: Not applicable.

Informed Consent Statement: Not applicable.

Data Availability Statement: The data used to support the findings of this study are available from the first author upon request. In this paper, some models and code used during the study are proprietary or confidential in nature and may only be provided with restrictions, such as the UMAT subroutine and python script in ABAQUS.

Conflicts of Interest: The authors declare that they have no known competing financial interests or personal relationships that could have appeared to influence the work reported in this paper.

References

1. Liu, X.; Zhang, X.; Wang, H.; Jiang, B. Laboratory Testing and Analysis of Dynamic and Static Resilient Modulus of Subgrade Soil under Various Influencing Factors. *Constr. Build. Mater.* **2019**, *195*, 178–186. [CrossRef]
2. Zhang, Y.; Ling, M.; Kaseer, F.; Arambula, E.; Lytton, R.L.; Martin, A.E. Prediction and Evaluation of Rutting and Moisture Susceptibility in Rejuvenated Asphalt Mixtures. *J. Clean. Prod.* **2022**, *333*, 129980. [CrossRef]
3. Deng, Y.; Zhang, Y.; Shi, X.; Hou, S.; Lytton, R.L. Stress–Strain Dependent Rutting Prediction Models for Multi-Layer Structures of Asphalt Mixtures. *Int. J. Pavement Eng.* **2022**, *23*, 2728–2745. [CrossRef]
4. Chu, X. A Review on the Resilient Response of Unsaturated Subgrade Soils. *Adv. Civ. Eng.* **2020**, *2020*, 7367484. [CrossRef]
5. Zhang, J.; Peng, J.; Zeng, L.; Li, J.; Li, F. Rapid Estimation of Resilient Modulus of Subgrade Soils Using Performance-Related Soil Properties. *Int. J. Pavement Eng.* **2021**, *22*, 732–739. [CrossRef]
6. Cary, C.E.; Zapata, C.E. Resilient Modulus for Unsaturated Unbound Materials. *Road Mater. Pavement Des.* **2011**, *12*, 615–638. [CrossRef]
7. Park, H.I.; Kweon, G.C.; Lee, S.R. Prediction of Resilient Modulus of Granular Subgrade Soils and Subbase Materials Using Artificial Neural Network. *Road Mater. Pavement Des.* **2009**, *10*, 647–665. [CrossRef]
8. Yao, Y.; Zheng, J.; Zhang, J.; Peng, J.; Li, J. Model for Predicting Resilient Modulus of Unsaturated Subgrade Soils in South China. *KSCE J. Civ. Eng.* **2018**, *22*, 2089–2098. [CrossRef]
9. Su, N.; Xiao, F.; Wang, J.; Amirkhanian, S. Characterizations of Base and Subbase Layers for Mechanistic-Empirical Pavement Design. *Constr. Build. Mater.* **2017**, *152*, 731–745. [CrossRef]
10. Jayakody, S.; Gallage, C.; Ramanujam, J. Performance Characteristics of Recycled Concrete Aggregate as an Unbound Pavement Material. *Heliyon* **2019**, *5*, e02494. [CrossRef]
11. El-Ashwah, A.S.; Awed, A.M.; El-Badawy, S.M.; Gabr, A.R. A New Approach for Developing Resilient Modulus Master Surface to Characterize Granular Pavement Materials and Subgrade Soils. *Constr. Build. Mater.* **2019**, *194*, 372–385. [CrossRef]
12. Naji, K. Resilient Modulus–Moisture Content Relationships for Pavement Engineering Applications. *Int. J. Pavement Eng.* **2018**, *19*, 651–660. [CrossRef]
13. Cui, Y.-J. Mechanical Behaviour of Coarse Grains/Fines Mixture under Monotonic and Cyclic Loadings. *Transp. Geotech.* **2018**, *17*, 91–97. [CrossRef]
14. Zheng, J.; Liu, S.; Hu, H. The calculation theory of humidity for subgrade:A perspective review. *J. China Foreign Highw.* **2023**, *43*, 1–10. (In Chinese)
15. Ling, J.; Chen, H.; Qian, J.; Zhou, D. Dynamic resilient modulus for unsaturated clay soils considering effect of limited moisture content fluctuation. *J. Jilin Univ. (Eng. Tech. Ed.)* **2018**, *50*, 613–620. (In Chinese)
16. Tan, W. Study on dynamic resilience modulus prediction model of subgrade fine-grained soil based on physical property parameters. *J. China Foreign Highw.* **2023**, *43*, 36–42. (In Chinese)
17. Xu, M.; Kang, Y.; Ma, S.; Zhang, H. Subgrade resilient modulus prediction with XGBoost model based on Bayesian optimization. *J. Highw. Transp. Res. Dev.* **2023**, *40*, 51–60. (In Chinese)
18. Ghadimi, B.; Nikraz, H. A Comparison of Implementation of Linear and Nonlinear Constitutive Models in Numerical Analysis of Layered Flexible Pavement. *Road Mater. Pavement Des.* **2017**, *18*, 550–572. [CrossRef]
19. Wang, H.; Li, M. Evaluation of Effects of Variations in Aggregate Base Layer Properties on Flexible Pavement Performance. *Transp. Res. Rec. J. Transp. Res. Board* **2015**, *2524*, 119–129. [CrossRef]
20. Wang, H.; Al-Qadi, I.L. Importance of Nonlinear Anisotropic Modeling of Granular Base for Predicting Maximum Viscoelastic Pavement Responses under Moving Vehicular Loading. *J. Eng. Mech.* **2013**, *139*, 29–38. [CrossRef]
21. Al-Qadi, I.L.; Wang, H.; Tutumluer, E. Dynamic Analysis of Thin Asphalt Pavements by Using Cross-Anisotropic Stress-Dependent Properties for Granular Layer. *Transp. Res. Rec. J. Transp. Res. Board* **2010**, *2154*, 156–163. [CrossRef]
22. Han, Z.; Vanapalli, S.K. State-of-the-Art: Prediction of Resilient Modulus of Unsaturated Subgrade Soils. *Int. J. Geomech.* **2016**, *16*, 04015104. [CrossRef]
23. Liang, R.Y.; Rabab'ah, S.; Khasawneh, M. Predicting Moisture-Dependent Resilient Modulus of Cohesive Soils Using Soil Suction Concept. *J. Transp. Eng.* **2008**, *134*, 34–40. [CrossRef]
24. Qian, J.; Li, J.; Zhou, D.; Ling, J. Prediction model of resilient modulus for unsaturated clay soils considering the effect of matric suction. *Rock Soil Mech.* **2018**, *39*, 6. [CrossRef]
25. Gu, F.; Sahin, H.; Luo, X.; Luo, R.; Lytton, R.L. Estimation of Resilient Modulus of Unbound Aggregates Using Performance-Related Base Course Properties. *J. Mater. Civ. Eng.* **2015**, *27*, 04014188. [CrossRef]
26. Zhang, J.; Peng, J.; Liu, W.; Lu, W. Predicting Resilient Modulus of Fine-Grained Subgrade Soils Considering Relative Compaction and Matric Suction. *Road Mater. Pavement Des.* **2021**, *22*, 703–715. [CrossRef]
27. Liu, J.; Niu, J.; Cui, M. Classification and comparison of modulus of resilience calculation models of subgrade unsaturated soil. *Sci. Technol. Eng.* **2021**, *21*, 13491–13496. (In Chinese)
28. Narzary, B.K.; Ahamad, K.U. Equivalent Modulus for Fine-Grained Subgrade Soil. *J. Transp. Eng. Part B Pavements* **2020**, *146*, 04020004. [CrossRef]
29. Gkyrtis, K. Pavement Analysis with the Consideration of Unbound Granular Material Nonlinearity. *Designs* **2023**, *7*, 142. [CrossRef]

30. Khasawneh, M.A.; Al-jamal, N.F. Modeling Resilient Modulus of Fine-Grained Materials Using Different Statistical Techniques. *Transp. Geotech.* **2019**, *21*, 100263. [CrossRef]
31. Tan, Z.; Wang, L. Equivalent resilient modulus of subgrade based on principle of flexural-tensile stress equivalence. *J. Highw. Transp. Res. Dev.* **2015**, *32*, 46–50. (In Chinese)
32. Yao, Y.; Qian, J.; Li, J.; Zhang, A.; Peng, J. Calculation and Control Methods for Equivalent Resilient Modulus of Subgrade Based on Nonuniform Distribution of Stress. *Adv. Civ. Eng.* **2019**, *2019*, 6809510. [CrossRef]
33. Li, J.; Zheng, J.; Yao, Y.; Zhang, J.; Peng, J. Numerical Method of Flexible Pavement Considering Moisture and Stress Sensitivity of Subgrade Soils. *Adv. Civ. Eng.* **2019**, *2019*, 7091210. [CrossRef]
34. Nam, B.; An, J.; Murphy, M.R. Improvements to the AASHTO Subgrade Resilient Modulus (M_R) Equation. In Proceedings of the Geo-Congress 2014, Atlanta, GA, USA, 24 February 2014; Geo-Congress 2014 Technical Papers. American Society of Civil Engineers: Reston, VA, USA, 2014; pp. 2414–2425.
35. Sarkhani Benemaran, R.; Esmaeili-Falak, M.; Javadi, A. Predicting Resilient Modulus of Flexible Pavement Foundation Using Extreme Gradient Boosting Based Optimised Models. *Int. J. Pavement Eng.* **2023**, *24*, 2095385. [CrossRef]
36. Sadrossadat, E.; Heidaripanah, A.; Osouli, S. Prediction of the Resilient Modulus of Flexible Pavement Subgrade Soils Using Adaptive Neuro-Fuzzy Inference Systems. *Constr. Build. Mater.* **2016**, *123*, 235–247. [CrossRef]
37. Fredlund, D.G.; Xing, A.; Xing, A. Equations for the Soil-Water Characteristic Curve. *Can. Geotech. J.* **1994**, *31*, 521–532. [CrossRef]
38. Wang, W.; Wang, L.; Xiong, H.; Luo, R. A Review and Perspective for Research on Moisture Damage in Asphalt Pavement Induced by Dynamic Pore Water Pressure. *Constr. Build. Mater.* **2019**, *204*, 631–642. [CrossRef]
39. Wang, W.; Zhao, K.; Li, J.; Luo, R.; Wang, L. Characterization of Dynamic Response of Asphalt Pavement in Dry and Saturated Conditions Using the Full-Scale Accelerated Loading Test. *Constr. Build. Mater.* **2021**, *312*, 125355. [CrossRef]

Disclaimer/Publisher's Note: The statements, opinions and data contained in all publications are solely those of the individual author(s) and contributor(s) and not of MDPI and/or the editor(s). MDPI and/or the editor(s) disclaim responsibility for any injury to people or property resulting from any ideas, methods, instructions or products referred to in the content.

Article

Evaluation of Fatigue Behavior of Asphalt Field Cores Using Discrete Element Modeling

Min Xiao [1], Yu Chen [2,*], Haohao Feng [2,3], Tingting Huang [2], Kai Xiong [1] and Yaoting Zhu [4]

[1] Jiangxi Provincial Communications Investment Group Co., Ltd., Project Construction Management Company, Nanchang 330200, China; xmkkxx2024@163.com (M.X.); xiongkai9996@163.com (K.X.)
[2] Hubei Highway Engineering Research Center, School of Transportation and Logistics Engineering, Wuhan University of Technology, Wuhan 430063, China; fenghaohao@whut.edu.cn (H.F.); huangtingting@whut.edu.cn (T.H.)
[3] Zhejiang Communications Construction Group Co., Ltd., Design Institute Branch, Hangzhou 310051, China
[4] Jiangxi Transportation Institute Co., Ltd., Nanchang 330200, China; zhuyt7538@163.com
* Correspondence: yu.chen@whut.edu.cn

Abstract: Fatigue cracking is one of the primary distresses of asphalt pavements, which significantly affects the asphalt pavement performance. The fatigue behavior of the asphalt mixture observed in the laboratory test can vary depending on the type of fatigue test and the dimension and shape of the test specimen. The variations can make it difficult to accurately evaluate the fatigue properties of the field asphalt concrete. Accordingly, this study proposed a reliable method to evaluate the fatigue behavior of the asphalt field cores based on discrete element modeling (DEM). The mesoscopic geometric model was built using discrete element software PFC (Particle Flow Code) and CT scan images of the asphalt field cores. The virtual fatigue test was simulated in accordance with the semi-circular bending (SCB) test. The mesoscopic parameters of the contacting model in the virtual test were determined through the uniaxial compression dynamic modulus test and SCB test. Based on the virtual SCB test, the displacement, contact forces, and crack growth were analyzed. The test results show that the fatigue life simulated in the virtual test was consistent with that of the SCB fatigue test. The fatigue cracks in the asphalt mixture were observed in three stages, i.e., crack initiation, crack propagation, and failure. It was found that the crack propagation stage consumes a significant portion of the fatigue life since the tensile contact forces mainly increase in this stage.

Keywords: fatigue behavior; asphalt field cores; discrete element method (DEM); semi-circular bending (SCB) test

Citation: Xiao, M.; Chen, Y.; Feng, H.; Huang, T.; Xiong, K.; Zhu, Y. Evaluation of Fatigue Behavior of Asphalt Field Cores Using Discrete Element Modeling. *Materials* 2024, 17, 3108. https://doi.org/10.3390/ma17133108

Academic Editor: Simon Hesp

Received: 3 May 2024
Revised: 16 June 2024
Accepted: 20 June 2024
Published: 25 June 2024

Copyright: © 2024 by the authors. Licensee MDPI, Basel, Switzerland. This article is an open access article distributed under the terms and conditions of the Creative Commons Attribution (CC BY) license (https://creativecommons.org/licenses/by/4.0/).

1. Introduction

Fatigue cracking is one of the most common distresses of asphalt pavement. Crack initiation and propagation can reduce the bearing capacity of the pavement structure and allow the penetration of water into the structure, resulting in other distresses such as water damage [1]. Fatigue cracking is mainly caused by repeated traffic loading. The repeated loading can lead to the micro-cracks forming and merging, eventually propagating into macro-cracks, i.e., alligator cracking, which can weaken the overall structural capacity of the asphalt pavement and deteriorate the bonding between asphalt and aggregates [2]. Therefore, it is of significance to evaluate the fatigue behavior of the asphalt mixture, which could help prolong the fatigue life.

Most of the existing studies have primarily analyzed the fatigue behavior of the asphalt mixture using laboratory tests and summarized the mechanism of fatigue cracking [3–5]. However, most of the previous studies focused on the properties of lab-fabricated asphalt mixture specimens and were limited when evaluating the properties of the core samples, which can lead to inefficient utilization of the core samples. Moreover, in practical pavement applications, asphalt pavement is constructed through three phases of compaction and ages under in-service conditions, which are subject to environmental fluctuations, such as temperature variation, precipitation, and ultraviolet rays. Thus, there is much difference in the internal structural distribution and aging between the lab-fabricated asphalt mixture specimens and those in the field. Consequently, many researchers have shifted focus to the asphalt field cores, studying the fracture properties of the cores drilled from the in-service pavement over various durations.

There are many lab tests to study the fatigue properties of the asphalt mixture, such as the indirect tensile (IDT) test, direct tensile (DT) test, overlay test (OT), four-point bending test, and semi-circular bending (SCB) test. Barman et al. conducted the IDT test to characterize the fatigue resistance of the asphalt mixture and proposed a simple data analysis approach [6]. Luo et al. used the controlled-strain repeated direct tensile test to evaluate the fatigue cracking [7]. Gu et al. conducted the overlay test to investigate the fracture properties of the field-aged asphalt concrete and found that the cracking resistance of the field reduced from 1st month to the 9th month [8]. Kim et al. performed a four-point bending test to investigate the fatigue life of a total of ten asphalt mixtures, including hot-mix asphalt (HMA) and warm-mix asphalt (WMA) with different amounts of reclaimed asphalt pavement (RAP) and recycled asphalt shingles (RAS) [9]. Du et al. performed the SCB test on the layer core samples drilled from five expressways to analyze the sensitivity of fracture energy to factors such as the equivalent single axle load, air void, service age, etc. [10]. These studies demonstrate that the field core samples can reflect the asphalt pavement conditions, and the test results of core samples can be effectively used for the decision-making related to pavement maintenance actions.

Based on the results of the fatigue test of the asphalt mixture, the fatigue cracking models of the asphalt mixtures were developed. The fracture mechanics and the dissipated energy approach are most widely used to evaluate the fatigue resistance of the asphalt mixtures [11]. However, the fatigue test results cannot describe the crack propagation of asphalt mixture at the mesoscopic level. Moreover, it is hard to validate the fatigue cracking models with only a limited number of core samples.

In recent years, researchers have attempted to use computer technology to simulate the fracture evolution of the asphalt mixture and investigate the various influencing factors on the fatigue behavior. The discrete element method (DEM) has been widely used in pavement engineering since discrete elements can reflect the discontinuous and non-uniform structural characteristics of asphalt mixtures. It can also help reveal the internal structural deformation, cracking, and other mechanical behaviors of asphalt mixtures. Ma et al. built a virtual specimen based on the DEM to estimate the fatigue life of an asphalt mixture and investigated the influence of air void on fatigue life [12]. Xue et al. developed a new approach combining algorithmic techniques and DEM to perform a heterogeneous fracture simulation, and the study proved that the DEM could provide a valid understanding of the fracture behavior of materials so as to be used to diminish the need for numerous laboratory tests [13]. Peng et al. adopted Python language and DEM to generate irregular particles and establish a three-dimensional (3D) discrete element model of asphalt surface to study the mechanical response under different working conditions [14]. However, due to the limitation of obtaining the raw material parameters, most of the existing studies depended on lab-fabricated specimens and were limited to discrete element simulation of the fractures of the core samples. Moreover, it is unclear whether the calibration of the mesoscopic parameters for the simulation through the lab tests can be effectively applied to the limited number of core samples.

The objective of this study is to propose a reliable method to evaluate the fatigue behavior of the asphalt field cores based on discrete element modeling and to conduct mesoscopic contact parameter calibration through lab tests, including the uniaxial compression dynamic modulus test, SCB test, and SCB fatigue test, which can enhance the utilization efficiency of core samples and provide a reliable representation of the fatigue behavior of the core samples.

This paper is organized as follows. The following section presents the test samples and lab tests conducted to determine the mesoscopic contact parameters in the discrete element modeling. The next section describes the establishment of the virtual specimen using CT scanning of the asphalt field cores and image processing technologies, as well as mesoscopic contact parameter calibration and virtual fatigue tests using discrete element modeling. The fatigue life obtained from the lab test and virtual fatigue test is compared, and the virtual test results of force chains, crack evolution, and displacement are discussed in the following section. The final section summarizes the findings of this study.

2. Laboratory Test

The field cores were drilled from the in-service asphalt pavement and used to conduct laboratory tests, including the uniaxial compression dynamic modulus test, SCB test, and SCB fatigue test, to evaluate the material properties of the field cores. The lab test results can be used to calibrate the mesoscopic parameters required in the discrete element modeling.

2.1. Asphalt Field Core

The asphalt field cores used in this study were taken from the Hubei sections of the G4 Beijing–Hong Kong–Macao Expressway, which opened to traffic in 2002. The Hubei sections have an asphalt surface course with a thickness of 16 cm in total. Even though the Hubei sections have been in service for over 20 years, pavement rehabilitation has been undertaken several times, and the originally designed structure of the asphalt surface course has still been chosen. The asphalt surface course are composed of three asphalt layers, i.e., the asphalt surface layer, asphalt middle layer, and asphalt bottom layer, as detailed in Table 1. The asphalt mixture composition was sourced from the original design documents. In the asphalt surface layer, basalt was used, while limestone was used for the asphalt middle and bottom layers. The modified asphalt binder with anti-stripping additives was used, and asphalt content in the mixtures ranged between 3% and 4%.

Table 1. Structure of the asphalt surface course.

No.	Directions	Materials	Thickness (cm)
1	Asphalt surface layer	SUP-12.5	4
2	Asphalt middle layer	AC-20I	6
3	Asphalt bottom layer	AC-20S	6

The field cores were drilled as cylinders near the wheel path. The cylinder of the field cores is 150 mm in diameter and 30 cm in thickness. The drilled cylinders were first cut into slices 50 mm in thickness and then cut into semi-cylinders as the SCB test specimen in the laboratory. Figure 1 shows the process of the drilled field cores into the SCB test specimen.

Figure 1. Core sample processing procedure.

2.2. Uniaxial Compression Dynamic Modulus Tests

The uniaxial compression dynamic modulus test was conducted by the multifunctional test system (MTS), as shown in Figure 2. This test was conducted by following the Chinese specification JTG E20-2011 at the temperatures of 5 °C, 20 °C, 35 °C, and 50 °C and frequencies of 0.1 Hz, 0.5 Hz, 1 Hz, 5 Hz, 10 Hz, and 25 Hz [15]. From low to high temperature and from high to low frequency, the sinusoidal load was applied to the specimen under the condition of no side limit, and the dynamic modulus and phase angle of each structural layer of the asphalt mixture were calculated according to the obtained stress–strain data and hysteresis time to quantify the linear viscoelastic mechanical properties of the asphalt mixture. The test results can be used to obtain the parameters of the viscoelastic contact model (e.g., Burgers model) in discrete element simulation.

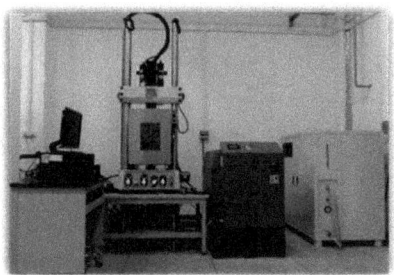

Figure 2. Multifunctional test system.

2.3. SCB Test

The semi-circular bending (SCB) test was conducted by the MTS machine and followed the control mode of the constant displacement rate loading according to the method of TP105-13 [16]. The test was conducted at a constant loading rate of 2 mm/min and a temperature of 20 °C. The dimensions of the semi-circular specimen are 150 mm in diameter, 75 mm in height, and 40 mm in thickness, which align with the thickness of the asphalt surface layer (4 cm). The notch was placed 10 mm in length and 1 mm in width at the center of the specimen to ensure the occurrence of crack initiation and propagation at the notch tip during the test. The test specimens were positioned on two rollers with a distance of 120 mm, which is 0.8 times the diameter of the specimen. Figure 3 shows the dimension of the semi-circular specimen. The SCB test results were obtained to determine the parameters of the parallel bonding model in the discrete element modeling.

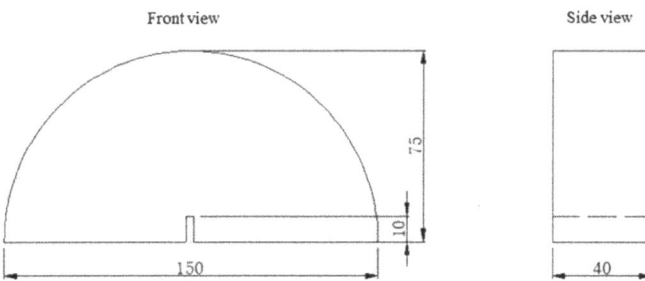

Figure 3. Size of SCB specimen (unit: mm).

2.4. SCB Fatigue Test

The SCB fatigue test was conducted in the loading mode of the stress control with a temperature of 20 °C. The loading mode of the semi-sinusoidal wave at the frequency of 10 Hz was adopted. Considering the cracking self-healing properties of the asphalt mixture, continuous loading without the intermittent time was selected. Since the stress ratio can greatly affect the fatigue life of the asphalt mixture, this study selected the stress ratio of

0.4, 0.5, and 0.6 based on the pre-test results to ensure that fatigue life could fall into the range between thousands and tens of thousands of cycles.

3. Discrete Element Modeling

3.1. The Establishment of Mesoscopic Virtual Specimen

The virtual specimen was established by CT scanning and the Digital Image Process (DIP) method. The CT scanning of the asphalt field cores was first conducted, and the scanned image of the field cores was processed by using Image-Pro Plus software (version 7.0). It was found that the air void content of the field cores is higher at both ends compared to the middle, ranging between 3% and 5%. Hence, the median value of 4% was set as the air void content of the virtual specimen for discrete element modeling. The CT-scanned images were proportionally cut by using Photoshop to ensure the geometry of the virtual specimen corresponded to that of the actual specimen. The distribution of aggregate and mortar in the actual specimen was further obtained by using the DIP method, which includes image enhancement, image noise reduction, threshold segmentation, feature extraction, and object recognition. The edge extraction was used to identify the edges of coarse aggregate [17]. The extracted outlines were converted into DXF file format, as shown in Figure 4.

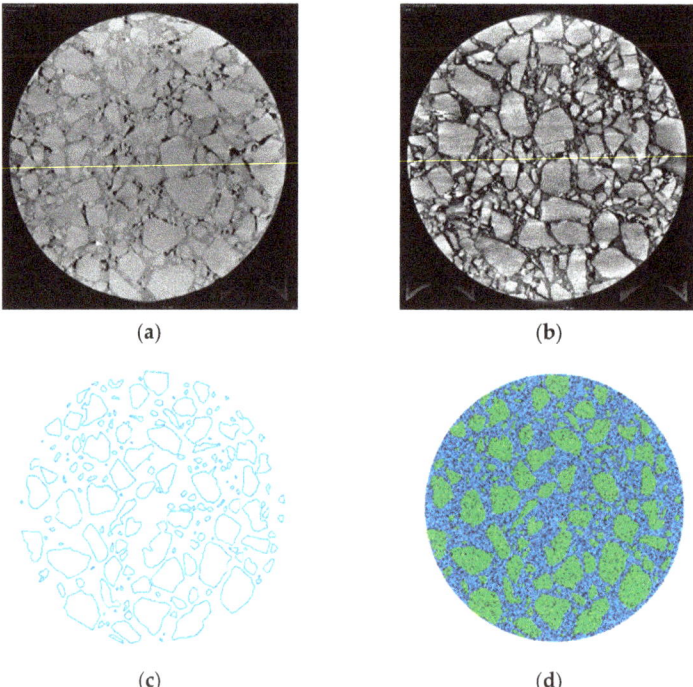

Figure 4. Image processing process. (**a**) Original image. The grayscale image shows the original CT scan of the asphalt core sample. (**b**) The enhanced image. (**c**) Image edge recognition. The blue lines delineate the boundaries of coarse aggregates. (**d**) Complete geometric model. The green area represents the aggregates and blue area represents the asphalt mortar.

Based on the CT scanning, the images at the center of the asphalt field cores with distinct structures of asphalt layers were selected and then processed. The distribution of coarse aggregates in the field cores was obtained using Photoshop software (version 21.2.11). To simulate the prefabricated notch, the particles at the center bottom of the virtual specimen were deleted. The wall was added at the top of the virtual specimen as a loading plate, and circular rigid walls were added at the bottom as constraints. The loading mode applied on the virtual specimen was consistent with that used in the lab test. The virtual specimens for the asphalt surface, middle, and bottom layers are shown in Figure 5.

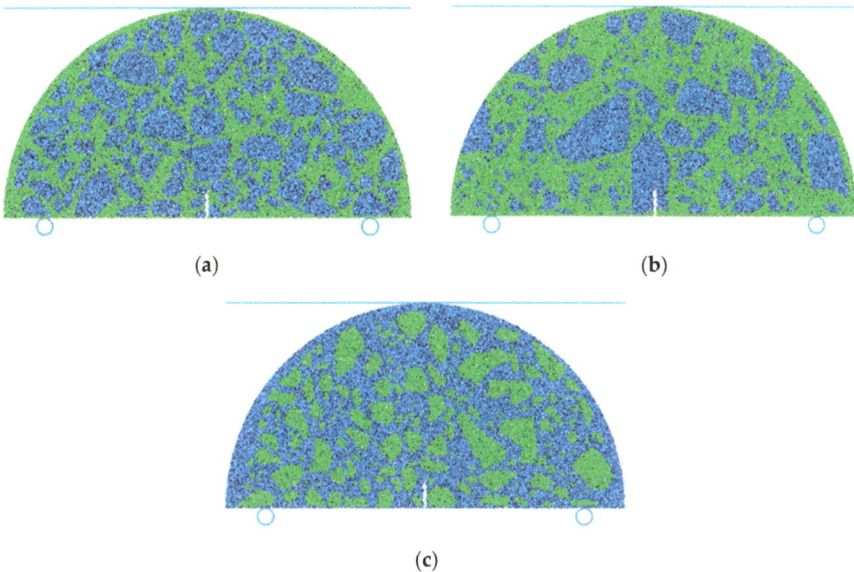

Figure 5. Virtual specimen for virtual SCB fatigue test (**a**) at the asphalt surface layer, (**b**) at the asphalt middle layer, and (**c**) at the asphalt bottom layer. The blue circles at the bottom represent the circular rigid walls as constraints. The blue line at the top represents the wall as a loading plate.

3.2. Contact Models and Parameter Calibration

The discrete element modeling was performed using the PFC (Particle Flow Code) software (version 5.0). The PFC software has four basic contact models, including the stiffness model, slipping model, parallel bonding model, and viscoelastic contact model. To simplify the contact between the asphalt mixture, the different contact models between the particle elements were selected, as shown in Table 2.

Table 2. Selection of contact model between particle elements.

Particle Elements	Contact Model Selection
Between coarse aggregate particle units	Linear stiffness model
Between coarse aggregate and asphalt mortar	Burgers model + parallel connection model
Between asphalt mortar	Burgers model + parallel connection model
Between the particle unit and the wall	Linear stiffness model

In this study, the viscoelastic behavior of the asphalt mixture is characterized using the Burgers model, which combines the Maxwell model and the Kelvin model, which act in series but in normal and shear directions, as depicted in Figure 6. The Maxwell and Kelvin models both include a spring with stiffness parameters and a dashpot with viscosity parameters. In the Maxwell model, these components are connected in series, while in the Kelvin model, they are connected in parallel. Hence, the Burgers model is divided into normal and shear directions at the mesoscopic level, resulting in eight parameters of the contact model between aggregate and asphalt mortar. These parameters can be calculated by the following equations.

$$C_{mn} = \eta_1 L K_{mn} = E_1 L C_{kn} = \eta_2 L K_{kn} = E_2 L \qquad (1)$$

$$K_{ms} = \frac{E_1 L}{2(1+\nu)} C_{ms} = \frac{\eta_1 L}{2(1+\nu)} K_{ks} = \frac{E_2 L}{2(1+\nu)} C_{ks} = \frac{\eta_2 L}{2(1+\nu)} \qquad (2)$$

where K_{mn} and C_{mn} are the stiffness and viscosity parameters of the Maxwell model, and K_{kn} and C_{kn} are the stiffness and viscosity parameters of the Kelvin model in the normal direction; since Burgers model can sustain tensile stress, K_{ms} and C_{ms} are the stiffness and viscosity parameters of the Maxwell model, and K_{ks} and C_{ks} are the stiffness and viscosity parameter of the Kelvin model in the shear direction. These parameters of stiffness and viscosity in the Burgers model determine the creep behavior of the virtual specimen. L is the length of the two-contacting discrete particle element for aggregate, i.e., the sum of two contacting particle radii; E_1, η_1, E_2, and η_2 are the macro-parameters of the Burgers model.

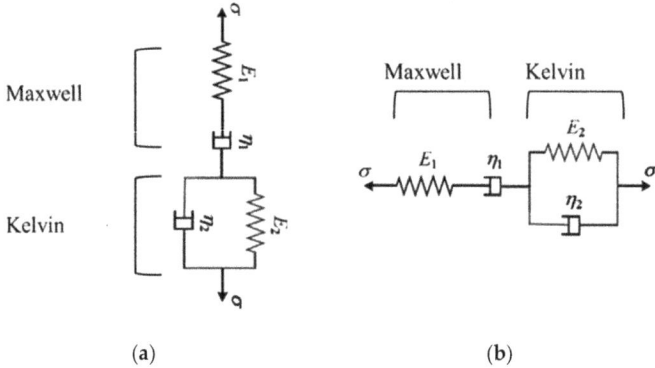

Figure 6. Schematic diagram of Burgers model. (**a**) Normal. (**b**) Shear.

The macro-parameters of the Burgers model (E_1, η_1, E_2, η_2) were converted from the uniaxial compression dynamic modulus test results of the field core specimen, as shown in Table 3. The macro-parameters of the Burgers model can be calculated by the following equations:

$$E_1 = [|E^*|]_{\omega=\omega_{max}} \qquad (3)$$

$$\eta_1 = \left[\frac{|E^*|}{\omega}\right]_{\omega=\omega_{min}} \qquad (4)$$

$$\frac{1}{|E^*|} = \sqrt{\frac{1}{E_1^2} + \frac{1}{\eta_1^2 \omega^2} + \frac{1 + 2(E_2/E_1 + \eta_2/\eta_1)}{E_2^2 + \eta_2^2 \omega^2}} \qquad (5)$$

$$\tan\varphi = \frac{E_1 [E_2^2 + \eta_2(\eta_1 + \eta_2)\omega^2]}{\eta_1 \omega (E_2^2 + E_1 E_2 + \eta_2^2 \omega^2)} \qquad (6)$$

where ω_{max} and ω_{min} are, respectively, the maximum and minimum values of angular frequency in the laboratory test. Using 20 °C as an example, the four macro-parameters of the asphalt surface, middle, and bottom layers of the pavement core sample are shown in Table 4.

Table 3. Dynamic modulus test results of each structural layer of the core sample at 20 °C.

Frequency	Asphalt Surface Layer		Asphalt Middle Layer		Asphalt Bottom Layer	
	Dynamic Modulus, MPa	Phase Angle	Dynamic Modulus, MPa	Phase Angle	Dynamic Modulus, MPa	Phase Angle
25 Hz	16,174	12.58	12,128	15.33	12,669	12.87
10 Hz	13,199	13.51	10,478	16.89	11,551	14.45
5 Hz	11,503	14.72	9152	18.52	10,172	15.64
1 Hz	8214	19.83	6199	24.15	7030	20.41
0.5 Hz	6871	21.90	5119	26.72	5855	22.60
0.1 Hz	4567	23.23	3177	30.66	3844	25.58

Table 4. Macroscopic parameters of the contact Burgers model at 20 °C.

Model Parameter	Asphalt Surface Layer	Asphalt Middle Layer	Asphalt Bottom Layer
E_1/GPa	14.96	12.46	12.88
E_2/GPa	10.72	7.08	8.98
η_1/(GPa·s)	23.89	11.93	17.70
η_2/(GPa·s)	1.22	0.82	0.99

Based on the conversion from the macro-parameters (E_1, η_1, E_2, η_2) to the mesoscopic parameters using Equations (1) and (2), the mesoscopic parameters of the contact models were calculated at 20 °C, as shown in Table 5.

Table 5. Calibration results of meso-contact parameters in Burgers model at 20 °C.

Meso-Structure Parameter	Asphalt Surface Layer	Asphalt Middle Layer	Asphalt Bottom Layer
K_{mn}	1.70×10^7	1.25×10^7	1.50×10^7
C_{mn}	2.10×10^7	1.10×10^7	1.50×10^7
K_{kn}	9.00×10^6	8.00×10^6	1.20×10^7
C_{kn}	1.45×10^6	9.50×10^5	1.10×10^6
K_{ms}	6.00×10^6	5.00×10^6	5.20×10^6
C_{ms}	1.00×10^7	1.20×10^7	7.00×10^6
K_{ks}	4.20×10^6	2.80×10^6	3.60×10^6
C_{ks}	5.00×10^5	3.50×10^5	4.00×10^5

3.3. Contact Parameter Calibration of Parallel Bonding Model

The parameters of the parallel bonding model were determined based on the SCB test results of the asphalt field cores and discrete element simulation. Firstly, the initial values and range of the parameters were selected based on the previous studies, as shown in Table 6. The parameters of the parallel bonding model were further determined by the trial-and-error approach based on the effect of parameters on the stress–strain curve in the discrete element simulation. Table 7 presents the contact parameters of the parallel bonding model calibrated and validated through the SCB test.

Table 6. Initial values of main parameters of the parallel bonding model.

Parameter Label	Parameter	Unit	Initial Value	Value Range
pb_emod	Parallel bond modulus	Pa	6.00×10^5	$6.00 \times 10^4 \sim 8.00 \times 10^6$
pb_ten	Strength of extension	Pa	8.00×10^5	$8.00 \times 10^4 \sim 8.00 \times 10^6$
pb_coh	Bonding force	Pa	4.00×10^5	$4.00 \times 10^4 \sim 4.00 \times 10^6$
pb_krat	Stiffness ratio	/	2.0	1~3
pb_fa	Internal friction angle	°	35	–
pb_rad	Parallel bond radius	mm	0.5	–

25

Table 7. Contact parameter values of parallel bonding model.

Parameter Label	Asphalt Surface Layer	Asphalt Middle Layer	Asphalt Bottom Layer
pb_emod	3.20×10^5	7.40×10^5	3.20×10^6
pb_ten	8.40×10^5	2.20×10^6	3.80×10^6
pb_coh	1.00×10^6	7.50×10^5	6.50×10^5
pb_krat	2.0	1.7	1.5
pb_fa	35	35	35
pb_rad	0.5	0.5	0.5

3.4. Virtual Fatigue Test

The virtual fatigue test was performed to simulate the SCB fatigue test with the loading mode consistent with that in the laboratory SCB fatigue test. In the virtual fatigue test, the fatigue behavior of the asphalt mixture is characterized by the deterioration at contacts within the asphalt mortar and between the aggregate and asphalt mortar. Hence, the virtual fatigue test of the asphalt mixture was conducted to investigate the deterioration of the mechanical properties of the parallel bonding model in the discrete element simulation. The micro-mechanical fatigue damage model was determined by the following equation [18,19].

$$\frac{d\overline{D}}{dt} = \beta_1 \left(\frac{\sigma}{\sigma_c}\right)^{\alpha_1} t^{\beta_2 (\frac{\sigma}{\sigma_c})^{\alpha_2}} \quad (7)$$

where D is the bonding diameter at contacts; t is the loading time; σ_c is the ultimate tensile strength; σ is the tensile stress between particle elements; and β_1, β_2, α_1, and α_2 are the coefficients of the fatigue damage model. The coefficients were initialized based on the existing studies [18,19] and then calibrated through the iterative virtual fatigue test to correspond with the SCB fatigue test results. Table 8 presents the calibrated coefficients of the fatigue damage model.

Table 8. Coefficients of the fatigue damage model.

Structural Layer	α_1	α_2	β_1	β_2
Asphalt surface layer	−1.420	0.053	2.20×10^6	−1.066
Asphalt middle layer	−1.350	0.050	2.50×10^6	−1.710
Asphalt bottom layer	−1.435	0.042	3.10×10^6	−1.790

4. Results and Discussion

4.1. Comparison between Laboratory and Virtual Fatigue Test

Table 9 presents the comparison of the results of fatigue life between the laboratory test and discrete element modeling at different stress ratios of 0.4, 0.5, and 0.6 for each asphalt structural layer. It was found that fatigue life decreases with the increase in the stress ratio and the depth of asphalt layers. The results from both lab and virtual tests indicate that the fatigue life of the asphalt surface layer was roughly twice as long as that of the asphalt bottom layer. The comparison, as shown in Table 9, also shows that the simulated fatigue life was shorter than the fatigue life from the lab test, which may result from the self-healing properties of the asphalt mortar. The variation in distribution and air void content may contribute to the differences in fatigue life between laboratory and virtual fatigue tests. However, the trend of the result of the virtual fatigue test was consistent with that of the lab test, and the error was below 20%. Compared to the benchmark established in the literature [12], the error range is considered acceptable. The comparison result indicates the virtual fatigue test simulated by discrete element modeling is reliable and acceptable.

Table 9. Comparison of fatigue life between lab and virtual fatigue test.

Asphalt Structure Layer	Stress Ratio	Test Fatigue Life	Simulated Fatigue Life	Error (%)
Asphalt surface layer	0.6	4310	3587	16.77
	0.5	6867	5388	21.54
	0.4	24,478	19,566	18.47
Middle layer	0.6	2155	1816	15.73
	0.5	4458	3653	18.06
	0.4	12,135	9585	21.01
Bottom layer	0.6	1751	1950	11.36
	0.5	4274	3653	17.00
	0.4	10,097	8086	19.91

4.2. Force Chain Evolution Process

The force chain, formed by the interaction between contacting particles, was analyzed to study the stress distribution of the virtual specimen since the evolution of the force chain reflects the variation in the mechanical response of the virtual specimen during loading. The process of force chain at three stages, i.e., the early, middle, and final stages (i.e., the crack initiation, crack propagation, and failure), is illustrated in Figure 7. The blue and green denote compressive and tensile stress, respectively, and the line size of the force chain reflects the stress level between particle contacts. It is observed from Figure 7 that in the early stage, the compressive force chains are primarily located near the aggregate particles. In the middle stage, the compressive force chains were concentrated to the loading point at the top and two bottom supports. The tensile force chains mainly appear near the notch cracks at the bottom center of the virtual specimen. In the final stage, the virtual specimen eventually failed due to fracture damage. It was found that crack propagation is mainly caused by concentrated tensile stress during loading.

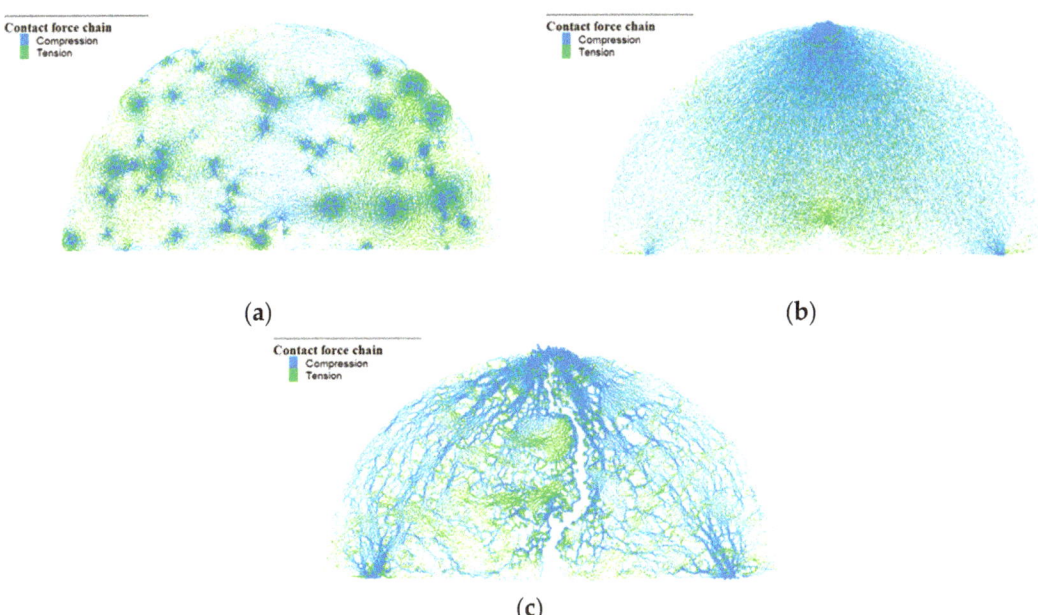

Figure 7. Evolution process of force chain: (a) early stage; (b) middle stage; (c) final stage.

Table 10 presents the distribution of the force chain under different stress ratios. We utilized the "Contact force chain" command in the PFC (Particle Flow Code) software to record the force chain evolution under loading. The term "proportion of the tensile force chain" in Table 10 refers to the percentage of the tensile force chains to all force chains. The variation in the proportion of the tensile/compressive force chains can reflect the mechanical response of the virtual specimen under loading, which can help to evaluate the fatigue behavior of the virtual specimen. It is observed that with the increase in the stress ratio, the tensile and compressive force chains increase slightly, and the proportion of the tensile force chains rises from 37.4% to 40.41%. The peak values of the force chains were observed as the tensile force chain of 5.18×10^4, 6.62×10^4, and 7.54×10^4 at the stress ratio of 0.4, 0.5, and 0.6, respectively. This indicates that crack propagation is closely associated with the growth of the tensile chains, which suggests that the crack propagation stage could consume a significant portion of the fatigue life since tensile force mainly increases in this stage.

Table 10. Distribution of force chains under different stress ratios.

Stress Ratio	Tensile Force Chain	Tensile Force Chain Proportion (%)	Compressive Force Chain	Compressive Force Chain Proportion (%)
0.4	11,843	37.40	19,819	62.60
0.5	13,536	40.22	20,121	59.78
0.6	13,688	40.41	20,185	59.59

4.3. Crack Evolution Process

Figure 8 shows the crack path of both virtual and lab specimens. It was found that the two crack paths are relatively consistent, and both of them first appear at the prefabricated notch and then gradually propagate to the top along the middle of the specimen. In the virtual test, the crack path turned close to a straight line, and the internal crack of the aggregate interface and asphalt mortar interface happened simultaneously. By contrast, in the lab test, the crack path seemed to wiggle and mostly grew along the aggregate interface.

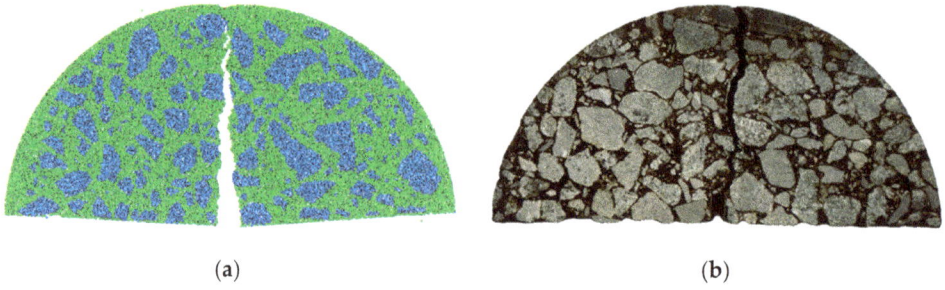

(a)   (b)

Figure 8. Crack propagation path: (**a**) virtual path. (**b**) test path.

Figure 9 shows the crack quantities and direction at three stages (i.e., the early stage, middle stage, and final stage) under loading. In Figure 9, it can be seen that in the early stage, cracks mainly grow vertically, and the direction of crack growth ranges between 90 and 110 degrees. At this stage, the cracks have not yet propagated upward at the prefabricated notch. In the middle stage, the crack growth deviates from the vertical direction and ranges from 60 to 100 degrees. In the final stage, the crack gradually propagates to the top of the specimen, and the direction of crack growth extends to a range between 50 and 130 degrees. The extension of the crack growth angle results in the formation of micro-crack branches.

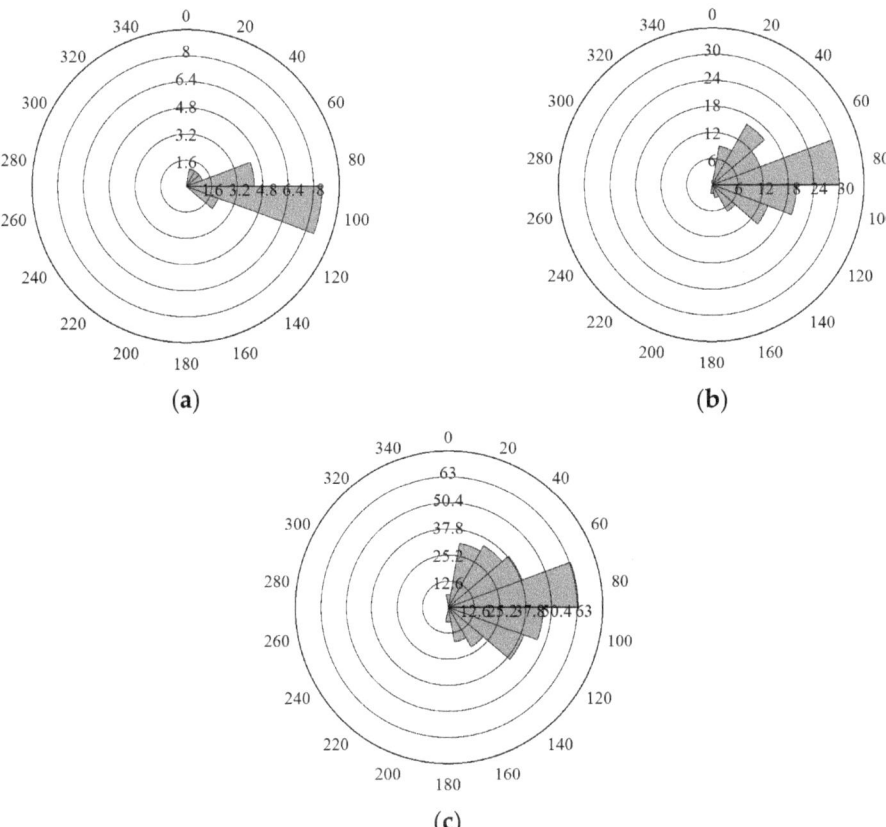

Figure 9. Crack quantities and directions in (**a**) early stage, (**b**) middle stage, and (**c**) final stage.

4.4. Displacement Evolution Process

The displacement field in the discrete element modeling is crucial to studying the movement of particle elements under loading and further evaluating fatigue behavior. Figure 10 shows the displacement field in the early, middle, and final stages of loading. It was found that the horizontal and vertical displacement fields of particles seem asymmetrical and close to zero, which indicates that the relative motion between particles is small. In the middle stage, the displacement between particles is symmetrically distributed with the central axis of the virtual specimen. In the horizontal direction (X direction), the relative displacement at the bottom near the prefabricated notch is the largest, about 0.25 mm. In the vertical direction (Y direction), the displacement at the loading head seems largest, about 0.43 mm. In the final stage, the displacement near the notch reaches about 2 mm in the horizontal direction. In the vertical direction, the displacement near the crack path appears consistent, yet the displacement field seems tortuous along the symmetry axis of the central axis due to the variation of the air void distribution. The vertical displacement near the loading head is about 2.2 mm, which shows a good agreement with the results of the SCB fatigue test.

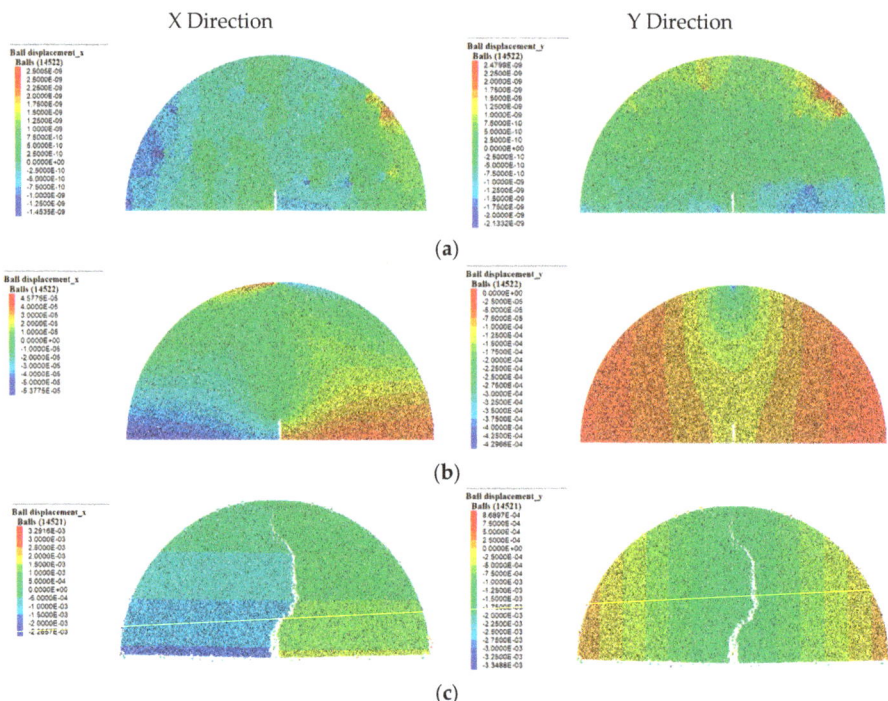

Figure 10. Displacement field of virtual specimen in (**a**) early stage, (**b**) middle stage, and (**c**) final stage.

5. Conclusions

In this study, the virtual SCB fatigue test was simulated by using discrete element modeling to evaluate the fatigue behavior of the asphalt field cores. The CT scan test was conducted to build the mesoscopic geometric model of the asphalt field cores. Additionally, the uniaxial compression dynamic modulus test and SCB test were performed to determine the parameters of the contact model in the virtual fatigue test. Based on the virtual SCB fatigue test, the displacement and contact forces, as well as crack growth, were analyzed. The main findings of this study can be drawn as follows.

(1) The evaluation methodology of fatigue behavior of the asphalt field cores based on the discrete element simulation was developed and can be used to enhance the effective usage of the field cores, which can help with the decision-making of pavement maintenance actions.

(2) The fatigue life simulated in the virtual fatigue test was consistent with that of the laboratory SCB fatigue test. The error between the simulated and test fatigue life was below 20%, which shows that the virtual fatigue test result is acceptable and reliable.

(3) It was found from the analysis of the force chain evolution process that concentrated tensile stress during loading can lead to crack initiation and propagation, ultimately resulting in material failure.

(4) The fatigue cracks in the asphalt mixture were observed as the three stages, i.e., crack initiation, crack propagation, and failure. It was found that the crack propagation stage consumes a significant portion of the fatigue life since tensile contact force mainly increases in this stage.

In this study, the discrete element modeling was restricted within 2D simulation due to the limited computational power. In future work, the 3D discrete element simulation will be performed to evaluate the fatigue behavior of the asphalt field cores, which could further improve the simulation accuracy. Additionally, future work will compare the asphalt field cores with the different aging times and investigate the difference in the fatigue behavior among them.

Author Contributions: Conceptualization, Y.C. and H.F.; methodology, Y.C., H.F., T.H., K.X. and Y.Z.; software, H.F.; validation, H.F.; writing—original draft preparation, Y.C., H.F. and M.X.; writing—review and editing, T.H.; funding acquisition, M.X., K.X. and Y.Z. All authors have read and agreed to the published version of the manuscript.

Funding: This work was financially supported by the Science and Technology Project of Jiangxi Provincial Department of Transportation of China (Project No. 2021C0008) and the Ganpo Talents Support Program High-level and High-skill Leading Talents Cultivation Project. This work was also supported by Supported by the Fundamental Research Funds for the Central University of China (WUT: 21302008).

Institutional Review Board Statement: Not applicable.

Informed Consent Statement: Not applicable.

Data Availability Statement: Data are contained within the article.

Conflicts of Interest: Authors Min Xiao, Haohao Feng, Kai Xiong and Yaoting Zhu were employed by the Jiangxi Provincial Communications Investment Group Co., Ltd., Project Construction Management, Zhejiang Communications Construction Group Co., Ltd. Design Institute Branch and Jiangxi Transportation Institute Co., Ltd. respectively. The remaining authors declare that the research was conducted in the absence of any commercial or financial relationships that could be construed as a potential conflict of interest.

References

1. Seo, Y.; Baek, C.; Kim, Y.R. Fatigue crack assessment of asphalt concrete pavement on a single span highway bridge subjected to a moving truck. *J. Test. Eval.* **2012**, *40*, 983–997. [CrossRef]
2. Zhang, Z.; Oeser, M. Residual strength model and cumulative damage characterization of asphalt mixture subjected to repeated loading. *Int. J. Fatigue* **2020**, *135*, 105534. [CrossRef]
3. Zhang, Y.; Gao, Y. Predicting crack growth in viscoelastic bitumen under a rotational shear fatigue load. *Road Mater. Pavement Des.* **2021**, *22*, 603–622. [CrossRef]
4. Luo, X.; Zhang, Y.; Lytton, R.L. Implementation of pseudo J-integral based Paris' law for fatigue cracking in asphalt mixtures and pavements. *Mater. Struct.* **2016**, *49*, 3713–3732. [CrossRef]
5. Zhang, Y.; Gu, F.; Birgisson, B.; Lytton, R.L. Viscoelasticplastic–fracture modeling of asphalt mixtures under monotonic and repeated loads. *Transp. Res. Rec.* **2017**, *2631*, 20–29. [CrossRef]
6. Barman, M.; Ghabchi, R.; Singh, D.; Zaman, M.; Commuri, S. An alternative analysis of indirect tensile test results for evaluating fatigue characteristics of asphalt mixes. *Constr. Build. Mater.* **2018**, *166*, 204–213. [CrossRef]
7. Luo, X.; Luo, R.; Lytton, R.L. Characterization of Asphalt Mixtures Using Controlled-Strain Repeated Direct Tension Test. *J. Mater. Civ. Eng.* **2013**, *25*, 194–207. [CrossRef]
8. Gu, F.; Luo, X.; Zhang, Y.; Lytton, R.L. Using overlay test to evaluate fracture properties of field-aged asphalt concrete. *Constr. Build. Mater.* **2015**, *101*, 1059–1068. [CrossRef]
9. Kim, M.; Mohammad, L.N.; Jordan, T.; Cooper, S.B. Fatigue performance of asphalt mixture containing recycled materials and warm-mix technologies under accelerated loading and four point bending beam test. *J. Clean. Prod.* **2018**, *192*, 656–664. [CrossRef]
10. Du, H.; Ni, F.; Ma, X. Crack resistance evaluation for In-service asphalt pavements by using SCB tests of layer-core samples. *J. Mater. Civ. Eng.* **2021**, *33*, 04020418. [CrossRef]
11. Luo, X.; Luo, R.; Lytton, R.L. Energy-based mechanistic approach to characterize crack growth of asphalt mixtures. *J. Mater. Civ. Eng.* **2013**, *25*, 1198–1208. [CrossRef]
12. Ma, T.; Zhang, Y.; Zhang, D.; Yan, J.; Ye, Q. Influences by air voids on fatigue life of asphalt mixture based on discrete element method. *Constr. Build. Mater.* **2016**, *126*, 785–799. [CrossRef]
13. Xue, B.; Pei, J.; Zhou, B.; Zhang, J.; Li, R.; Guo, F. Using random heterogeneous DEM model to simulate the SCB fracture behavior of asphalt concrete. *Constr. Build. Mater.* **2020**, *236*, 117580. [CrossRef]
14. Peng, Y.; Xia, S.; Xu, Y.-R.; Lu, X.-Y.; Li, Y.-W. Mechanical response of asphalt surfaces under moving traffic loads using 3D discrete element method. *J. Transp. Eng. Part B Pavements* **2022**, *148*, 04022006. [CrossRef]
15. *JTG E20-2011*; Standard Test Methods of Bitumen and Bituminous Mixtures for Highway Engineering. Ministry of Transportation: Beijing, China, 2011.

16. *TP105-13*; Standard Method of Test for Determining the Fracture Energy of Asphalt Mixtures Using the Semicircular Bend Geometry (SCB). AASHTO: Washington, DC, USA, 2013.
17. Chen, X.; Zhang, J.; Wang, X.; Zhu, Y.; Guo, H.; Xu, Y. State-of-the-art review on asphalt mixture distribution uniformity based on digital image processing technology. *IOP Conf. Ser. Earth Environ. Sci.* **2021**, *638*, 012074. [CrossRef]
18. Potyondy, D.O. Simulating stress corrosion with a bonded-particle model for rock. *Int. J. Rock Mech. Min. Sci.* **2007**, *44*, 677–691. [CrossRef]
19. Nguyen, N.H.; Bui, H.H.; Kodikara, J.; Arooran, S.; Darve, F. A discrete element modelling approach for fatigue damage growth in cemented materials. *Int. J. Plast.* **2019**, *112*, 68–88. [CrossRef]

Disclaimer/Publisher's Note: The statements, opinions and data contained in all publications are solely those of the individual author(s) and contributor(s) and not of MDPI and/or the editor(s). MDPI and/or the editor(s) disclaim responsibility for any injury to people or property resulting from any ideas, methods, instructions or products referred to in the content.

Article

Effect of Basalt Fiber Diameter on the Properties of Asphalt Mastic and Asphalt Mixture

Bo Li [1], Minghao Liu [1], Aihong Kang [1,2,*], Yao Zhang [1] and Zhetao Zheng [1]

[1] College of Civil Science and Engineering, Yangzhou University, Yangzhou 225100, China; libo@yzu.edu.cn (B.L.); lmhanksuper@163.com (M.L.); yaozhang@yzu.edu.cn (Y.Z.); zzt451873798@163.com (Z.Z.)

[2] Research Center for Basalt Fiber Composite Construction Materials, Yangzhou 225127, China

* Correspondence: ahkang@yzu.edu.cn; Tel.: +86-0514-8797-9418

Abstract: In this study, basalt fiber having two types of diameters (16 μm and 25 μm) was selected and added to asphalt mastic and asphalt mixtures using different fiber proportions. The influences of fiber diameters and proportions on the properties of asphalt mastic and mixtures were studied. The adhesion behavior of the fiber-asphalt mastic (FAM) interface was evaluated by a monofilament pullout test, and the rheological properties of FAM were evaluated by temperature sweep, linear amplitude sweep, and bending beam rheological tests. In addition, the high-temperature stability, intermediate and low-temperature cracking resistance, and water stability of fiber-modified mixtures were studied by wheel tracking, ideal cracking, a low-temperature bending beam, and a water-immersed Marshall test. The results showed that the interface adhesion behavior between 16 μm fiber and asphalt mastic was more likely in the fiber failure mode at both −12 °C and 25 °C. Adding basalt fiber can significantly improve the high-temperature and fatigue properties of asphalt mastics. Moreover, 16 μm fiber had a better modifying effect on asphalt mastic than 25 μm fiber. The same enhancement trend can be observed in asphalt mixtures. Basalt fibers with 16 μm diameters can improve the high-temperature performance of asphalt mixtures more significantly. In addition, 16 μm fiber could sharply enhance the cracking performance of the mixtures at intermediate and low temperatures, while the enhancing effect of 25 μm fiber on the mixture is insignificant, though both diameters of the fibers have a minor effect on the water stability.

Keywords: basalt fibers; diameter; asphalt mastic; monofilament pullout test; asphalt mixture

1. Introduction

The traditional asphalt mixture is a composite material that is mainly obtained by mixing aggregates, asphalt, and fillers. Due to its excellent performance, it is widely used in different types of road engineering fields [1–3]. The service function of asphalt pavement has drastically deteriorated due to the increasing traffic volume and the effects of climate change. And the pavement structure suffers distress such as cracking, aging, rutting, water seepage, etc. [1]. Therefore, searching for a more durable, safe, and environmentally friendly pavement material has become a research hotspot in global transportation construction [4–7], such as incorporating fibers and anti-rutting agents.

Since the 1990s, lignin fibers have been added to Stone Mastic Asphalt to increase its stability [8]. Due to their fluffy structure and rough surface, lignin fibers can better adsorb free asphalt and increase the stability of the asphalt mixture [9]. Other types of fibers, such as glass, polyester, and polyacrylonitrile fibers, are also frequently employed in asphalt mixtures and can vary in their ability to enhance the overall performance of asphalt mixtures [10–12]. As an environmentally friendly material [13], basalt fiber has been widely studied as an additive because of its unique advantages of high-temperature resistance, high strength, and stable chemical properties [11]. Two levels—asphalt mastic and asphalt mixture—can be used to examine the modification impact of basalt fiber on asphalt materials. Although asphalt mastic has a smaller proportion in the mixture compared with that of aggregates, its role in the asphalt mixture is irreplaceable. The mastic's composition, microstructure, and rheological properties directly affect the performance of asphalt mixtures [14–17]. Using fiber materials is an effective way to improve the properties of asphalt mastic, and fiber types, characteristics, and content also have other enhancement effects on asphalt mastic [18,19]. Xing et al. observed the impact of basalt fibers on the asphalt material's performance. The results indicated that basalt fibers can improve the toughness of asphalt [20]. Qin et al. investigated the enhancing effect of fiber types on asphalt mastic properties. They found that basalt fiber presented the best overall performance, and basalt fiber with a length of 6 mm had the best-enhancing effect on the cracking resistance [21].

Studies have shown that basalt fibers can significantly enhance the asphalt mixture's performance. Ramesh et al. determined that the RAP material of the AC mixture accounted for 30% by weight of the total aggregate and the chopped basalt fiber accounted for 6% by weight of the binder, which met the target AC mixture cracking standard, using the standard semicircular bending (SCB) test. [22]. Song et al. combined an indoor SCB test and digital image correlation method to analyze the fracture characteristics of DTC phase-change asphalt mix reinforced by basalt fiber. The research results revealed that basalt fiber improved the fracture toughness of SMA and increased its resistance to low-temperature cracking [23]. Lou et al. investigated the reinforcing ability of basalt fibers of various lengths on asphalt mixtures and found that mixed-length basalt fibers further improved mix crack resistance [24]. Pirmohammad et al. analyzed the impacts of the content and length of basalt fibers on the fracture toughness of the mixture through SCB experiments. They found that 0.3% basalt fiber with a length of 4 mm had the best-improving effect on fracture toughness [25]. Zhang et al. used the Hamburg rutting experiment to test an OGFC asphalt mixture with different contents of basalt fiber. They found that 0.15% basalt fiber had the best effect on mixture performance [26]. Lou et al. used uniaxial penetration tests to find that the permeability of the ultra-thin wearing course (UTWC) reached the maximum value of AC-10 when the fiber content of 6 mm fiber was 0.3% [27]. Luo et al. discussed how fiber type and dose affected the mixture's high-temperature stability at the Micro-surfacing, showing that the impact of fiber on the enhancement of rutting resistance at the Micro-surfacing was relatively small [28]. Lou et al. explored the impact of fiber length on various gradation mixtures. They found it significantly affected the hot mix asphalt's fatigue, intermediate, and low-temperature cracking resistance. In addition, they had no discernible impact on water stability or high-temperature deformation performance [29]. The asphalt mixture's performance at low temperatures first increased and then decreased as the amount of basalt fiber in the asphalt mixture increased. The results revealed that the asphalt mixture performance at low temperatures was best at the particular amount of basalt fiber [30]. This fiber agglomeration phenomenon in the mixture will have detrimental effects. To increase the mixture's water stability, the optimal amount and length of fibers can create a stable three-dimensional network structure [28]. This is primarily due to the fibers' even distribution throughout the asphalt mixture, which creates a three-dimensional network structure. It can decrease the fluidity of asphalt close to the fiber and then enhance the performance of asphalt mastic and asphalt mixture, resulting in prolonging the pavement's service life.

In summary, fiber's properties are the primary indexes causing the reinforcing effect on the asphalt mixture. The diameter of basalt fiber directly affects the amount of fiber monofilaments and the specific surface area of fiber at a given weight content. Hence, its influence on the mixture is also crucial. There are, however, few investigations on the enhancing ability of fiber diameter on the characteristics of asphalt mixtures and even fewer on various grades of asphalt mixtures. Due to the limitations of the production process, the most widely used diameters of basalt fiber are 7 μm, 16 μm, and 25 μm. The price of basalt fiber with a diameter of 7 μm is much higher than the other two, and it is mainly used in the textile industry. Meanwhile, basalt fiber with diameters of 16 μm and 25 μm is mainly used in road engineering. In addition, there might be the phenomenon of mixed diameters in the same batch of fibers produced by the same manufacturer. Subsequently, the suitability of using this mixed fiber type in engineering must be further explored. As the research objects, two distinct gradations (AC-13 and AC-20) and two different diameters of basalt fibers (16 μm and 25 μm) were chosen. And multiple experiments were conducted to investigate the impact of basalt fibers on the properties of fiber-asphalt mastic and fiber-asphalt mixtures. The findings can provide guidelines for future studies and technical applications of basalt fibers in asphalt mixtures.

2. Materials and Methods

2.1. Materials

2.1.1. Asphalt

Jiangsu Nantong Tongsha Asphalt Technology Co., Ltd. (Nantong, China) supplied us with SBS-modified asphalt (PG76-22), and its properties are listed in Table 1.

Table 1. Performance indicators and test results of SBS-modified asphalt.

Properties		Specification	Test Results	Test Method [31]
	Penetration (25 °C)/0.1 mm	40~70	56	T0604
	Penetration index PI	−0.4~1.0	0.6	T0604
	Softening point/°C	≮80	84.7	T0606
	Ductility (5 cm/min, 5 °C)/cm	≮30	46	T0605
	Segregation (softening spreads)/°C	≯2.5	1.5	T0661
	Resilient recovery (25 °C)/%	≮65	77	T0662
Residue after RTFOT	Quality changes/%	±1.0	−0.09	T0610
	Penetration ratio/%	≮60	88	T0604
	5 °C residual ductility/cm	≮20	39	T0605

2.1.2. Fiber

The basalt fibers used in this research are short-chopped and have two different diameters (16 μm and 25 μm). Jiangsu Tianlong Basalt Continuous Fiber Co., Ltd. (Yangzhou, China). supplied the fibers. Basalt fiber is solid, can resist corrosion and high temperatures, and has excellent performance indicators (shown in Table 2). Figure 1 displays the macro-micro diagram of the basalt fibers used.

Table 2. Performance indices and test results of basalt fiber.

Properties	Test Results	
	16 μm	25 μm
Breaking force/cN	32.69	38.61
Elongation/%	4.68	5.06
Modulus/GPa	88	79
Density/g·cm^{-3}	2.72	2.66
Water content/%	0.1	0.1

2.1.3. Filler

The mastic in the experiment is made by SBS asphalt and the mineral powder filler using a weight ratio of 1:1 [32], and the mineral powder in the experiment is selected from Zhenjiang high-capital limestone ore powder; the property indices of the mineral powder filler are shown in Table 3.

Figure 1. Macroscopic and microscopic diagrams of basalt fiber morphology; (**a**) 16 μm Basalt fiber; (**b**) 25 μm Basalt fiber; (**c**) Microscopic diagrams of 16 μm basalt fiber; (**d**) Microscopic diagrams of 25 μm basalt fiber.

Table 3. Technical indicators of limestone ore powder.

Properties	Specification	Test Results	Test Method [33]
Water content/%	≤1.0	0.43	Drying method
Relative density	≥2.50	2.708	T0352
Hydrophilic coefficient	<1	0.66	T0353
Particle size range: <0.6 mm	100	100	T0351
Particle size range: <0.15 mm	90~100	93.4	T0351
Particle size range: <0.075 mm	75~100	81.6	T0351

2.1.4. Aggregate and Gradation

Limestone and basalt aggregates were used in this investigation. Table 4 displays the findings for determining apparent density and gross bulk density. The fiber-reinforced asphalt mixture is graded as AC-13 and AC-20, and the data for this gradation is displayed in Tables 5 and 6.

Table 4. Test results of aggregate density.

Type		Apparent Relative Density (g/cm³)	Relative Density of Gross Volume (g/cm³)	Asphalt Mixture
Limestone	1#	2.785	2.731	AC-13
	2#	2.781	2.722	
Basalt	3#	2.900	2.850	
	4#	2.861	2.796	
Limestone	1#	2.718	2.686	AC-20
	2#	2.779	2.731	
Basalt	3#	2.776	2.719	
	4#	2.855	2.791	

Table 5. AC-13 gradation.

Screen Aggregate	Percentage of Passing through a Square Hole Sieve (mm)/%									
	16	13.2	9.5	4.75	2.36	1.18	0.6	0.3	0.15	0.075
1#	100	91.1	33.5	0.2	0.2	0.2	0.2	0.2	0.2	0.2
2#	100	98.3	83.5	4	0.1	0.1	0.1	0.1	0.1	0.1
3#	100	100	100	92.9	6.2	2.3	2	1.9	1.8	1.6
4#	100	100	100	97.7	66.9	35.5	21.1	9.9	5.5	3.9

Table 6. AC-20 gradation.

Screen Aggregate	Percentage of Passing through a Square Hole Sieve (mm)/%										
	26.5	19	16	13.2	9.5	4.75	2.36	1.18	0.6	0.3	0.15	0.075
1#	100	31.9	12.8	1.9	0.5	0.5	0.5	0.4	0.4	0.3	0.2	0.1
2#	100	100	100	91.1	33.5	0.2	0.2	0.2	0.2	0.2	0.2	0.2
3#	100	100	100	98.3	83.5	4	0.1	0.1	0.1	0.1	0.1	0.1
4#	100	100	100	100	100	97.7	66.9	35.5	21.1	9.9	5.5	3.9

2.2. Specimen Preparation

Before the experiment, SBS-modified asphalt, mineral powder, and basalt fiber were heated in the oven at 175 °C for 2 h to make the asphalt flow and to remove the water in the mineral powder and fiber. Based on the previous research, the fiber content in asphalt mastic and asphalt mixture was 5% and 0.3% [32], respectively.

First, the weight ratio of mineral powder and asphalt was 1.0. Then, it was mixed with asphalt mastic at 600 r/min until the mineral powder was evenly distributed in the asphalt mastic. Finally, 5% fiber to the total weight of asphalt mastic was added to the material three times, and it was stirred with a stirring bar until the fiber was completely dispersed in the asphalt.

As to the preparation of asphalt mixture samples, aggregate, fiber, mineral powder, and asphalt were heated at 175 °C. The aggregate was then mixed with fiber and stirred for 90 s until the fiber was evenly dispersed between the aggregates. Then, asphalt and aggregates were stirred evenly, and finally, the mineral powder was added for stirring to obtain the fiber asphalt mixture. Six sets of samples were fabricated: samples without doped fibers (blank); only 16 μm basalt fiber specimens (16 μm); only 25 μm basalt fiber specimens (25 μm); and samples with mass ratios of 16 μm to 25 μm basalt fiber of 1:1 (1:1), 1:2 (1:2), and 2:1 (2:1), respectively.

2.3. Test Methods

In this study, the fiber-asphalt interface adhesion behavior was evaluated using the monofilament fiber-asphalt pullout instrument, which was self-developed. The experimental process is shown in Figure 2. As there is presently no approach for investigating asphalt mastic, the experimental standards for the fiber asphalt mastic in this study were the same as the asphalt experimental standards. Fiber asphalt mastic was tested with DSR and BBR instruments, while UTM-25 was used to test fiber asphalt mixtures.

2.3.1. Monofilament Fiber Pullout Test

In this study, the monofilament fiber-asphalt pullout instrument was used. The following were the preparation steps: (1) the monofilament fiber was first peeled off with metal tweezers; (2) the monofilament was passed through the soft rubber mold with a diameter of 2 cm in the round groove, and then the asphalt mastic was poured; (3) the excessive fiber monofilament at the bottom was removed after cooling; (4) the specimen was moved to the wire pullout die and poured to the required depth for the experiment, as shown in Figure 3; After the specimen was placed in the instrument for 2 h, the pull-out test was carried out, and the loading rate was set to be 1 mm/s. The maximum tensile load and fiber tensile displacement curves at −12 °C and 25 °C were tested. The calculation formulas are shown in Equations (1)–(3).

$$S = \pi \times d \times L \tag{1}$$

$$\tau_m = \frac{F_{max}}{S} \tag{2}$$

$$W_a = \int_0^{x_a} F d(x) \tag{3}$$

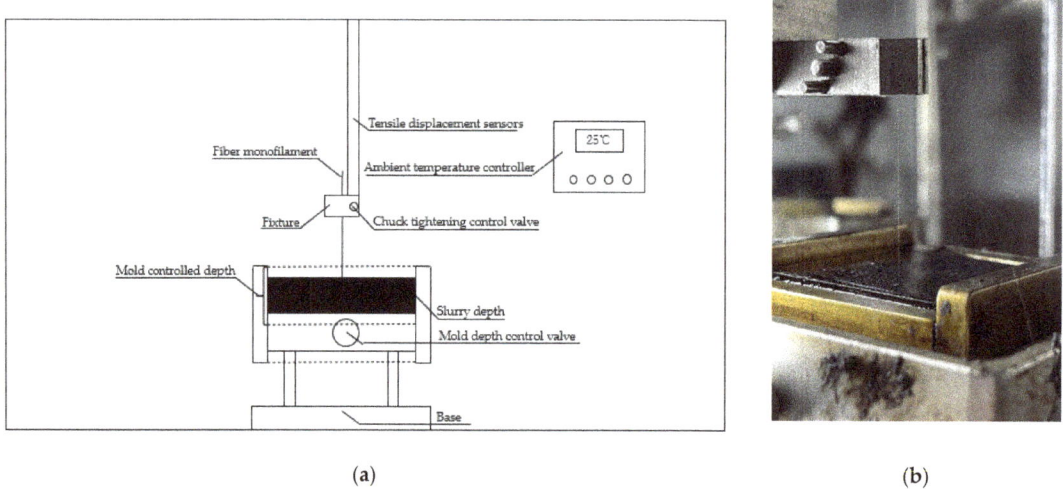

Figure 2. Fiber pullout test: (**a**) Schematic diagram of a monofilament-asphalt pullout device; (**b**) Actual test diagram.

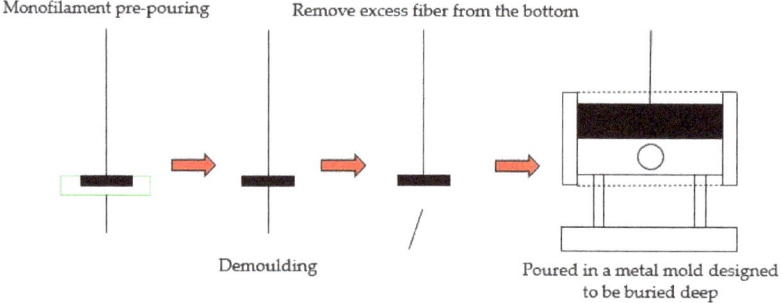

Figure 3. Schematic diagram of monofilament fiber-asphalt pullout specimen production.

Here F is the pulling force in the pulling period, cN; S is the cohesion area between asphalt and fiber, mm²; d is the diameter of the fiber monofilament, mm; x is the real-time displacement during the pullout process, mm; x_a is the displacement corresponding to the maximum pulling force, mm; L is the fiber embedding depth, mm; W_a is the energy generated before the fiber-asphalt interface begins to debond, which is the integral of the tensile force and the displacement of the fiber before the moment of complete debonding, 10^{-2}·mJ.

2.3.2. Temperature Sweep Test of Asphalt Mastic

The DSR temperature sweep test was conducted using the dynamic shear rheometer from TA Instruments according to the AASHTO T 315-12 specification [34], as shown in Figure 4 with the temperature set to 52 °C~82 °C. The fiber asphalt mastic was injected into an elastic rubber mold with a diameter of 25 mm and a thickness of 2 mm. After the sample was cooled, it was removed and set on the fixture, the excess material was tripped off, and the test data were collected at intervals of every 6 °C. The final data were the average value of three duplicates.

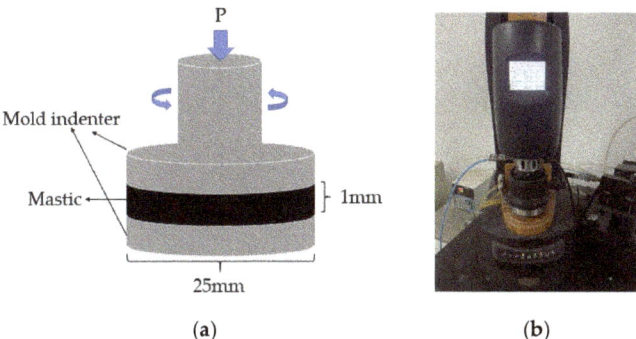

Figure 4. Temperature sweep test: (**a**) experimental model; (**b**) actual test diagram.

2.3.3. Linear Amplitude Sweep Test of Asphalt Mastic

The fatigue performance of asphalt was assessed using the linear amplitude sweep (LAS) technique following the AASHTO TP 101-12 guidelines [35]. The test uses three strain levels of 2.5%, 5.0%, and 10% at a temperature of 25 °C as shown in Figure 5. The diameter of the mold indenter is 8 mm, and the gap is 2 mm. The fatigue life N_f was calculated in Equation (5), and the coefficients A_{35} and B were calculated by Equations (6) and (7). The final data were averaged across three samples.

$$G'(\omega) = |G^*|(\omega) \times \cos \delta(\omega) \quad (4)$$

$$N_f = (\gamma_{\max})^{-B} \quad (5)$$

$$A_{35} = \frac{f(D_f)^k}{k(\pi I_D C_1 C_2)^\alpha} \quad (6)$$

$$B = 2\alpha \quad (7)$$

where γ_{max} is the maximum strain based on the applied load; f is the frequency of loading, which is 10 Hz; k is the coefficient and the formula is $k = 1 + (1 - C_2) \cdot \alpha$; C_1, C_2, D_f, α is calculated rheological parameter; I_D is the original value of $|G^*|$ from the 1.0 percent applied strain interval, (MPa); $|G^*|$ is the dynamic modulus in the test, (MPa); $\delta(\omega)$ is the phase angle of each frequency, (°); $G'(\omega)$ is the converted storage modulus.

2.3.4. Bending Beam Rheometer Test of Asphalt Mastic

The low-temperature characteristics of fiber-modified asphalt were investigated using a TE-BBR-F type bending beam rheometer according to the AASHTO TP 125-16 specification [36]. The test plot is shown in Figure 6. The test temperatures are from −6 °C to −24 °C, setting 6 °C as the interval. The sample was poured into a stainless-steel mold of 125 × 12.5 × 6.25 mm, and a water bath in ethanol was placed to maintain the experimental temperature before testing. The creep stiffness (S) and creep rate (m) of the beam sample were used to assess the fiber mastic's low-temperature performance. Formulas (8)–(10) display the calculation mechanism for m and S.

$$S(t) = \frac{1}{D} = \frac{\sigma}{\varepsilon} = \frac{Pl^3}{4bh^3 \delta(t)} \quad (8)$$

$$\log S'(t) = A + B[\log(t)] + C[\log(t)]^2 \tag{9}$$

$$|m(t)| = \frac{d[\log S'(t)]}{d[\log(t)]} = B + 2C[\log(t)] \tag{10}$$

where $S(t)$ is the real-time creep stiffness of the beam at 60 s (MPa); $m(t)$ is the real-time creep rate; P is a constant load of 100 g; L is the length of the asphalt mastic beam, (mm); b is the sample width of the asphalt mastic beam, (mm); h is the sample height of the asphalt mastic beam, (mm); $\delta(t)$ is the real-time deflection of the specimen at 60 s, $S'(t)$ is an estimate of the creep stiffness of flexural over time (MPa), and A, B, and C are the regression coefficients.

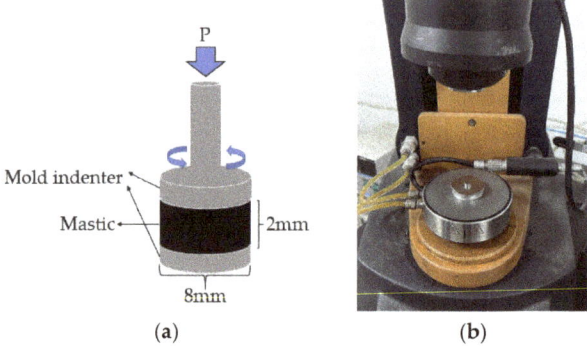

Figure 5. Linear amplitude sweep test: (**a**) experimental model; (**b**) mold indenter.

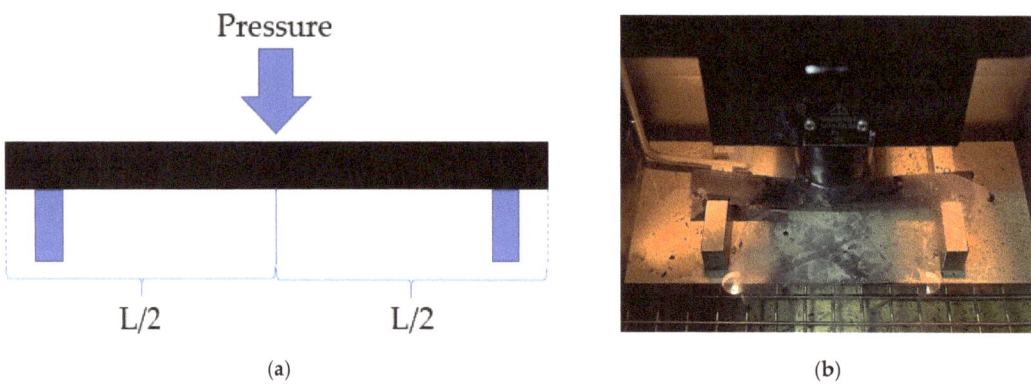

Figure 6. BBR test: (**a**) BBR experimental model; (**b**) actual test diagram.

2.3.5. High-Temperature Stability Test of the Mixture

The mixture was formed into a cube specimen with dimensions of 300 × 300 × 50 mm by the rutting plate roller described in detail in JTG E20-2011 [31]. After the specimen was formed, it was first put into the experimental cabin for heat preservation, and the experiment was carried out after the specimen reached 60 °C, as shown in Figure 7. The rutting resistance of asphalt mixtures is assessed using the dynamic stability index, calculated by Equation (11).

$$DS = \frac{(t_2 - t_1) \times N}{d_2 - d_1} \times C_1 \times C_2 \tag{11}$$

where DS is the dynamic stability of the asphalt mixture (times/mm); t_1 and t_2 are the times of the experiment, usually 45 min and 60 min; d_1 and d_2 are the related deformations of t_1 and t_2 (mm); C_1 is the correction coefficient of the experimental instrument, and it is set as 1; C_2 is the correction coefficient of the specimen, and it is set as 1; N is the rolling number of wheels per minute, which is 42 times/min.

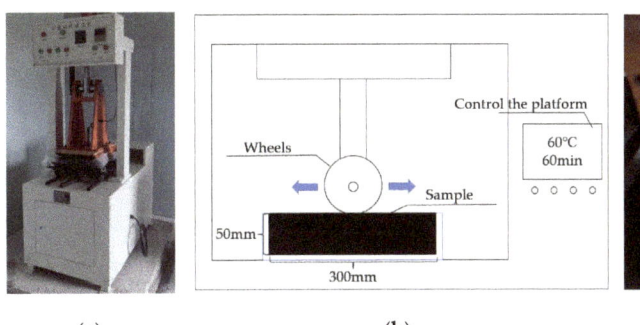

(a)   (b)   (c)

Figure 7. Rutting test: (**a**) QCX-4 Rut plate former; (**b**) Rutting experimental model; (**c**) Actual test diagram.

2.3.6. IDEAL-CT Test of Mixture

The intermediate-temperature cracking resistance of fiber asphalt mixtures was studied using the "ideal cracking test" (IDEAL-CT). The 150 mm diameter and 62 mm height specimens were fabricated by the Ict-Rotary compactor. The specimens and the indenter were insulated to 25 °C and tested, as shown in Figure 8c. The CT_{index} was obtained, and the crack initiation energy G_r (the work done in the area corresponding to points 0–3 in Figure 8b) and the total fracture energy G_f (the work done in the region corresponding to points 0-endpoint in Figure 8b over the cracking area) were adopted to assess the anti-crack ability at intermediate temperature. Equations (12) and (13) display the indicator computation.

$$CT_{index} = \frac{G_f}{|m_{75}|} \times \frac{l_{75}}{D} \tag{12}$$

$$|m_{75}| = \left| \frac{(p_{85} - p_{65})}{(l_{85} - l_{65})} \right| \tag{13}$$

where CT_{index} is the index of the cracking test; $|m_{75}|$ is the absolute value of the gradient of 75% of the post-peak section; G_f is the energy of fracture, (J/m^2); l_{75} is the displacement of 75% of the post-peak section, (mm); D is the diameter of the specimen, (mm).

2.3.7. Low-Temperature Bending Beam Test of the Mixture

The fiber asphalt mixture was made into a 30 × 35 × 250 mm prismatic beam specimen and tested after 2 h of insulation at −10 °C, as shown in Figure 9. Then, it was determined what the bending stiffness modulus S_B, maximum bending strain ε_B, and bending tensile strength R_B were. Equations (14)–(16) display the calculating formula.

$$R_B = \frac{3LP_B}{2bh^2} \tag{14}$$

$$\varepsilon_B = \frac{6hd}{L^2} \tag{15}$$

$$S_B = \frac{R_B}{\varepsilon_B} \tag{16}$$

where b is the actual width of the beam, (mm); h is the sample height of the beam, (mm); L is the length of the beam, (mm); P_B is the maximum load (N); d is the mid-span deflection (mm).

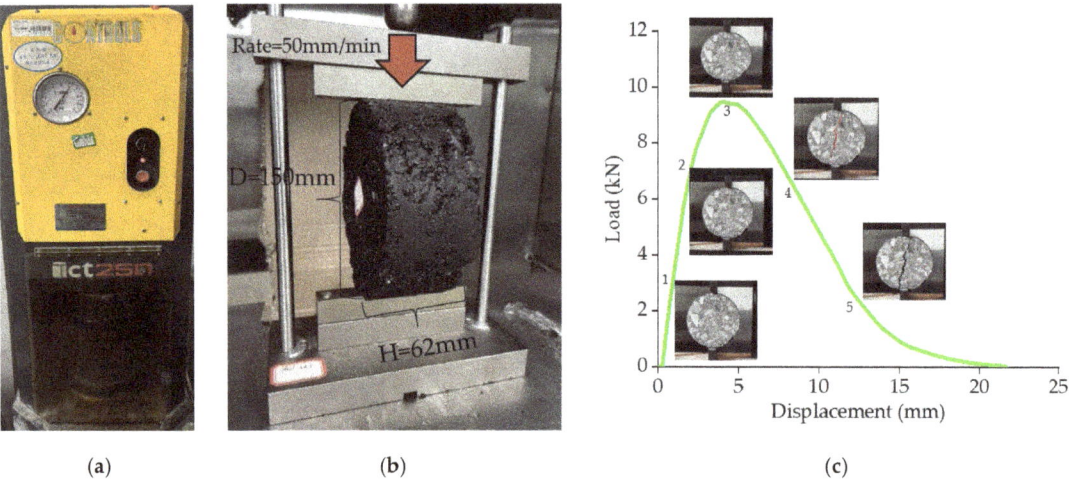

Figure 8. IDEAL-CT test: (**a**) Ict-Rotary compactor; (**b**) Actual test diagram; (**c**) IDEAL-CT test load-displacement curve.

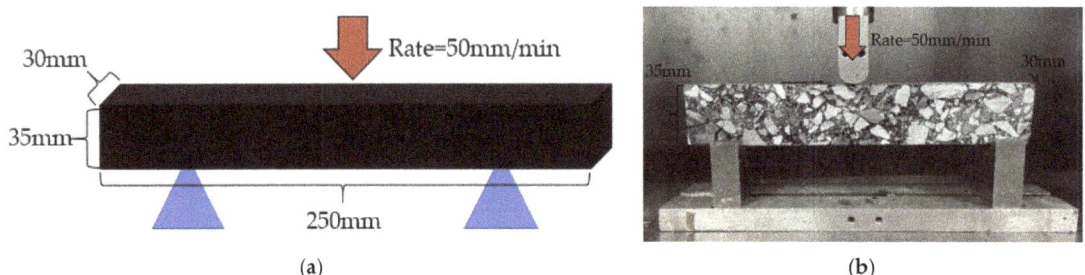

Figure 9. Bending beam test: (**a**) Model of test; (**b**) Actual test diagram.

2.3.8. Water Stability Test of the Mixture

Cylindrical asphalt mixture specimens fabricated by gyratory compaction were closer to those compacted in the field by a roller compactor [37]. Since the water-immersion Marshall test was conducted according to the specification of JTG E20-2011, the cylindrical specimen with a diameter of 101.6 mm and height of 63.5 mm was fabricated by the Marshall compactor, as shown in Figure 10. The specimens for the water immersion Marshall test were divided into two batches and immersed entirely in water at 60 °C; one batch was removed for the stability test after two hours, while the other batch was removed and tested after 48 h. The stability was expressed by Equation (17).

$$MS_0 = \frac{MS_1}{MS} \times 100 \tag{17}$$

where MS_0 is the residual stability of the specimen, (%); MS_1 is the Marshall stability of the specimen after 48 h of water bath, (kN); MS is the Marshall stability after 2 h of water bathing (kN).

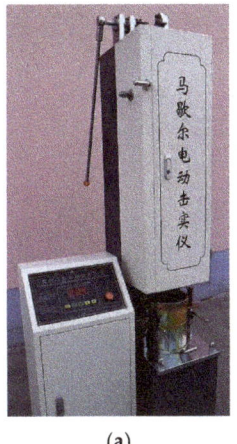 

(a) (b)

Figure 10. Water stability test: (**a**) Automatic Marshall compactor; (**b**) Schematic for water stability test.

3. Results and Discussion
3.1. Fiber-Asphalt Interface Adhesion Characteristics

This experiment aims to analyze the bonding characteristics between fiber monofilaments and asphalt mastic at $-12\ °C$ and $25\ °C$. The pulling force-displacement curves are shown in Figures 11 and 12.

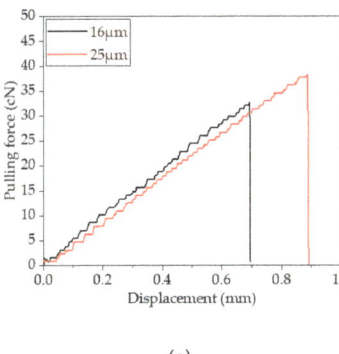 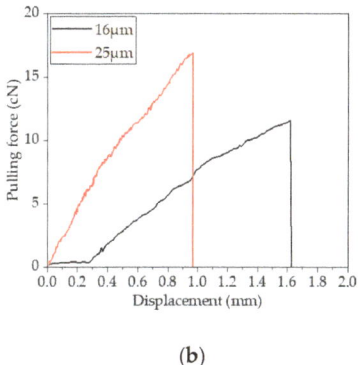

(a) (b)

Figure 11. Fiber-SBS-modified asphalt mastic monofilament pullout curve: (**a**) $-12\ °C$; (**b**) $25\ °C$.

Figure 11a demonstrates that, at $-12\ °C$, the pulling force of the two fibers increases roughly linearly as displacement increases. When the pulling force reached a certain level, it dropped sharply to 0, and the fiber was broken. This shows that at $-12\ °C$, the failure is more likely in fiber fracture mode. Experiments showed that the cohesion ability between asphalt mastic and fiber was very strong at low temperatures. Since the asphalt mastic hardened at lower temperatures and there was almost no viscous flow state, the pulling force mainly acted on the fibers, so the fibers broke. In contrast, the failure displacement of 25 μm basalt fibers was about 0.9 mm, which was greater than 0.7 mm for the 16 μm basalt fibers. This demonstrated that the maximum tensile force and elongation at the breaking point the fiber could withstand at low temperatures were slightly higher than those of 16 μm basalt fiber because 25 μm basalt fiber had a slightly larger diameter than 16 μm basalt fiber.

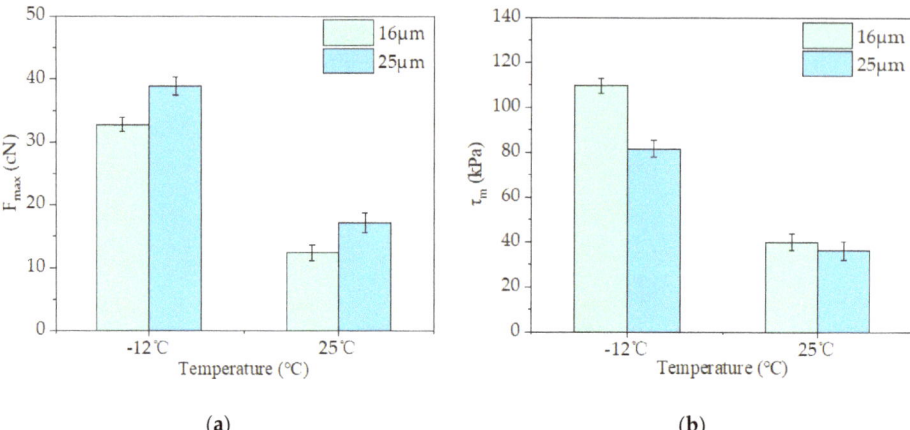

Figure 12. Fiber-asphalt interface adhesion parameters (the error bars illustrate the standard deviations of testing results): (a) F_{max}; (b) τ_m.

At 25 °C, the tensile force of two fibers with different diameters gradually increased with the increase in tensile displacement in the stretching process. The interface damage between them and asphalt mastic eventually displayed a violent fracture, showing that the fiber fracture mode was still present. As shown in Figure 11b, the displacement of the 16 μm basalt fiber when it was pulled off was close to 1.6 mm at the breaking point, which was greater than 0.97 mm of the 25 μm fiber. Because at 25 °C, asphalt mastic presented a viscoelastic state, the restraining effect on the fiber was smaller than that at −12 °C. Due to the large contact area between basalt fiber monofilament with a diameter of 25 μm and asphalt mastic, the adhesion area between asphalt mastic and the fiber was large [11]. Therefore, more pulling force was needed to make the same displacement; basalt fiber with a diameter of 25 μm had a much higher maximum pulling force than basalt fiber with a diameter of 16 μm.

Since the interface failure of two different diameters of fibers and asphalt mastic was manifested as fiber fracture mode at both −12 °C and 25 °C, there is no interface debonding work (W_a was 0).

Figure 12a shows that the pulling force F_{max} of both fibers was higher at −12 °C than at 25 °C, and the fiber's fracture strength parameters were in line with its size. The pulling force of 25 μm fiber in asphalt mastic was larger than that of 16 μm fiber at the same temperature. Figure 12b demonstrates that, nevertheless, the adhesion strength between fiber and asphalt mastic increases with the decreasing diameter of the fiber. At −12 °C, the interfacial adhesion strength of 16 μm basalt fibers increased by 34.18% compared to 25 μm fibers, and at 25 °C, this value increased only by 11.01%. This was due to the different diameters of fiber, and its cohesion area with the asphalt was also different. From Equation (2), it can be seen that in the case of F_{max}, it was not much different; the larger the S, the smaller the interface adhesion strength.

3.2. High-Temperature Performance Analysis of Fiber Asphalt Mastic

The relative amount of elastic deformation in asphalt could be characterized using the phase angle index, which increases as the phase angle decreases and makes the deformation caused by the load easier to recover. The complex modulus could be used to express the deformation resistance of a material, which is the ratio of the shear stress and strain. The rutting factor is usually used to characterize the rutting resistance of asphalt. The test results are illustrated in Figures 13–15.

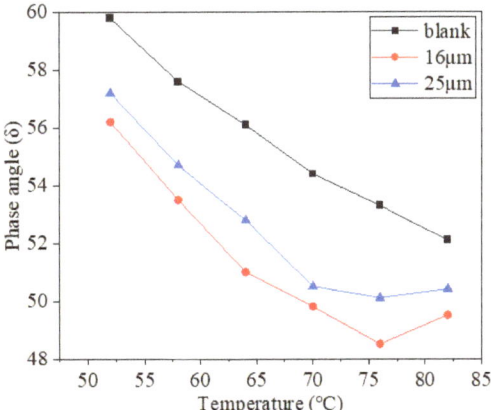

Figure 13. Test results for phase angles.

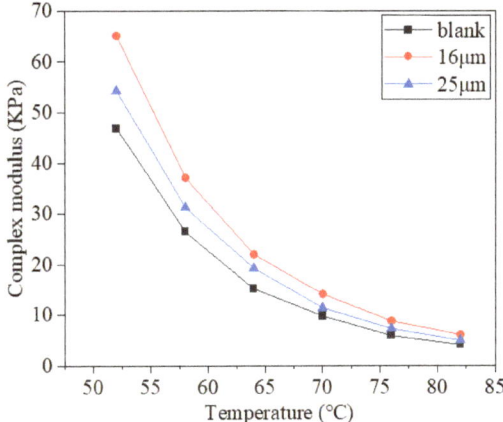

Figure 14. Test results for complex modulus.

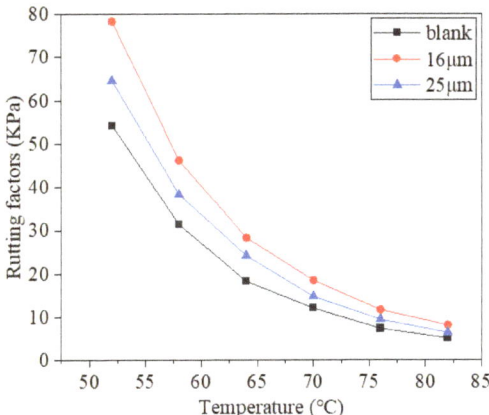

Figure 15. Test results of the rutting factor.

As shown in Figure 13, as the temperature increases, the phase angle generally indicates a downward trend, and the overall change is significant. The experimental findings demonstrated that as the temperature rose, the viscosity of asphalt mastic decreased while the elasticity component decreased [16]. At the same temperature, the phase angle of basalt fiber asphalt mastic was smaller than that of asphalt mastic without fiber. Compared to the asphalt mastic, the phase angle of the fiber asphalt mastic dropped by 4.8° and 3.2°, respectively, after adding 16 μm and 25 μm basalt fiber, according to the phase angle data at 76 °C. Because the fibers are covered with one another to shape a "bracket" and assumed to be part of a "network" in the asphalt mastic [38,39], a structural asphalt interfacial layer with a strong adhesion force formed on the fiber's surface after the asphalt made contact with it, improving the bonding performance of asphalt while also limiting the mastic's ability to deform around the fiber and lowering its fluidity [20]. Larger specific surface area fibers can absorb more asphalt and exhibit greater elastic properties. At the same weight content, 16 μm basalt fiber had more monofilaments and specific surface area than 25 μm, resulting in improved elastic characteristics in asphalt mastic, which is in accordance with the findings in reference [20].

Complex modulus test results are shown in Figure 14. As the temperature rose, the complex modulus decreased [20]. This was mainly due to the deformation and thermal effect of asphalt molecules under loading, and the motility of asphalt molecules was enhanced. Deformation was more likely to occur, which was manifested by the decrease in modulus. Results showed that its resistance to deformation when subjected to traffic loads and temperature changes was constantly weakened. Asphalt mastic with two fibers of different diameters had a larger complex modulus than that of the blank asphalt mastic at the same temperature. Following the addition of 16 μm and 25 μm basalt fibers, the complex modulus of mastic increased by 49.15% and 23.73%, respectively, when the testing temperature reached 76 °C. These outcomes demonstrated that fibers might be added to the SBS-modified asphalt to increase its deformation resistance further [32]. The complex modulus of 25 μm basalt fiber asphalt mastic was smaller than that of 16 μm fiber, indicating that 16 μm mastic presented stronger resistance to deformation, which is consistent with the research results of the references [40,41].

As the temperature rose, the rutting factor values of all three asphalt mastics decreased, indicating that the high-temperature environment weakened its resistance to rutting (Figure 15). At the same temperature, blank asphalt mastic had a lower rutting factor than the other two asphalt mastics with different diameter fibers. Taking the rutting factor data at 76 °C as an example, after adding 16 μm and 25 μm basalt fibers, the rutting factor values of fiber asphalt mastic increased by 58.11% and 28.38%, respectively. The outcomes demonstrated that fiber addition could enhance rutting resistance at high temperatures. At the same temperature, the rutting factor of 16 μm basalt fiber asphalt mastic was greater than that of 25 μm fiber, indicating smaller fiber diameter guarantees greater resistance to rutting at high temperatures.

In conclusion, basalt fiber may be added to asphalt mastic to improve its high-temperature stability, and 16 μm basalt fiber performed better than 25 μm fiber. Because of the larger specific surface area of 16 μm basalt fiber and the higher amount of monofilament fibers under the same fiber content compared to 25 μm basalt fiber, the adsorption capacity of fibers was proportional to their surface area, so fibers with a larger specific surface area can adsorb more asphalt, and fibers with smaller diameters in asphalt mastic showed better elasticity effects. Asphalt mastic had a three-dimensional network structure made of layers of evenly spaced-out fibers [38,39]. After the asphalt and fiber made contact, a layer of asphalt with a strong adhesion force formed on the fiber's surface, improving the bonding performance between asphalt and fiber and preventing the asphalt mastic from deforming around the fiber monofilament [20]. This decreased the fluidity of asphalt mastic and increased its deformation resistance.

3.3. Mid-Temperature Performance Analysis of Fiber Asphalt Mastic

To describe the fatigue properties of asphalt, the LAS experiment was conducted using the viscoelastic continuum damage theory (VECD model). Table 7 and Figure 16 display the experimental findings.

Table 7. The result of the LAS test.

Specimen	Model A	Model B	α	2.5%N_f	5%N_f	10%N_f
SBS	70,400	4.57	2.28	1067.157	44.323	1.985
SBS-16 μm	6,154,000	3.42	1.71	268,315.212	25,089.889	2346.64
SBS-25 μm	1,336,000	3.02	1.51	83,716.875	10,298.677	1269.985

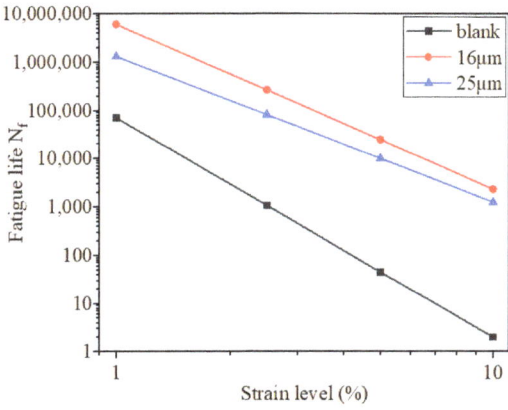

Figure 16. Logarithmic plot of fatigue life vs. shear strain level.

Figure 16 shows that the fatigue resistance of asphalt mastic weakens as strain increases. The fatigue life of fiber asphalt mastic with different diameters was higher than that of the blank asphalt mastic under the same strain level. It demonstrated that fiber addition could enhance the fatigue property of SBS asphalt [19,42].

At the same strain level, 16 μm basalt fiber asphalt mastic had a higher fatigue life N_f than 25 μm, indicating stronger fatigue resistance could be observed in 16 μm basalt fiber asphalt mastic. When compared to asphalt mastic with 25 μm of fiber, the fatigue life N_f of asphalt mastic with 16 μm of basalt fiber rose by 220.50%, 143.62%, and 84.78% at strain levels of 2.5%, 5%, and 10%, respectively. Because under the same fiber content, the specific surface area and more monofilament of 16 μm basalt fiber could be seen than those of 25 μm basalt fiber. Therefore, 16 μm basalt fiber could absorb more asphalt to form structural asphalt [43], which slowed down the dissipation rate of the asphalt mastic to maintain good viscoelasticity for a long time and improve its fatigue life [40,41].

3.4. Low-Temperature Performance Analysis of Fiber Asphalt Mastic

The flexural creep stiffness modulus S in the BBR test, which characterizes the flexibility of asphalt mastic, was utilized to measure the rheological property at low temperatures. With a decline in the S value, flexibility increases. The ability of the sample to relieve stress was evaluated by the creep rate m. As m increases, the stress relaxation ability of asphalt improves. Figure 17a,b display the test findings.

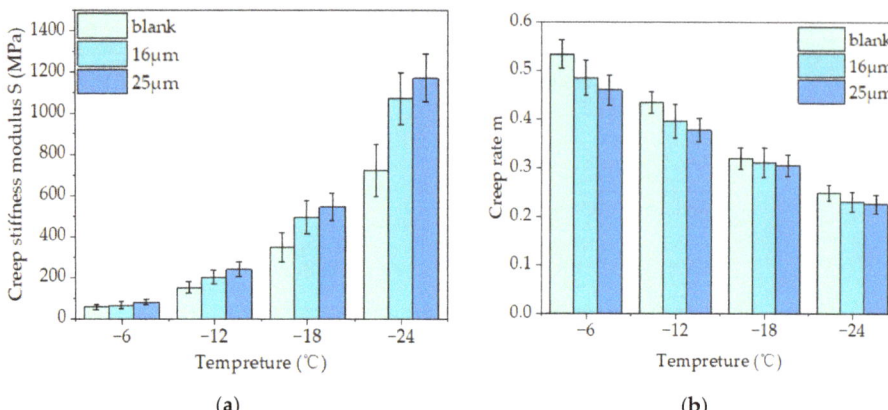

Figure 17. Test results of the BBR test: (**a**) creep stiffness modulus; (**b**) creep rate.

Figure 17a illustrates the test results for creep stiffness. As the temperature drops, the creep stiffness of the sample increases. Asphalt mastic had a lower creep stiffness modulus than basalt-reinforced mastic at the same temperature. Compared to the blank asphalt mastic, the creep stiffness values of the 16 μm and 25 μm fiber asphalt mastics increased by 32.89% and 57.24%, respectively, at −12 °C. A high creep stiffness modulus of 25 μm fiber asphalt mastic was observed at all four testing temperatures, indicating the inferior low-temperature performance of 25 μm fiber asphalt mastic compared to that of 16 μm.

As shown in Figure 17b, the creep rate of the test sample reduces with temperature drops, showing a decrease in the capacity of asphalt to release stresses at low temperatures. The creep rate of blank asphalt mastic was higher than that of the other two basalt fiber asphalt mastics at the same temperature.

In addition, the creep rate values of 16 μm and 25 μm fiber-modified asphalt mastic decreased by 9.87% and 13.13%, respectively, compared to the blank asphalt mastic at −12 °C. The findings indicated that fibers hindered the capacity of asphalt mastic to relax under stress at low temperatures [19,20]. The creep rate of 25 μm fiber-modified asphalt mastic was lower than that of 16 μm fiber-modified asphalt mastic and blank asphalt mastic at the same temperature, indicating that its low-temperature stress relaxation capacity had significantly deteriorated.

Fibers increased the creep stiffness and decreased the creep rate of asphalt mastic in the BBR test, demonstrating that fibers reduced their low-temperature rheological qualities [16]. However, the evaluation of the modified agent and fiber-modified asphalt in the BBR test was inaccurate because the test specification did not account for the material's inherent strength [44]. Analytical methodologies should also be pursued from alternative perspectives.

3.5. High-Temperature Performance Analysis of Fiber Asphalt Mixture

The impact of fibers on the dynamic stability of mixtures is depicted in Figure 18.

It can be seen from Figure 18 that, compared to the blank mixture, asphalt mixtures with 16 μm and 25 μm basalt fiber had dynamic stability of 43.00% and 22.26% higher for the AC-13 gradation. In addition, the dynamic stability of the mixture rose by 32.30%, 28.35%, and 36.28% when the mixing scheme (16 μm: 25 μm) was 1:1, 1:2, and 2:1, respectively. The dynamic stability enhancement trend of the fiber-modified AC-20 mixture was comparable to that in the AC-13 gradation. The dynamic stability of the corresponding fiber asphalt mixture increased by 40.64%, 20.88%, 29.55%, 24.98%, and 33.62% over the blank sample. This was attributed to the fact that the viscosity of the asphalt mixture might be improved by basalt fibers' ability to absorb some of the free asphalt and inhibit the flow ability of asphalt, as reflected in the DSR test results in Section 2.3.2. High-temperature stability can be improved by adding fibers to asphalt mastic, enhancing the properties of the asphalt mixture. The fibers in the mixture were equally dispersed and created a three-dimensional network structure simultaneously [45]. This construction could better reinforce the aggregate skeleton and strengthen the mixture. Thus, the rutting resistance performance of the mixture was improved.

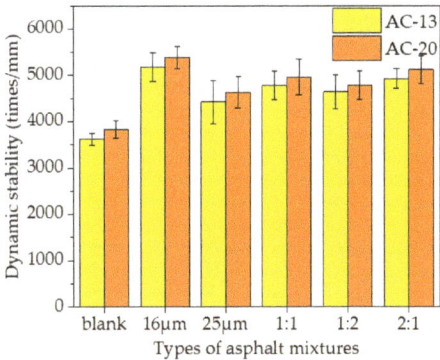

Figure 18. Dynamic stability results of asphalt mixtures with different fiber dosages.

Compared to asphalt mixtures with different fiber diameters, dynamic stability increased with decreasing fiber diameters. Compared to the mixtures with 25 μm basalt fiber, the dynamic stability of the AC-13 mixture with 16 μm basalt fiber rose by 16.96%. The index of the mixed-diameter fiber asphalt mixture was 6~14% bigger than that of 25 μm fiber. After fiber was used to modify the AC-20 mixture, its dynamic stability change law and improvement degree matched those of AC-13. In general, adding basalt fiber could boost dynamic stability by more than 20%. The dynamic stability values of the mixture in the remixing scheme were basically at the same level, and DS values were proportional to the content of 16 μm fiber.

3.6. Mid-Temperature Performance Analysis of Fiber Asphalt Mixture

The results of the load-displacement curves from the IDEAL-CT test are illustrated in Figure 19.

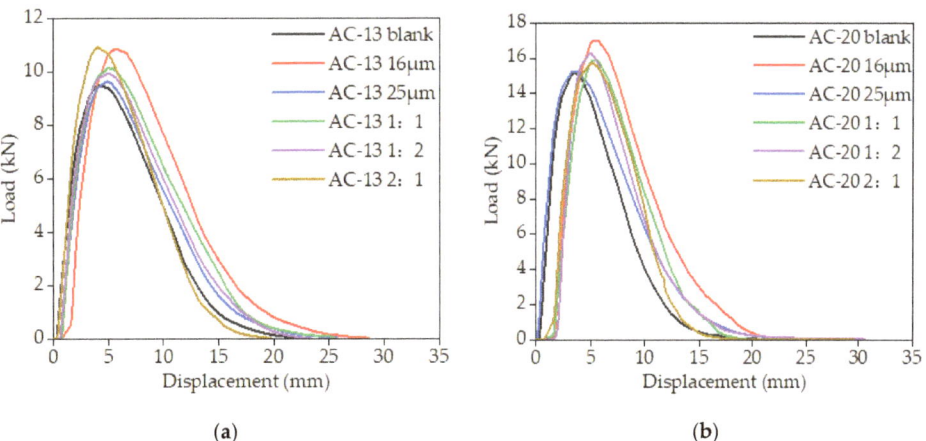

Figure 19. IDEAL-CT test load-displacement curve of asphalt mixture: (a): AC-13; (b): AC-20.

It can be seen from Figure 19a,b that after the fiber was incorporated, the post-peak section of the curve was smoother, and the post-peak slope was also reduced. It can be explained that adding fiber could increase the toughness of the mixture, resulting in a delay in crack expansion, an extension of the cracking time of the asphalt mixture, and more energy being consumed in the cracking process to achieve the aim of enhancing crack resistance. Additionally, incorporating fiber might increase the peak load of cracking in asphalt mixtures, indicating that the fiber could strengthen the bond between aggregate and mastic and increase the energy required for crack initiation [24].

The load-displacement curve was processed to calculate the indicators such as G_r, G_f, and CT_{index}, as shown in Figures 20 and 21. The crack initiation energy G_r and total fracture energy G_f were used, and the crack resistance increased with G_r and G_f values. CT_{index} was used to assess the resistance of asphalt mixtures to crack extension, and the higher the CT_{index} value, the slower the crack extension and propagation, and the stronger the crack resistance achieved.

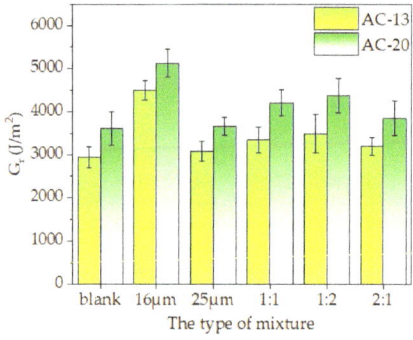

Figure 20. The crack initiation energy of the asphalt mixture.

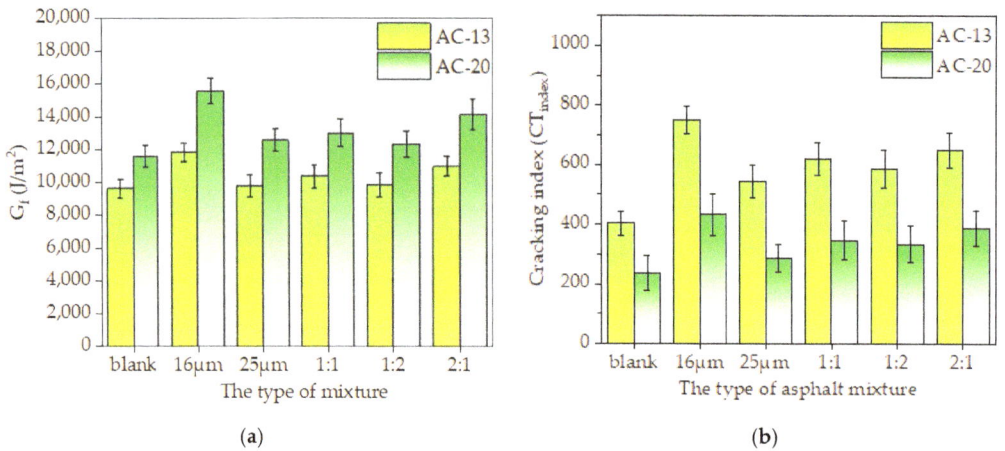

Figure 21. The results of the IDEAL-CT test: (**a**): Fracture energy; (**b**): Cracking index.

It can be seen from Figure 20 that the crack resistance of the asphalt mixture with basalt fiber was significantly higher than that of the blank asphalt mixture. The crack initiation energy (G_r) of a single-doped asphalt mixture with a diameter of 16 μm or 25 μm increased by 52.97% and 4.94%, respectively, compared to the blank mixture of AC-13, while the G_r values increased by 13.89%, 18.82%, and 8.61% when the compounding scheme (16 μm: 25 μm) was 1:1, 1:2, and 2:1, respectively. In addition, the rupture energy (G_r) of the AC-20 followed the same change law as the AC-13 mixture. The strength and anti-crack ability of the asphalt mixture were improved by weaving a large number of basalt fibers into a three-dimensional grid structure [11]. When the mixture starts cracking, the fiber might effectively stop the crack from spreading and help the asphalt mixture withstand crack germination. This was consistent with Wu's findings [11].

Figure 21a,b show that, as contrasted with the blank mixture of AC-13, single-doped asphalt mixtures with diameters of 16 μm or 25 μm had fracture energies G_f that were 22.87% and 1.67% higher, while with the cracking indices CT_{index} of 86.17% and 34.67% higher. The fracture energies G_f of the mixture increased by 7.66%, 2.40%, and 13.94%, respectively, when the mixing scheme (16 μm: 25 μm) was 1:1, 1:2, and 2:1 asphalt mixture. The cracking index CT_{index} values of the corresponding mixtures also increased by 53.33%, 44.94%, and 60.58%, respectively. The fracture energy G_f of the AC-20 mixture followed a similar change law to that of the AC-13 mixture. Additionally, the anti-crack ability at intermediate temperatures was significantly influenced by the fiber diameter. According to the results, when fiber diameter decreased, the fracture energy (G_f) and cracking index (CT_{index}) of the asphalt mixture increased. As the AC-13 asphalt mixture with 16 μm basalt fiber was combined, the fracture energy G_f and cracking index CT_{index} rose by 20.8% and 37.4%, respectively, compared to the mixture with 25 μm fiber. However, the fracture energy G_f and the CT_{index} of the asphalt mixture mixed with two fibers of different diameters simultaneously were not much different. Still, the anti-crack ability of the asphalt mixture increased with an increase in the fiber ratio of 16 μm.

In general, the inclusion of 16 μm basalt fiber can significantly increase the anti-crack and anti-propagate abilities of the asphalt mixture [43]. However, at intermediate temperatures, the addition of 25 μm basalt fiber had minimal impact on the asphalt mixture's resistance to cracking. The findings of the monofilament pullout test in Section 3.1 further demonstrated that at 25 °C, 16 μm basalt fiber had a higher interfacial adhesion strength than 25 μm fiber, which can further enhance the anti-crack ability of the asphalt mixtures.

3.7. Low-Temperature Performance Analysis of Fiber Asphalt Mixture

One of the most significant road features is the asphalt mixture's endurance to crack at low temperatures. Asphalt mixtures are temperature-sensitive materials that are prone to cracks when they are in a low-temperature environment. In this study, a bending beam test was adopted to assess the impact of basalt fiber on the properties of the mixture at low temperatures. The results are shown in Figure 22.

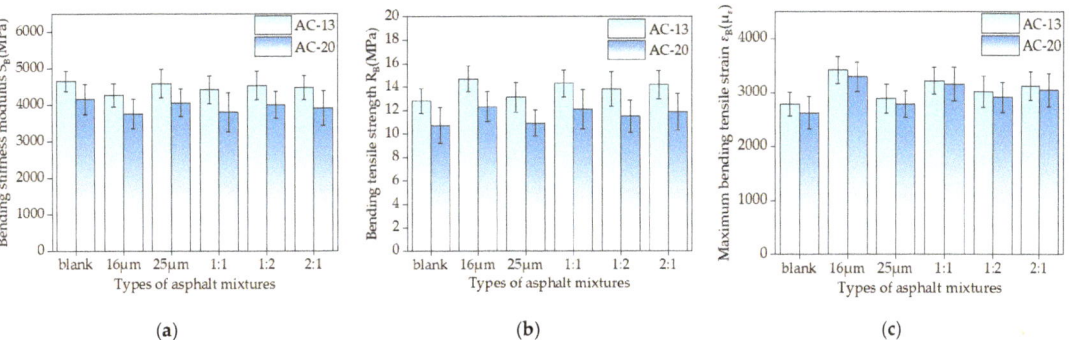

Figure 22. The results of low-temperature beam experiments: (**a**): Bending stiffness modulus; (**b**): Bending tensile strength; (**c**): Maximum bending tensile strain.

As can be seen in Figure 22a–c, the addition of basalt fiber increased the bending tensile strength and maximum bending tensile strain of the mixture of the two gradations while decreasing the bending stiffness modulus. It demonstrates how the inclusion of basalt fiber could boost the flexibility of the asphalt mixture, acting as a strengthening agent and enhancing the endurance of the asphalt mixture to crack at low temperatures. The fibers were woven into the mixture and were able to adhere to the asphalt, which served to toughen the fibers. This toughening effect successfully avoided the formation of cracks, consumed some of the stress concentration, and increased the resistance of material to cracks at low temperatures [30]. Figure 22b shows that the maximum bending strain of the single-doped 16 μm or 25 μm basalt fiber asphalt mixtures was raised by 23.07% and 3.81%, respectively, in comparison to the blank AC-13 asphalt mixture. The maximum bending strain increased by 15.63%, 8.55%, and 12.25% when the asphalt mixture with a fiber ratio of 1:1, 1:2, and 2:1 was mixed. The maximum bending strain of asphalt mixtures with single-doped 16 μm and 25 μm basalt fibers increased by 25.67% and 6.37%, respectively, compared to those without fibers in the AC-20 asphalt mixtures. The maximum bending strain increased by 20.67%, 11.10%, and 16.25%, respectively, when the mixing ratio was 1:1, 1:2, and 2:1.

When fiber asphalt mixtures of various diameters were compared, it was found that the fiber diameter had a significant influence on the low-temperature crack resistance of the mixture. The smaller the fiber diameter, the lower the bending stiffness of the corresponding mixture and the higher the failure strain. This means that a smaller fiber diameter possesses better low-temperature performance. The results of the monofilament pullout test in Section 3.1 also show that under the low-temperature condition of −12 °C, the interfacial adhesion strength of 16 μm basalt fiber is 34.18% higher than that of 25 μm fiber, and 16 μm fiber could better increase the anti-crack ability of asphalt mixture [43]. When fibers of two different diameters were combined with an asphalt mixture, the properties of the corresponding mixture at low temperatures improved when the content of 16 μm fiber increased [40,41]. Generally speaking, the effect of fiber diameter on mixture cracking at low temperatures was comparable to the impact on cracking resistance at intermediate temperatures.

3.8. Water Stability Analysis of Fiber Asphalt Mixture

It is crucial to investigate how fiber diameter affects the water stability of the mixture because the pavement is vulnerable to rain, freezing, and other types of water erosion while in use. The outcomes and standard deviations are displayed in Figure 23.

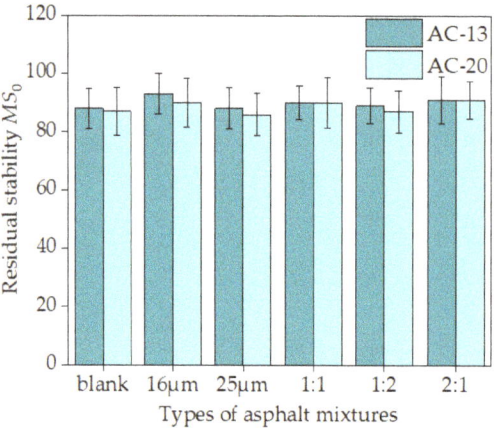

Figure 23. The results of the residual stability MS_0 of the mixture.

Figure 23 shows that AC-13 or AC-20 asphalt mixtures with 16 μm or 25 μm diameters had only about 4% higher residual stability MS_0 in water immersion than the corresponding blank asphalt mixtures. When the remixing scheme was 1:1, 1:2, and 2:1, the residual stability of water immersion in the mixture increased by about 3%. It is clear that the water stability was not significantly affected by the fiber diameter [29].

4. Conclusions

In this research, two basalt fibers with diameters of 16 μm and 25 μm were chosen to make fiber asphalt mastics and asphalt mixtures (AC-13 and AC-20 grades). This study examined how well fiber sticks to the asphalt mastic and the properties of fiber-modified asphalt mastic. It also looked into how changing fiber diameter affects the stability of asphalt mixtures at high, intermediate, and low temperatures and in water baths. Within the scope of this research, the findings are as follows:

(1) The test showed that the cohesiveness between 16 and 25 μm basalt fiber and asphalt mastic was in fiber failure mode at both temperatures of −12 °C and 25 °C. At the same temperature, the interface bonding strength of 16 μm fiber with asphalt mastic is larger than that of 25 μm, which increased by 34.18% and 11.01% at −12 °C and 25 °C, respectively.
(2) Adding basalt fiber improved the SBS-modified asphalt's rutting resistance and deformation recovery ability at high temperatures. The improvement effect of 16 mm basalt fiber was superior to that of 25 mm fiber.
(3) A linear amplitude scanning test showed that basalt fiber could improve the fatigue resistance of asphalt mastic. At strain levels of 2.5%, 5.0%, and 10%, the fatigue life N_f of 16 μm basalt fiber asphalt mastic increased by 220.50%, 143.62%, and 84.78%, respectively, compared with that of 25 μm basalt fiber.
(4) According to the low-temperature bending beam rheological test, basalt fiber would lower the mastic's low-temperature performance, which conflicts with the low-temperature performance of the mixtures, indicating the bending beam rheological test may not be suitable for fiber asphalt mastics.
(5) Adding fiber may raise the rutting stability of the asphalt mixture by 20.88% to 43.00%. As the fiber diameter shrank, the high-temperature stability improved.
(6) The addition of 16 μm basalt fiber significantly increased the resistance of the mixture to crack initiation and propagation (18~85%), whereas the addition of a single doped 25 μm basalt fiber did not considerably increase the crack resistance at intermediate-temperature and low-temperature (about 5%). The diameter of the fiber asphalt mixture was also compounded, and the more 16 μm fiber there was in the mixture, the better resistance to cracking was achieved.
(7) The water stability of the basalt fiber mixture could meet the specifications, and the fiber diameter had no appreciable impact on the water stability.

Author Contributions: Conceptualization, A.K. and Y.Z.; methodology, B.L. and M.L.; validation, B.L. and M.L.; formal analysis, B.L. and M.L.; investigation, M.L.; resources, Z.Z. and B.L.; data curation, A.K. and M.L.; writing—original draft preparation, M.L. and B.L.; writing—review and editing, M.L. and B.L.; project administration, Y.Z. and B.L.; funding acquisition, A.K. and Y.Z. All authors have read and agreed to the published version of the manuscript.

Funding: This study is financially supported by the National Natural Science Foundation of China (Grant No. 52178439) and the Yangzhou Government-Yangzhou University Cooperative Platform Project for Science and Technology Innovation (No. YZ2020262).

Institutional Review Board Statement: Not applicable.

Informed Consent Statement: Not applicable.

Data Availability Statement: Not applicable.

Conflicts of Interest: The authors declare no conflict of interest.

References

1. Jafarifar, N.; Pilakoutas, K.; Bennett, T. Moisture transport and drying shrinkage properties of steel–fiber-reinforced-concrete. *Constr. Build. Mater.* **2014**, *73*, 41–50. [CrossRef]
2. Turbay, E.; Martinez-Arguelles, M.; Navarro-Donado, T.; Sánchez-Cotte, E.; Polo-Mendoza, R.; Covilla-Valera, E. Rheological Behaviour of WMA-Modified Asphalt Binders with Crumb Rubber. *Polymers* **2022**, *14*, 4148. [CrossRef] [PubMed]
3. Park, B.; Cho, S.; Rahbar-Rastegar, R.; Nantung, T.E.; Haddock, J. Prediction of critical responses in full-depth asphalt pavements using the falling weight deflectometer deflection basin parameters. *Constr. Build. Mater.* **2022**, *318*, 126019. [CrossRef]
4. Ghani, U.; Zamin, B.; Bashir, M.; Ahmad, M.; Sabri, M.; Keawsawasvong, S.; Anupam, B.R.; Sahoo, U.; Chandrappa, A. A methodological review on self-healing asphalt pavements. *Constr. Build. Mater.* **2022**, *321*, 126395.
5. Ghani, U.; Zamin, B.; Bashir, M.; Ahmad, M.; Sabri, M.; Keawsawasvong, S. Comprehensive Study on the Performance of Waste HDPE and LDPE Modified Asphalt Binders for Construction of Asphalt Pavements Application. *Polymers* **2022**, *14*, 3673. [CrossRef] [PubMed]
6. Perez, S.P.M.; Maicelo, P.A.A.O. Use of recycled asphalt as an aggregate for asphalt mixtures: Literary review. *Innov. Infrastruct. Solut.* **2021**, *6*, 146. [CrossRef]
7. Hamedi, G.H.; Tahami, S.A. The effect of using anti-stripping additives on moisture damage of hot mix asphalt. *Int. J. Adhes. Adhes.* **2018**, *81*, 90–97. [CrossRef]
8. Stempihar, J.J.; Souliman, M.; Kaloush, K.E. Fiber-Reinforced Asphalt Concrete as Sustainable Paving Material for Airfields. *Transp. Res. Rec. J. Transp. Res. Board* **2012**, *2266*, 60–68. [CrossRef]
9. Abdelsalam, M.; Yue, Y.; Khater, A.; Luo, D.; Musanyufu, J.; Qin, X. Laboratory Study on the Performance of Asphalt Mixes Modified with a Novel Composite of Diatomite Powder and Lignin Fiber. *Appl. Sci.* **2020**, *10*, 5517. [CrossRef]
10. Park, K.S.; Shoukat, T.; Yoo, P.J.; Lee, S.H. Strengthening of hybrid glass fiber reinforced recycled hot-mix asphalt mixtures. *Constr. Build. Mater.* **2020**, *258*, 118947. [CrossRef]
11. Wu, B.; Pei, Z.; Xiao, P.; Lou, K.; Wu, X. Influence of fiber-asphalt interface property on crack resistance of asphalt mixture. *Case Stud. Constr. Mater.* **2022**, *17*, e01703. [CrossRef]
12. Slebi-Acevedo, C.J.; Lastra-González, P.; Castro-Fresno, D.; Bueno, M. An experimental laboratory study of fiber-reinforced asphalt mortars with polyolefin-aramid and polyacrylonitrile fibers. *Constr. Build. Mater.* **2020**, *248*, 118622. [CrossRef]
13. Fiore, V.; Scalici, T.; Bella, G.D.; Valenza, A. A review on basalt fibre and its composites. *Compos. B Eng.* **2015**, *74*, 74–79. [CrossRef]
14. Taherkhani, H.; Kamsari, S. Evaluating the properties of zinc production wastes as filler and theireffects on asphalt mastic. *Constr. Build. Mater.* **2020**, *265*, 120748. [CrossRef]
15. Mohamed, A.; Han, Y.; Cao, Z.; Xu, X.; Xiao, F. Investigating bonding mechanism of rubberized RAP asphalt mastic from rheological and physiochemical perspectives. *J. Clean. Prod.* **2023**, *404*, 136978. [CrossRef]
16. Das, A.; Singh, D. Evaluation of fatigue performance of asphalt mastics composed of nano hydrated lime filler. *Constr. Build. Mater.* **2021**, *269*, 121322. [CrossRef]
17. Veropalumbo, R.; Oreto, C.; Viscione, N.; Russo, F. Investigating the environmental and mechanical properties of sustainable asphalt mastic solutions for road flexible pavements. *Transp. Res. Procedia.* **2023**, *69*, 225–232. [CrossRef]
18. Bieliatynskyi, A.; Yang, S.; Pershakov, V.; Shao, M.; Ta, M. Study of crushed stone-mastic asphalt concrete using fiber from fly ash of thermal power plants. *Case Stud. Constr. Mater.* **2022**, *16*, e00877. [CrossRef]
19. Wang, S.; Mallick, R.; Rahbar, N. Toughening mechanisms in polypropylene fiber-reinforced asphalt mastic at low temperature. *Constr. Build. Mater.* **2020**, *248*, 118690. [CrossRef]
20. Xing, X.; Chen, S.; Li, Y.; Pei, J.; Zhang, J.; Wen, Y.; Li, R.; Cui, S. Effect of different fibers on the properties of asphalt mastics. *Constr. Build. Mater.* **2020**, *262*, 120005. [CrossRef]
21. Qin, X.; Shen, A.; Guo, Y.; Li, Z.; Lv, Z. Characterization of asphalt mastics reinforced with basalt fibers. *Constr. Build. Mater.* **2018**, *159*, 508–516. [CrossRef]
22. Adepu, R.; Ramayya, V.; Mamatha, A.; Ram, V. Fracture studies on basalt fiber reinforced asphalt mixtures with reclaimed asphalt pavement derived aggregates and warm mix additives. *Constr. Build. Mater.* **2023**, *386*, 131548. [CrossRef]
23. Song, Y.; Sun, Y. Low-Temperature Crack Resistance of Basalt Fiber-Reinforced Phase-Change Asphalt Mixture Based on Digital-Image Correlation Technology. *J. Mater. Civ. Eng.* **2023**, *35*, 04023141. [CrossRef]
24. Lou, K.; Xiao, P.; Kang, A.; Wu, Z.; Li, B.; Lu, P. Performance evaluation and adaptability optimization of hot mix asphalt reinforced by mixed lengths basalt fibers. *Constr. Build. Mater.* **2021**, *292*, 123373. [CrossRef]
25. Pirmohammad, S.; Amani, B.; Shokorlou, Y.M. The effect of basalt fibres on fracture toughness of asphalt mixture. *Fatigue Fract. Eng. M.* **2020**, *43*, 1446–1460. [CrossRef]
26. Zhang, J.; Huang, W.; Zhang, Y.; Lv, Q.; Yan, C. Evaluating four typical fibers used for OGFC mixture modification regarding drainage, raveling, rutting and fatigue resistance. *Constr. Build. Mater.* **2020**, *253*, 119131. [CrossRef]
27. Lou, K.; Xiao, P.; Wu, B.; Kang, A.; Wu, X. Effects of fiber length and content on the performance of ultra-thin wearing course modified by basalt fibers. *Constr. Build. Mater.* **2021**, *313*, 125439. [CrossRef]
28. Luo, Y.; Zhang, K.; Xie, X.; Yao, X. Performance evaluation and material optimization of Micro-surfacing based on cracking and rutting resistance. *Const. Build. Mater.* **2019**, *206*, 193–200. [CrossRef]
29. Lou, K.; Xiao, P.; Kang, A.; Wu, Z.; Lu, P. Suitability of fiber lengths for hot mix asphalt with different nominal maximum aggregate size: A pilot experimental investigation. *Materials* **2020**, *13*, 3685. [CrossRef] [PubMed]

30. Pirmohammad, S.; Mengharpey, M. Influence of natural fibers on fracture strength of WMA (warm mix asphalt) concretes using a new fracture test specimen. *Constr. Build. Mater.* **2020**, *251*, 118927. [CrossRef]
31. JTG E20-2011; Standard Test Methods of Bitumen and Bituminous Mixtures for Highway Engineering. China Communications Press: Beijing, China, 2011.
32. Wu, B.; Pei, Z.; Luo, C.; Xia, J.; Chen, C.; Kang, A. Effect of different basalt fibers on the rheological behavior of asphalt mastic. *Constr. Build. Mater.* **2022**, *318*, 125718. [CrossRef]
33. JTG E42-2004; Test Methods of Aggregate for Highway Engineering. China Communications Press: Beijing, China, 2005.
34. AASHTO T 315-12; Standard Method of Test for Determining the Rheological Properties of Asphalt Binder Using a Dynamic Shear Rheometer (DSR). American Association of State Highway and Transportation Officials (AASHTO): Washington, DC, USA, 2016.
35. AASHTO TP101-12; Standard Method of Test for Estimating Fatigue Resistance of Asphalt Binders Using the Linear Amplitude Sweep. American Association of State Highway and Transportation Officials (AASHTO): Washington, DC, USA, 2016.
36. AASHTO TP 125-16; Standard Method of Test for Determining the Flexural Creep Stiffness of Asphalt Mixtures Using the Bending Beam Rheometer (BBR). American Association of State Highway and Transportation Officials (AASHTO): Washington, DC, USA, 2016.
37. Pasetto, M.; Baldo, N. Fatigue Performance of Recycled Hot Mix Asphalt: A Laboratory Study. *Adv. Mater. Sci. Eng.* **2017**, *10*, 4397957. [CrossRef]
38. Hussein, F.; Ismael, M.; Huseien, G. Rock Wool Fiber-Reinforced and Recycled Concrete Aggregate-Imbued Hot Asphalt Mixtures: Design and Moisture Susceptibility Evaluation. *J. Compos. Sci.* **2023**, *7*, 428. [CrossRef]
39. Serin, S.; Önal, Y.; Emiroğlu, M.; Demir, E. Comparison of the effect of basalt and glass fibers on the fracture energy of asphalt mixes using semi-circular bending test. *Constr. Build. Mater.* **2020**, *406*, 133460. [CrossRef]
40. Huang, C.; Gao, D.; Meng, T.; Yang, C. Investigation into Viscoelastic Properties of Fiber-Reinforced Asphalt Composite Concrete Based on the Burgers Model. *Buildings* **2023**, *13*, 499. [CrossRef]
41. Meng, F.; Gao, D.; Chen, F.; Huang, C. Fatigue Performance Test and Life Calculation of Fiber-Reinforced Asphalt Concrete. *Adv. World. Inform. Eng.* **2022**, *44*, 133–139. [CrossRef]
42. Davar, A.; Tanzadeh, J.; Fadaee, O. Experimental evaluation of the basalt fibers and diatomite powder compound on enhanced fatigue life and tensile strength of hot mix asphalt at low temperatures. *Constr. Build. Mater.* **2017**, *153*, 238–246. [CrossRef]
43. Pei, Z.; Lou, K.; Kong, H.; Wu, B.; Wu, X. Effects of fiber diameter on crack resistance of asphalt mixtures reinforced by basalt fibers based on digital image correlation technology. *Materials* **2021**, *14*, 7426. [CrossRef]
44. Cheng, Y.; Zhu, C.; Tan, G.; Lv, Z.; Yang, J.; Ma, J. Laboratory Study on Properties of Diatomite and Basalt Fiber Compound Modified Asphalt Mastic. *Adv. Mater. Sci. Eng.* **2017**, *10*, 4175167. [CrossRef]
45. Krayushkina, K.; Khymeryk, T.; Bieliatynskyi, A. Basalt fiber concrete as a new construction material for roads and airfields. *IOP Conf. Ser. Mater. Sci. Eng.* **2019**, *708*, 012088. [CrossRef]

Disclaimer/Publisher's Note: The statements, opinions and data contained in all publications are solely those of the individual author(s) and contributor(s) and not of MDPI and/or the editor(s). MDPI and/or the editor(s) disclaim responsibility for any injury to people or property resulting from any ideas, methods, instructions or products referred to in the content.

Article

Multi-Step Relaxation Characterization and Viscoelastic Modeling to Predict the Long-Term Behavior of Bitumen-Free Road Pavements Based on Polymeric Resin and Thixotropic Filler

Carina Emminger *, Umut D. Cakmak and Zoltan Major

Institute of Polymer Product Engineering, Johannes Kepler University Linz, Altenbergerstraße 69, 4040 Linz, Austria; umut.cakmak@jku.at (U.D.C.); zoltan.major@jku.at (Z.M.)
* Correspondence: carina.emminger@jku.at; Tel.: +43-732-2468-6654

Abstract: Asphalt pavements are fundamental to modern transportation infrastructure, requiring elasticity, firmness, and longevity. However, traditional asphalt, based on bitumen, faces several limitations. To improve pavement performance, polymer resins are being used to substitute bitumen and improve requirements. Therefore, a deep understanding of the material behavior is required. This study presents the analysis of the relaxation behavior of a poly(methyl methacrylate)-based pavement and the influence of mineral fillers. An approach using a linear elastic–viscoelastic material model was selected based on evidence and validated across the linear and nonlinear deformation range. The results reveal no influence of the mineral fillers on the relaxation behavior. The presented modification of the linear elastic and viscoelastic modeling reveals accurate results to predict long-term pavement performance. This approach offers a practical method for forecasting asphalt behavior. Further research is needed to incorporate deformation behavior into the model.

Keywords: bitumen-free asphalt pavement; linear elastic modeling; viscoelastic modeling; relaxation characterization; polymeric resin pavement; mineral filler; long-term asphalt behavior

1. Introduction

Asphalt pavements are critical components of modern transportation infrastructure, providing the necessary surface for vehicles to travel safely and efficiently. Traditionally, asphalt mixtures have relied on bitumen as a binder due to its adhesive and waterproofing properties [1]. However, bitumen-based asphalt presents several challenges, including limited processability at ambient temperatures [2,3] and significant environmental concerns due to its high global warming potential (GWP) [4]. These limitations have driven the search for alternative materials that can enhance the performance and sustainability of asphalt pavements.

In recent years, polymer resins have emerged as a promising substitute for bitumen in road pavements [1]. Polymeric materials such as styrene–butadiene–styrene (SBS) and ethylene–vinyl acetate (EVA) have proven to improve the durability and resistance of pavements while reducing their GWP [2,5–8]. Resin-based pavements offer a versatile foundation for customizing properties to meet specific needs. For example, their static friction can be adjusted to improve slip resistance in industrial areas. By incorporating reinforcing agents and other additives, the pavement's elastic and thermoelastic properties can be tailored to ensure it fulfills the required performance standards for various applications [9]. In addition, the incorporation of thixotropic fillers has been shown to optimize the rheological properties of pavements, addressing issues such as segregation and flow during processing [10,11].

Despite these advancements, a comprehensive understanding of the viscoelastic behavior of polymer-based pavements is essential for predicting their long-term performance [12–14].

Citation: Emminger, C.; Cakmak, U.D.; Major, Z. Multi-Step Relaxation Characterization and Viscoelastic Modeling to Predict the Long-Term Behavior of Bitumen-Free Road Pavements Based on Polymeric Resin and Thixotropic Filler. *Materials* **2024**, *17*, 3511. https://doi.org/10.3390/ma17143511

Academic Editors: Yao Zhang, Haibo Ding, Yu Chen and Meng Ling

Received: 3 June 2024
Revised: 10 July 2024
Accepted: 12 July 2024
Published: 15 July 2024

Copyright: © 2024 by the authors. Licensee MDPI, Basel, Switzerland. This article is an open access article distributed under the terms and conditions of the Creative Commons Attribution (CC BY) license (https://creativecommons.org/licenses/by/4.0/).

Moreover, the influence of processing and filler additives and how these can be used to modify pavements are of interest too (cf. [9,15,16]). The behavior of asphalt under various loads can be categorized into three distinct ranges: linear viscoelastic, nonlinear viscoelastic, and destructive [17]. Characterizing the viscoelastic properties, particularly the relaxation behavior, is crucial for developing reliable material models that can forecast the performance and longevity of asphalt pavements under real-world conditions. The standardized viscoelastic characterization is performed with dynamic experimental tests, which are specified by guidelines such as the ASHTO or ASTM specifications [18]. All define the dynamic mechanical complex modulus tests at specified temperatures and frequencies. A master curve can be created from these results, and material parameters can be derived from it for finite element (FE) calculations [19,20]. Bitumen-based and bitumen-free pavements exhibit viscoelastic behavior [15,21–23]. Viscoelastic modeling plays a pivotal role in pavement condition assessment, enabling the prediction of material responses to dynamic loads and environmental variations [24,25]. Bai et al. [26] used a viscoelastic model to predict the stress–strain response of asphalt pavements under nonuniform-distributed tire-pavement contact pressure. Asim and Khan [27] used uniaxial tensile stress-relaxation tests to model the viscoelastic behavior of asphalt concrete, whereas Ban et al. [28] conducted creep tests to obtain viscoelastic material properties of pavements. However, all material models were generated in the linear viscoelastic range at small deformations [19,21,24,25,27–29]. If higher deformations were considered, the models were extended to viscoplastic [30] or nonlinear viscoelastic behavior [24], which requires further experimental investigations. In this study, a linear elastic–viscoelastic approach for bitumen-free pavements is presented, generated from three-step relaxation tests, within and outside the linear elastic range.

The objective of this study was to gain insights into the relaxation behavior of bitumen-free asphalt pavements and, specifically, the influence of three mineral fillers (i.e., basalt sand (BS), silica sand (SS), and silica dust (SD)). The methodological framework is presented in Figure 1. An experimental three-step compression relaxation measurement was performed (Experimental Characterization). From these results, the influences of the three mineral fillers on the relaxation behavior, as well as on the compression set, were investigated. Moreover, the experimental data were used to determine parameters for a material model, which can be used in numerical analysis (Material Modeling).

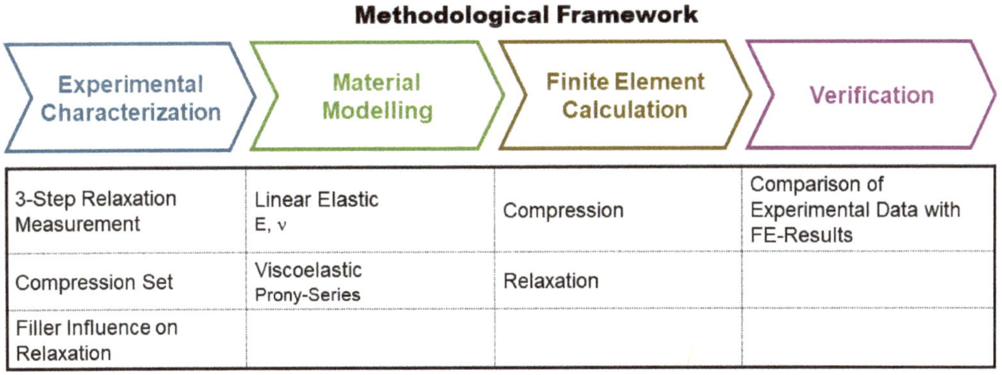

Methodological Framework

Experimental Characterization	Material Modelling	Finite Element Calculation	Verification
3-Step Relaxation Measurement	Linear Elastic E, ν	Compression	Comparison of Experimental Data with FE-Results
Compression Set	Viscoelastic Prony-Series	Relaxation	
Filler Influence on Relaxation			

Figure 1. Methodological framework.

Regarding the material model, a linear elastic–viscoelastic approach was used. The linear elastic model was chosen for its simplicity, requiring only two parameters. The viscoelastic behavior was modeled with a Prony series with five parameters, according to the experimental relaxation measurement. The parameters for modeling the long-term behavior were generated and validated across the linear and nonlinear deformation range.

2. Materials and Methods

2.1. Materials

Within this study, seven different material formulations were selected to investigate the influence of three different mineral fillers on the relaxation behavior of bitumen-free asphalt pavements and determine parameters of a Prony series to model the linear viscoelasticity based on five elements of the generalized Maxwell model. A poly(methyl methacrylate) (PMMA)-based resin from Silikal GmbH (Mainhausen, DEU) was used and filled with a thixotropy agent (hydrophilic fumed silica with a specific surface area of 200 m^2/g), a binder, a catalysator, and a color pigment to improve processability. The three mineral fillers were silica dust (SD), silica sand (SS), and basalt sand (BS). The difference between SD and SS lies in the particle size. Previous works [15,31] reveal an increase in compressive strength with a higher amount of SD, due to the smaller particle size. The total amount of mineral fillers was 80 wt% but with systematically varying composition for each formulation. The different formulations are presented in Table 1, and the bulk density [32–34] of each filler and processing agent is shown in Table 2. F2, F6, and F7 refer to dust-dominant formulations, whereas the remaining are named sand-dominant formulations.

Table 1. Material formulations with the amount of fillers.

Material Formulation	Silica Dust (SD) in wt%	Silica Sand (SS) in wt%	Basalt Sand (BS) in wt%	Ratio Dust–Sand
F1	0	0	80	0:80
F2	20	0	60	20:60
F3	0	20	60	0:80
F4	5	5	70	5:75
F5	0	10	70	0:80
F6	10	0	70	10:70
F7	15	15	50	15:65

Table 2. Fillers and processing agents and their bulk density.

Filler/Processing Agent	Bulk Density [g/cm^3]
Silica dust	2.65
Silica Sand	2.65
Basalt sand	2.71
Thixotropy agent	2.2
Pigment	4.6
Catalysator	0.62
Binder	0.98

The viscoelastic behavior of these materials and the optimum amount of filler were characterized in a previous study [15]. The materials were provided by RoadPlast Mohr GmbH (Vorarlberg, AUT), and the specimens were cast cylinders with ⌀15 mm × 15 mm.

2.2. Methods

2.2.1. Experimental Relaxation Test

For the characterization of the relaxation behavior, three-step relaxation tests were performed. The measurement procedure is shown in Figure 2. The specimens were loaded while controlling displacement with 0.1 mm/s under compression at three different states: −1 mm, −2 mm, and −4 mm. Each position was held for 60 s, and after the relaxation time, the specimen was fully unloaded (back to 0 mm) within 30 s. The experiment was performed with a servo-hydraulic test system (MTS 852, MTS System Corporation, Eden Prairie, MN, USA) under isothermal conditions at 20 °C. The force was recorded with a 10 kN load cell (661.19F-02 Force Transducer-10kN, MTS System Corporation, Eden Prairie, MN, USA). The specimens had a cylindrical shape (⌀15 mm × 15 mm). From the experiment, the relaxation behavior at different compression states, the compression set (CS), and the relaxation slope (k) of each material formulation were measured and analyzed.

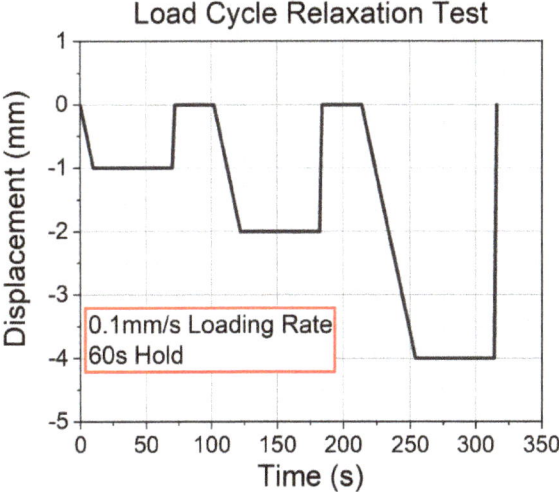

Figure 2. The procedure of the three-step relaxation test.

The CS [%] was calculated according to Equation (1), where h_0 is the original specimen height, h_i is the specimen height after testing, and h_n is the spacer thickness during the measurement. CS was calculated after the final compression step of 4 mm. So, $h_n = h_0 - 4$.

$$CS = \frac{h_0 - h_i}{h_0 - h_n} \times 100 \qquad (1)$$

The relaxation slope k [s^{-1}] was calculated according to Equation (2), where F_{Comp} is the load; t_{Comp} is the time after the compression at each compression level (-1 mm, -2 mm, and -4 mm); and $F(t_i)$ is the load after t_i, which was set to 2 s. For better comparison, k was normalized to F_{Comp}. The exact values are given in Section 3.

$$k = \frac{1}{F_{Comp}} \times \frac{\Delta F}{\Delta t} = \frac{1}{F_{Comp}} \times \frac{F(t_i) - F_{Comp}}{t_i - t_{Comp}} \qquad (2)$$

2.2.2. Numerical Implementation

Material Modeling for Prony Series Parameters

To predict the relaxation behavior of each material formulation, Prony parameters were determined using MCalibration (MCalibration 7.2.6, Ansys, Canonsburg, PA, USA). Therefore, the *Abaqus Linear Elastic–Viscoelastic* material model was chosen, due to the simplicity of the linear elastic model, requiring only two parameters. The linear elastic model is used to simulate the compressive deformation, whereby it is considered that the materials do not reveal ideal linear elastic material behavior. The focus of this research is on the determination of the relaxation behavior of the materials, which is modeled with the generalized Maxwell viscoelastic (phenomenological) model. A Prony series with five parameters (Maxwell elements) was chosen for the modeling of the viscoelastic behavior.

The parameters were generated by fitting the experimental data from the relaxation tests of -1 mm and -2 mm compression, as depicted in Figure 3. The figure shows the experimental data (red curves) and the modeled data (blue curves). The error (R2, coefficient of determination) was calculated to analyze how well the experimental results were reproduced by the model. The Poisson's ratio (ν) was set to 0.3 in all cases. According to the work of Aurangzeb et al. [35], ν was used in the range from 0.25 to 0.35 for asphalt pavements. Gonzalez et al. [36] also used 0.3 as the value for ν. Additionally, 0.3 was chosen because it aligns with commonly accepted values found in related studies, ensuring consistency and comparability of results. It is important to note that the value of ν set to 0.3 is only valid at 20 °C. It has to be adjusted accordingly due to the high-temperature dependence of ν [37,38]. The determined parameters for the *Abaqus Linear Elastic–Viscoelastic* model of the material formulation F1 are given in Table 3. The values for F2–F7 are given in Tables A1–A6 in Appendix A.

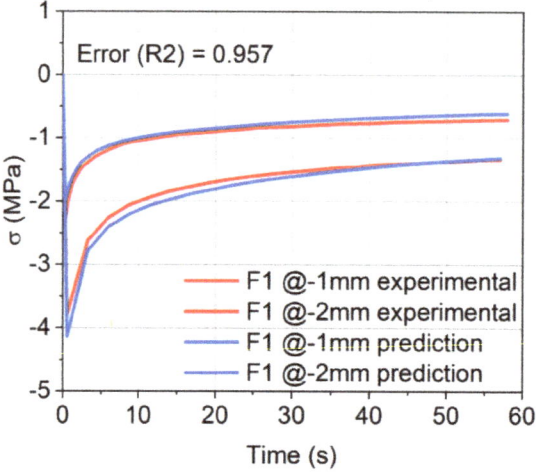

Figure 3. Linear elastic and Prony series fit of the experimental relaxation data at −1 mm and −2 mm compression, with error (R2), the coefficient of determination, and metric for understanding the proportion of variance fitted by the model.

Table 3. Prony parameters of F1 from the relaxation measurement using the FE software Abaqus 2020.

Parameters Units	g_i -	k_i -	$\tau_{i,t}$ s	E MPa	ν -
1	0.36044	0.3504	0.50	33.8	0.3
2	0.15096	0.1910	3.23		
3	0.10181	0.0952	8.62		
4	0.05211	0.0524	22.06		
5	0.11754	0.1163	46.30		

According to the results presented in Section 3, the material formulation is no longer linear at a deformation of −4 mm. This is due to the higher deformation compared to −1 mm or −2 mm. In contrast, the material was pre-stressed and compacted by the first two measurement cycles, leading to a higher compressive set. For the specified material models, these two reasons lead to large deviations in the calculation of F_{Comp}. To apply the linear elastic–viscoelastic model to the results of the −4 mm compression tests, Young's modulus has to be adjusted by a factor γ, as shown in Equation (3). The predicted moduli of each formulation, the adjusted moduli, and the adjusting factor γ are shown in Table 4. γ was retrieved by dividing the $E_{Adjusted}$ with E, and prior $E_{Adjusted}$ was fitted linear with the experimental data from −4 mm compression.

$$E_{Adjusted} = \gamma \times E \qquad (3)$$

Table 4. Young's moduli (E) and adjusted Young's moduli ($E_{Adjusted}$) of all material formulations.

Material Formulation	E [MPa]	$E_{Adjusted}$ [MPa]	Γ [-]
F1	33.80	18.00	0.53
F2	18.17	18.17	1.00
F3	29.50	14.75	0.50
F4	24.24	21.22	0.88
F5	33.46	15.00	0.45
F6	21.26	14.30	0.67
F7	25.44	20.00	0.79

Virtual Setup

The finite element simulation was performed using Abaqus 2020 (Abaqus CAE, Dassault Systems, FRA). A *solid* cylinder with dimensions of ⌀15 mm × 15 mm was created and meshed using an 8-node linear brick (C3D8) with a mesh size of 0.6 mm. The clamp was modeled as *discrete rigid* with a diameter of ⌀20 mm, using a 4-node 3D bilinear rigid quadrilateral (R3D4) mesh with a size of 2.0 mm. The material properties of the specimen were modeled as *Elastic* and *Viscoelastic*, with parameters as described previously. Prior to the performed simulations, a mesh-sensitivity analysis was performed to optimize the mesh size regarding the calculation time of the simulation.

To model the experimental setup, the cylinder was positioned between two clamps: one at the bottom, constrained with the boundary conditions encastre, and another at the top to apply the compressive deformations of −1 mm, −2 mm, and −4 mm. The interaction between the clamps and the cylinder was modeled as *surface-to-surface* contact, with the clamp as the *master surface* and the specimen as the *slave surface*. A tangential behavior with an estimated coefficient of friction of 0.3 was specified as the *contact interaction property*.

The simulation was executed in two sequential steps: first, compression was applied using a *static, general* procedure, and subsequently, relaxation was simulated for a duration of 60 s using *visco* analysis.

Numerical Verification

To validate the results of the FE simulation, three values of the experiment and the simulation were compared. The errors are shown in Figure 4. First, the load at the end of the compression (F_{Comp}) was determined. The calculation of F_{Comp} in the numerical analysis should be as accurate as possible because it is the beginning of the relaxation calculation. High deviations at this stage cause follow-up errors in the prediction of relaxation. In the second step, the load at the end of the relaxation (F_{Relax}), and in the third, the total relaxation ($A_{Relaxation}$) were determined, where the relaxation data were integrated over the entire relaxation time, and the deviation in area between the experimental and simulation data was determined. Since one of the objectives was to predict the relaxation behavior, a low deviation from the FE-calculated F_{Relax} from the experimentally measured F_{Relax} was targeted. Moreover, not only was the value at the end of the numerical analysis of interest but also the modeling of the whole relaxation process. Therefore, the $A_{Relaxation}$ of the experiment was compared with the $A_{Relaxation}$ of the FE calculation. Accurate modeling allows the relaxation to be calculated at any time within 60 s.

The relative error (ER) of all three parameters was calculated according to Equation (4), where x_{SIM} is a variable for $F_{Comp,SIM}$, $F_{Relax,SIM}$, or $A_{Relaxation,SIM}$ from the simulation results, and x_{EXP} is a variable for $F_{Comp,EXP}$, $F_{Relax,EXP}$, or $A_{Relaxation,EXP}$ from the experimental data.

$$ER\ [\%] = \frac{x_{SIM} - x_{EXP}}{x_{EXP}} \times 100 \qquad (4)$$

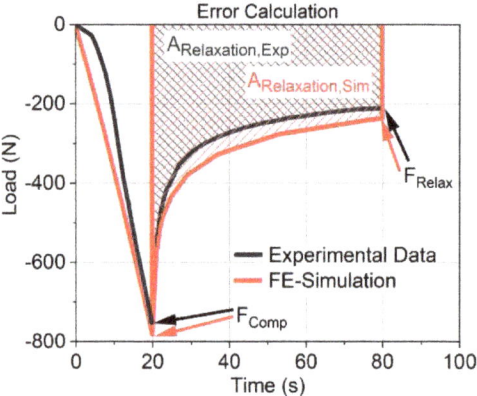

Figure 4. Illustration of the error calculation of the whole relaxation time.

3. Results

3.1. Results of the Experimental Relaxation Measurements

The experimental results of the three-step relaxation measurements are shown in Figure 5. Figure 5a reveals the material response of all seven material formulations at each compression step, and Figure 5b,c display the relaxation behavior of all formulations at one compression step from −1 mm to −4 mm.

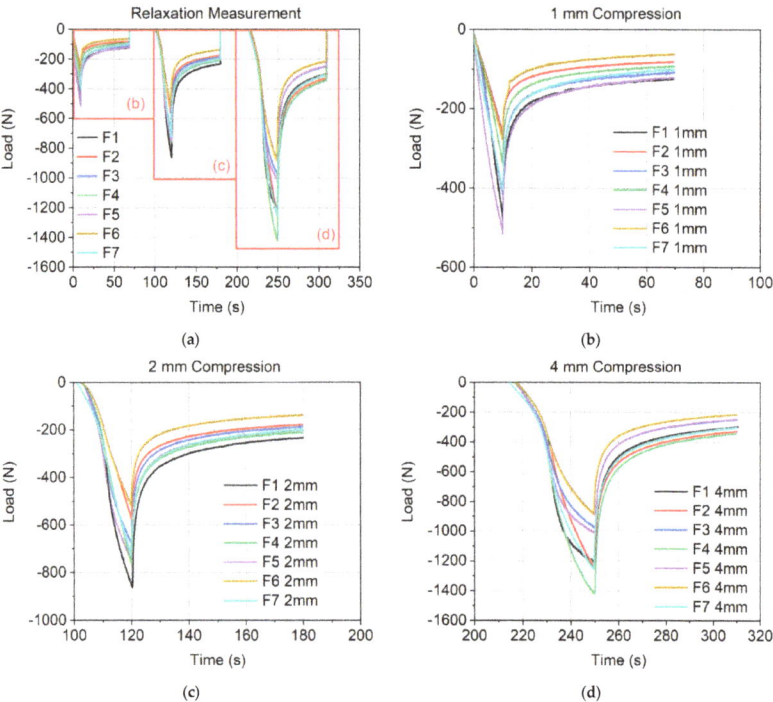

Figure 5. Results of the experimental relaxation measurement: (**a**) relaxation data of the whole measurement cycle; (**b**) relaxation data of all formulations at the first compression step at −1 mm; (**c**) relaxation data of all formulations at the second compression step at −2 mm; (**d**) relaxation data of all formulations at the third compression step at −4 mm.

Figure 6 shows the results of the compression set (CS) in order of a decreasing amount of SD and an increasing amount of sand, as well as a decreasing amount of SS.

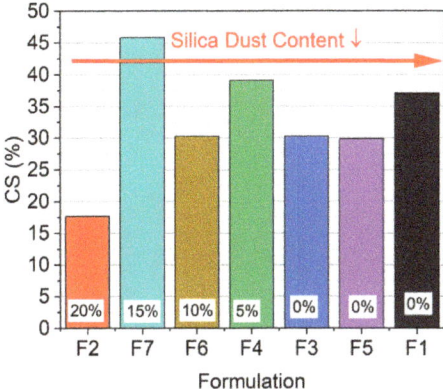

Figure 6. Compression set (CS) of all material formulations after 73.5% compression.

Figure 7 shows the averages of the relaxation slope k at all compression levels, including the ±standard deviation. Additionally, Table 5 lists the exact values of the results, the averages, and the standard deviation.

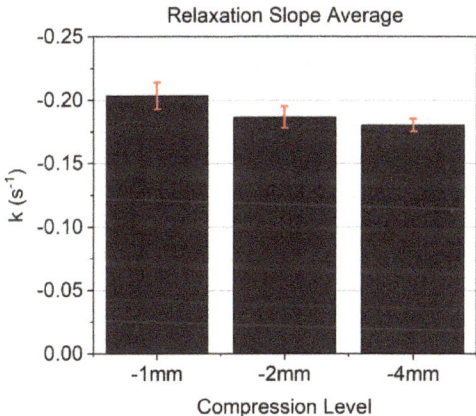

Figure 7. Average of the relaxation slope k of all three compression levels.

Table 5. Calculated relaxation slope k at all compression levels.

Material Formulation	k @−1 mm [s^{-1}]	k @−2 mm [s^{-1}]	k @−4 mm [s^{-1}]
F1	−0.2045	−0.1995	−0.1734
F2	−0.1864	−0.1747	−0.1727
F3	−0.2043	−0.1925	−0.1816
F4	−0.2017	−0.1888	−0.1830
F5	−0.2174	−0.1887	−0.1855
F6	−0.2144	−0.1857	−0.1846
F7	−0.1956	−0.1765	−0.1817
Average	−0.2035	−0.1866	−0.1804
Standard deviation	±0.0106	±0.0087	±0.0052

3.2. Results of the Numerical Analyses

Figure 8a–c illustrate the results of the finite element (FE) simulation at each compression state for material formulation F1. The results for F2–F7 are presented in Appendix A in Figure A1a–f. Figure 8c presents the results of the simulations with the same modulus used in Figure 8a,b and the adjusted modulus. Further, it highlights the necessity to adjust the moduli to generate good predictions. As shown in Table 4, the $E_{Adjusted}$ of F1 is only 53% of the original modulus E.

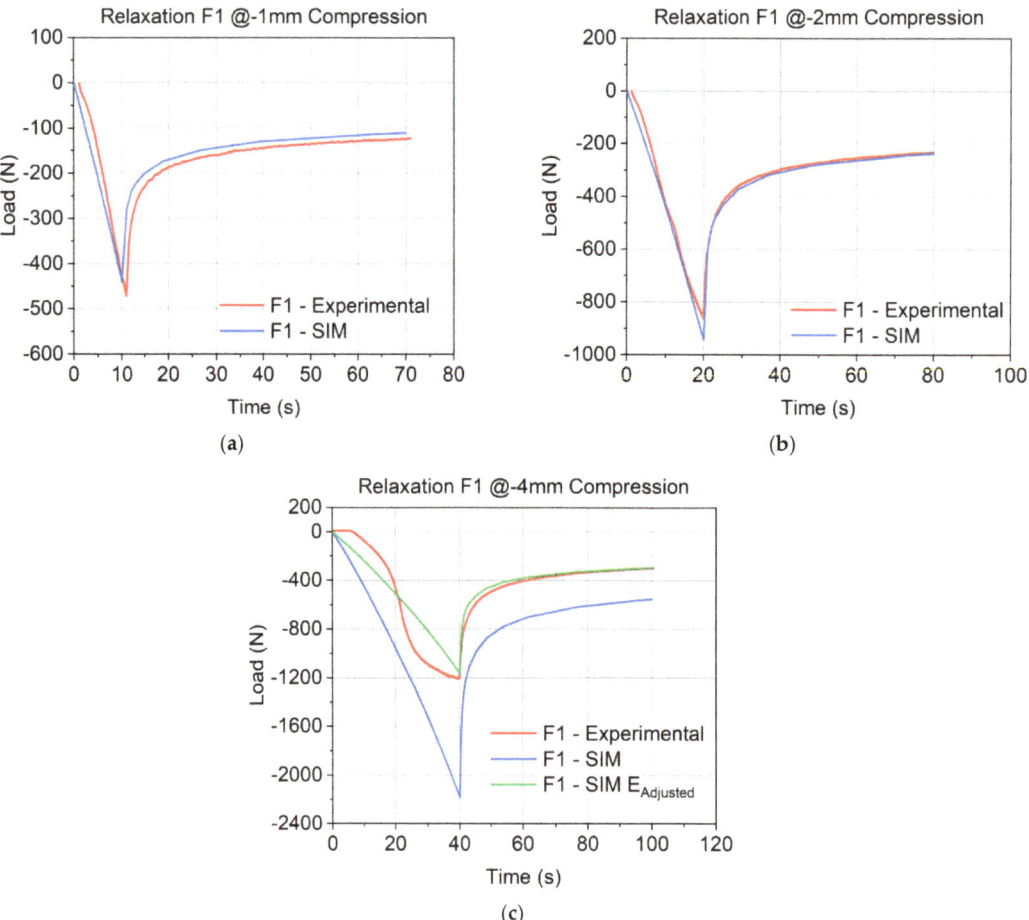

Figure 8. Results of the FE simulation of material formulation F1: (**a**) results at −1 mm compression; (**b**) results at −2 mm compression; (**c**) results at −4 mm compression; the blue curve shows the results with the same E as for −1 mm and −2 mm, and the green curve presents the results with the adjusted modulus.

3.3. Results of the Verification

The calculated relative errors of the compression load (ER of F_{Comp}) between the experimental and the FE-calculated load of all material formulations are presented in Figure 9a. For each material formulation, all ER values for each compression state (−1 mm, −2 mm, and −4 mm) are presented in graduated shades of gray. The errors presented at the −4 mm compression state were calculated with the results of the adjusted modulus ($E_{Adjusted}$). Figure 9b illustrates the high deviation of F_{Comp} of all materials at −4 mm between modulus E values, fitted for −1 mm and −2 mm compression and $E_{Adjusted}$. Due to the better results with $E_{Adjusted}$ at −4 mm compression, in what follows, the values only show the results obtained with $E_{Adjusted}$ and no longer refer to the results achieved with E.

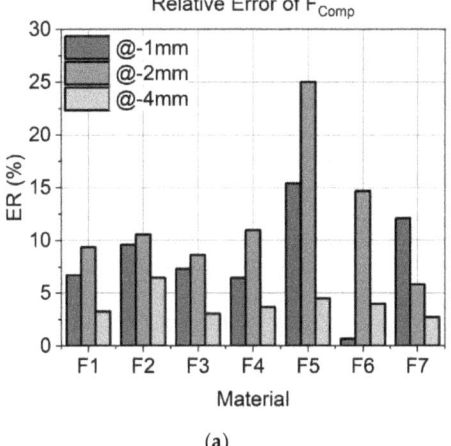

(a)

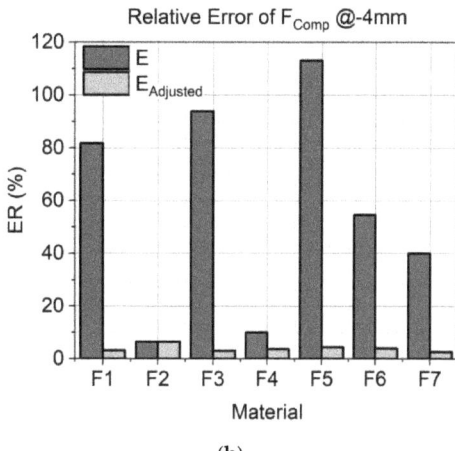

(b)

Figure 9. Error calculation (ER) of the experimentally determined load at the end of the compression and the FE calculated one for each material formulation: (**a**) error of F_{Comp} at all compression states of all material formulations with $E_{Adjusted}$ at −4 mm; (**b**) error of F_{Comp} at −4 mm in comparison of the modulus E, used for −1 mm and −2 mm, and $E_{Adjusted}$.

Figure 10a shows the calculated relative error (ER) of the relaxation load F_{Relax} between the experimental load and the FE-calculated load of all materials for the compression at −1 mm, −2 mm, and −4 mm in shades of gray. Furthermore, the relative error of the areas (ER of $A_{Relaxation}$) of the total relaxation time for all seven material formulations is presented in Figure 10b for all three compression states.

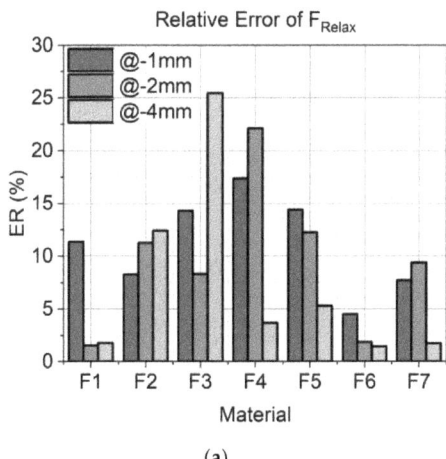

(a)

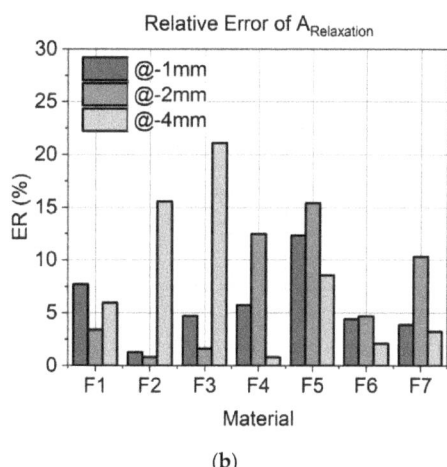

(b)

Figure 10. (**a**) Error calculation of the experimentally determined load at the end of the relaxation and the FE-calculated one for each material formulation at each compression state; (**b**) error calculation of the experimentally determined relaxation area $A_{Relaxation,Exp}$ and the FE-calculated one $A_{Relaxation,Sim}$ for each material formulation at each compression state.

4. Discussion

The results of the three-step relaxation measurement, shown in Figure 5a–d, reveal a similar relaxation behavior for all seven material formulations at each compression level (−1 mm, −2 mm, and −4 mm). This correlation is also shown in the mean values of the relaxation slope k in Figure 7, which leads to the expected conclusion that regarding the relaxation behavior of the investigated materials, their polymeric matrix is the origin for viscoelasticity, and the mineral fillers have no influence on the relaxation behavior. Furthermore, k exhibited a decrease with an increase in compression from −1 mm to −2 mm but stayed constant for −2 mm and −4 mm. The work of Liu et al. [39] shows that relaxation decreases after passing the yield point in glassy polymers in tension and compression. This states a decrease in k for the higher compression at −2 mm and −4 mm. However, the relaxation behavior of the materials was similar (see Figure 5), and the compression behavior showed a more nonlinear behavior with an increased compression state. As shown in Figure 5b, all material formulations show a linear behavior, whereas in Figure 5c, F5, F6, and F7 already show a nonlinear behavior. Proceeding to the higher compression level, as shown in Figure 5d, only F2 and F4 remain linear, whereas all other formulations show pronounced nonlinearity. The increase in nonlinearity with an increase in compression was already studied in [15], which revealed that the material formulations with a higher amount of SD (also referred to as dust-dominant formulation) show a stiffer behavior compared to sand-dominant formulations. Dust-dominant formulations reveal a more linear compression behavior too, whereas sand-dominant materials show a broader load-carrying plateau.

The examined compression set (CS), shown in Figure 6, points to the influence of mineral fillers. A significant difference in the CS was observed for all material formulations. The results indicated that formulations with greater variation in the filler particle size had greater variation in the CS. F7 and F4 were the only formulations that contained both SD and SS and, therefore, had the highest variation in particle size and the highest CS. This finding led to the conclusion that the more particles of the same size, the more constant the value of the CS. This was observed in F3 (20 wt% SD: 60 wt% BS), F5 (10 wt% SS: 70 wt% BS), and F6 (10 wt% SD: 70 wt% BS). The results also showed that a higher number of small particles led to a lower CS than larger particles (F2 (20 wt% SD: 60 wt% BS) vs. F1 (80 wt% BS)).

The fitted material models reveal good simulation results. Figure 8a–c show the experimental data in comparison to the results of the simulation for F1. For −1 mm (Figure 8a) and −2 mm (Figure 8b), the compression behavior as well as the relaxation behavior were reproduced with high accuracy. This is also shown in the low values of relative error of F_{Comp} in Figure 9a (6.7% at −1 mm and 9.3% at −2 mm), the relative error of F_{Relax} in Figure 10a (11.4% at −1 mm and 1.6% at −2 mm), and the error of $A_{Relaxation}$ in Figure 10b (7.7% at −1 mm and 3.4% at −2 mm). For higher compression states, the compression behavior was found to be far too stiff, as shown in Figure 8c. Hence, the implementation of γ to adjust modulus E was necessary to predict good results. With the adjusted modulus $E_{Adjusted}$, the prediction of F_{Comp} is possible with good quality, as shown in Figure 9b. Only F2 and F4 would have predicted good results without $E_{Adjusted}$, due to their highly linear behavior. This is also demonstrated by the high values of γ (F2 1.00 and F4 0.88), whereas high nonlinear formulations require low γ values like F1 and F5 (0.53 and 0.45). Comparing the error of F_{Comp} in Figure 9a, it can be inferred that all material formulations revealed low deviations between the experimental and the FE-calculated load values, except for F5, which showed high deviations for the results of −1 mm and −2 mm (15.4% and 25%). As F5 has a lower content of SD, a significant nonlinearity is observed at −2 mm (see Figure 5c), and the fitted modulus is an average of both compression states and, therefore, shows higher deviations. As mentioned above, F6 and F7 behaved nonlinearly at −2 mm too, which exhibited a higher deviation from −1 mm (0.7%) to −2 mm (14.7%) for F6. Interestingly, F7 showed a higher deviation for −1 mm (12.1%) compared to −2 mm (5.8%). The modeling fitted Young's modulus for both load cases in one step, and in this case, it fitted the second load case better than the first.

To conclude on the relaxation behavior, an exact calculation of F_{Comp} is required, because it is the beginning of the relaxation calculation and can cause follow-up errors. Figure 10a illustrates the error of F_{Relax} for all material formulations, while Figure 10b depicts the error of $A_{Relaxation}$. Interestingly, no direct correlation was observed between the error of $A_{Relaxation}$ and the error of F_{Relax}. Using the material parameters, the relaxation load F_{Relax} F_{Relax} could be predicted with a deviation of 15% from the experimental results for most materials and load cases. However, formulations F3 and F4 exhibited higher deviations. F3 had an error of 26% at −4 mm compression. The relaxation behavior was also modeled using experimental data from −1 mm and −2 mm compressions. Due to reduced relaxation at −4 mm, indicated by a decrease in the relaxation time constant with increased compression, the model requires further refinement to yield accurate predictions. F4 showed an error of 17% for −1 mm and 22% for −2 mm compressions. For −4 mm compression, the error was below 5%. According to the decreased relaxation time constant at higher compression levels, the calibrated model predicted too low relaxation for small deformations.

The deviation in $A_{Relaxation}$ was below 5% for more than half of the materials and load cases, with only 5 out of 21 cases revealing deviations higher than 10%. Notably, F2 exhibited a 16% deviation at −4 mm, and F3 showed a 22% deviation at −4 mm. These higher deviations are due to the variation in relaxation at higher compression levels, as previously discussed. F4, with a 13% deviation at −2 mm, reflected higher deviations in F_{Relax}, while F5, with deviations of 13% at −1 mm and 15% at −2 mm, indicated follow-up errors from the high deviations of F_{Comp} at −1 mm and −2 mm compressions. However, F_{Relax} results for these formulations still showed a deviation below 15%.

5. Conclusions

To conclude, the results of the experimental investigations revealed that mineral fillers have no influence on the relaxation behavior of the different material formulations but influence the stiffness of the compression behavior. They also affected the compression set (CS) according to particle sizes and their distribution; with higher content of smaller particles (SD), a lower CS was observed and vice versa. Further, in formulations with more different particles, the CS revealed higher values than for formulations with only one or two types of particles.

The investigated linear elastic–viscoelastic approach and the used material models reveal good results with respect to F_{Comp}, F_{Relax}, and $A_{Relaxation}$, which enables a prediction of the relaxation behavior and the long-term behavior of the different material formulations. The results highlight the applicability and the limitations of the linear elastic model. For low-compression deformations, the linear elastic–viscoelastic approach fits the compression behavior as well as the relaxation behavior. For higher deformations, the linear elastic model reveals stiff results, but the limitations can be extended with the adjusted modulus and the required adjusting factor γ. This enables an exact prediction of the compression force for higher nonlinear deformations. However, in the current study, the compression behavior was not exactly reproduced with the adjusted version.

These results highlight an understanding of the composition and influence of mineral fillers in bitumen-free asphalt pavements. The presented modeling approach enables an uncomplicated parameter fitting, which can be used for numerical analysis to predict the composition of material formulations for desired mechanical properties in the application. However, compressive deformation is only accurately modeled in the linear elastic range. Further investigations are required for a detailed simulation of compression in the nonlinear range. Moreover, the material model can be extended by damage hypotheses, but these require additional experimental studies.

Author Contributions: Conceptualization, U.D.C. and C.E.; methodology, U.D.C.; software, C.E.; validation, C.E. and U.D.C.; formal analysis, C.E. and U.D.C.; investigation, C.E. and U.D.C.; resources, Z.M.; data curation, C.E.; writing—original draft preparation, C.E. and U.D.C.; writing—review and editing, U.D.C. and Z.M.; visualization, C.E. and U.D.C.; supervision, U.D.C. and Z.M.; project administration, C.E. and U.D.C.; funding acquisition, Z.M. All authors have read and agreed to the published version of the manuscript.

Funding: This research received no external funding, and the APC was funded by the Johannes Kepler Open Access Publishing Fund and the federal state of Upper Austria.

Institutional Review Board Statement: Not applicable.

Informed Consent Statement: Not applicable.

Data Availability Statement: The original contributions presented in the study are included in the article, further inquiries can be directed to the corresponding author.

Acknowledgments: The authors acknowledge Emanuel Mohr (RoadPlast Mohr GmbH, Vorarlberg, Austria) for the preparation of the material formulations.

Conflicts of Interest: The authors declare no conflicts of interest.

Appendix A

In the following tables (Tables A1–A6), the parameters generated with MCalibration for the material formulations F2 to F7 are presented. The used model was the *Abaqus Linear Elastic–Viscoelastic*.

Table A1. Prony parameters of F2 from the relaxation measurement for the FE software Abaqus.

Parameters	g_i	k_i	$\tau_{i,t}$	E	ν
Units	-	-	s	MPa	-
1	0.3496	0.3498	1.79	18.17	0.3
2	0.0336	0.0762	4.95		
3	0.0313	0.0374	5.91		
4	0.1966	0.1797	49.69		
5	0.2019	0.1612	51.69		

Table A2. Prony parameters of F3 from the relaxation measurement for the FE software Abaqus.

Parameters	g_i	k_i	$\tau_{i,t}$	E	ν
Units	-	-	s	MPa	-
1	0.0242	0.8845	0.40	29.50	0.3
2	0.0025	0.0032	3.92		
3	0.0181	0.0014	28.55		
4	0.7255	0.0263	61.78		
5	0.2297	0.0550	118.15		

Table A3. Prony parameters of F4 from the relaxation measurement for the FE software Abaqus.

Parameters	g_i	k_i	$\tau_{i,t}$	E	ν
Units	-	-	s	MPa	-
1	0.2580	0.3458	0.70	24.24	0.3
2	0.2387	0.2394	3.26		
3	0.0683	0.0921	25.62		
4	0.1015	0.0840	36.68		
5	0.1558	0.1232	56.35		

Table A4. Prony parameters of F5 from the relaxation measurement for the FE software Abaqus.

Parameters	g_i	k_i	$\tau_{i,t}$	E	ν
Units	-	-	s	MPa	-
1	0.3441	0.3441	1.51	33.46	0.3
2	0.2069	0.2069	2.83		
3	0.1125	0.1125	11.94		
4	0.0583	0.0583	52.44		
5	0.1008	0.1008	60.17		

Table A5. Prony parameters of F6 from the relaxation measurement for the FE software Abaqus.

Parameters	g_i	k_i	$\tau_{i,t}$	E	ν
Units	-	-	s	MPa	-
1	0.2956	0.3558	0.3	21.26	0.3
2	0.2379	0.2364	2.8		
3	0.1619	0.1543	16.7		
4	0.0737	0.0117	48		
5	0.0235	0.0730	56		

Table A6. Prony parameters of F7 from the relaxation measurement for the FE software Abaqus.

Parameters	g_i	k_i	$\tau_{i,t}$	E	ν
Units	-	-	s	MPa	-
1	0.0065	0.7679	0.30	25.44	0.3
2	0.0024	0.0742	2.81		
3	0.0023	0.0854	15.46		
4	0.5398	0.0001	63.72		
5	0	0.0002	93.33		

In Figure A1a–f, the results of the FE calculations are presented for all three load cases and compared to the experimental data. The presented results for −4 mm compression were calculated with the adjusted modulus. The red curves present the experimental values, and the blue curves show the calculated data. The solid lines show the compression at −1 mm, the dashed line shows the compression at −2 mm, and the dash-dotted line shows the compression at −4 mm.

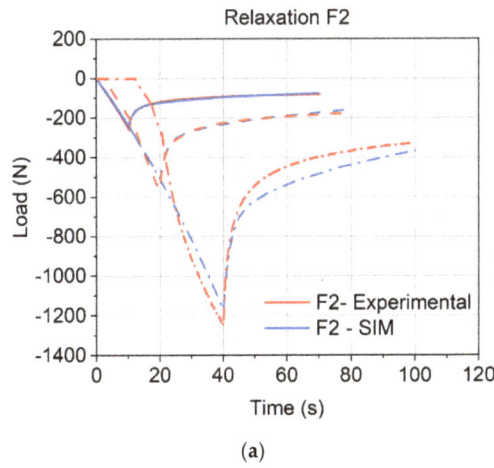

Figure A1. Cont.

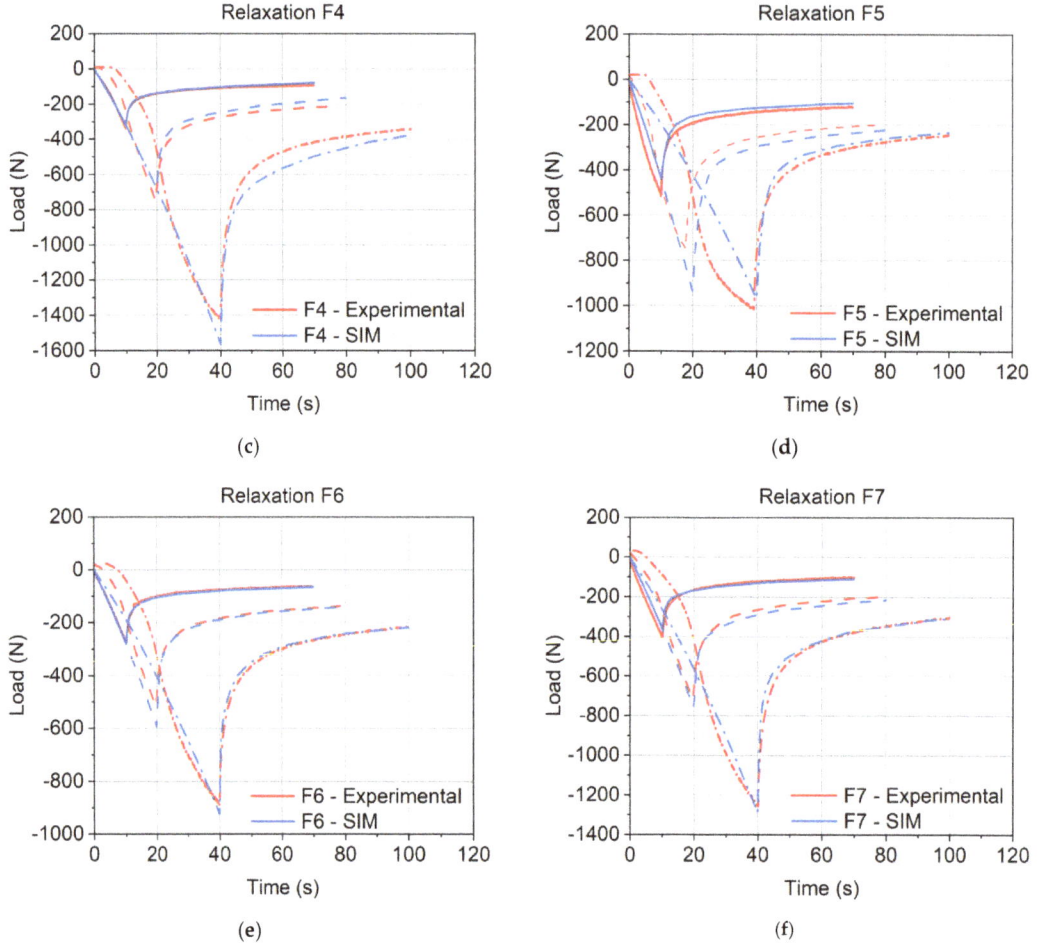

Figure A1. Results of the FE calculations for all three load cases. For the −4 mm compression, the adjusted modulus was used: (**a**) F2, (**b**) F3, (**c**) F4, (**d**) F5, (**e**) F6, and (**f**) F7. The solid lines represent the −1 mm, the dashed line the −2 mm, and the dash-dotted line the −4 mm compression.

References

1. Mallick, R.B.; El-Korchi, T. *Pavement Engineering: Principles and Practice*; CRC Press: Boca Raton, FL, USA, 2023.
2. Sienkiewicz, M.; Gnatowski, P.; Malus, M.; Grzegórska, A.; Ipakchi, H.; Jouyandeh, M.; Kucińska-Lipka, J.; Navarro, F.J.; Saeb, M.R. Eco-friendly modification of bitumen: The effects of rubber wastes and castor oil on the microstructure, processability and properties. *J. Clean. Prod.* **2024**, *447*, 141524. [CrossRef]
3. Yuliestyan, A.; Cuadri, A.A.; García-Morales, M.; Partal, P. Binder Design for Asphalt Mixes with Reduced Temperature: EVA Modified Bitumen and its Emulsions. *Transp. Res. Procedia* **2016**, *14*, 3512–3518. [CrossRef]
4. Saberi, F.; Fakhri, M.; Azami, A. Evaluation of warm mix asphalt mixtures containing reclaimed asphalt pavement and crumb rubber. *J. Clean. Prod.* **2017**, *165*, 1125–1132. [CrossRef]
5. Bao, B.; Liu, J.; Li, S.; Si, C.; Zhang, Q. Laboratory Evaluation of the Relationship of Asphalt Binder and Asphalt Mastic via a Modified MSCR Test. *Coatings* **2023**, *13*, 304. [CrossRef]
6. Dehouche, N.; Kaci, M.; Mokhtar, K.A. Influence of thermo-oxidative aging on chemical composition and physical properties of polymer modified bitumens. *Constr. Build. Mater.* **2012**, *26*, 350–356. [CrossRef]
7. Fan, S.; Zhu, H.; Lu, Z. Fatigue Behavior and Healing Properties of Aged Asphalt Binders. *J. Mater. Civ. Eng.* **2022**, *34*, 04022117. [CrossRef]

8. Mazzoni, G.; Stimilli, A.; Cardone, F.; Canestrari, F. Fatigue, self-healing and thixotropy of bituminous mastics including aged modified bitumens and different filler contents. *Constr. Build. Mater.* **2017**, *131*, 496–502. [CrossRef]
9. Partl, M.N. Towards improved testing of modern asphalt pavements. *Mater. Struct.* **2018**, *51*, 166. [CrossRef]
10. Mewis, J.; Wagner, N.J. Thixotropy. *Adv. Colloid Interface Sci.* **2009**, *147–148*, 214–227. [CrossRef] [PubMed]
11. Miglietta, F.; Tsantilis, L.; Baglieri, O.; Santagata, E. A new approach for the evaluation of time–temperature superposition effects on the self-healing of bituminous binders. *Constr. Build. Mater.* **2021**, *287*, 122987. [CrossRef]
12. Zhao, R.; Jing, F.; Wang, R.; Cai, J.; Zhang, J.; Wang, Q.; Xie, H. Influence of oligomer content on viscosity and dynamic mechanical properties of epoxy asphalt binders. *Constr. Build. Mater.* **2022**, *338*, 127524. [CrossRef]
13. Tan, G.; Wang, W.; Cheng, Y.; Wang, Y.; Zhu, Z. Master Curve Establishment and Complex Modulus Evaluation of SBS-Modified Asphalt Mixture Reinforced with Basalt Fiber Based on Generalized Sigmoidal Model. *Polymers* **2020**, *12*, 1586. [CrossRef] [PubMed]
14. Jing, F.; Wang, R.; Zhao, R.; Li, C.; Cai, J.; Ding, G.; Wang, Q.; Xie, H. Enhancement of Bonding and Mechanical Performance of Epoxy Asphalt Bond Coats with Graphene Nanoplatelets. *Polymers* **2023**, *15*, 412. [CrossRef] [PubMed]
15. Emminger, C.; Cakmak, U.D.; Lackner, M.; Major, Z. Mechanical Characterization of Asphalt Mixtures Based on Polymeric Resin and Thixotropic Filler as a Substitute for Bitumen. *Coatings* **2023**, *13*, 932. [CrossRef]
16. Antunes, V.; Freire, A.C.; Quaresma, L.; Micaelo, R. Influence of the geometrical and physical properties of filler in the filler–bitumen interaction. *Constr. Build. Mater.* **2015**, *76*, 322–329. [CrossRef]
17. Luo, X.; Luo, R.; Lytton, R.L. Characterization of Fatigue Damage in Asphalt Mixtures Using Pseudostrain Energy. *J. Mater. Civ. Eng.* **2013**, *25*, 208–218. [CrossRef]
18. Liu, H.; Zeiada, W.; Al-Khateeb, G.G.; Shanableh, A.; Samarai, M. A framework for linear viscoelastic characterization of asphalt mixtures. *Mater. Struct.* **2020**, *53*, 32. [CrossRef]
19. Liu, H.; Luo, R. Development of master curve models complying with linear viscoelastic theory for complex moduli of asphalt mixtures with improved accuracy. *Constr. Build. Mater.* **2017**, *152*, 259–268. [CrossRef]
20. Tschoegl, N.W. *The Phenomenological Theory of Linear Viscoelastic Behavior: An Introduction*; Springer: Berlin/Heidelberg, Germany, 1989.
21. Bai, T.; Hu, Z.; Hu, X.; Liu, Y.; Fuentes, L.; Walubita, L.F. Rejuvenation of short-term aged asphalt-binder using waste engine oil. *Can. J. Civ. Eng.* **2020**, *47*, 822–832. [CrossRef]
22. Zhang, Y.; Ma, T.; Ling, M.; Zhang, D.; Huang, X. Predicting Dynamic Shear Modulus of Asphalt Mastics Using Discretized-Element Simulation and Reinforcement Mechanisms. *J. Mater. Civ. Eng.* **2019**, *31*, 04019163. [CrossRef]
23. Blab, R.; Harvey, J.T. Modeling Measured 3D Tire Contact Stresses in a Viscoelastic FE Pavement Model. *Int. J. Geomech.* **2002**, *2*, 271–290. [CrossRef]
24. Khurshid, A.; Khan, R.; Khan, D.; Jamal, H.; Hasan, M.R.M.; Khedher, K.M.; Salem, M.A. Micromechanical modeling for analyzing non-linear behavior of flexible pavements under truck loading. *Case Stud. Constr. Mater.* **2024**, *20*, e02754. [CrossRef]
25. Keshavarzi, B.; Kim, Y.R. A viscoelastic-based model for predicting the strength of asphalt concrete in direct tension. *Constr. Build. Mater.* **2016**, *122*, 721–727. [CrossRef]
26. Bai, T.; Cheng, Z.; Hu, X.; Fuentes, L.; Walubita, L.F. Viscoelastic modelling of an asphalt pavement based on actual tire-pavement contact pressure. *Road Mater. Pavement Des.* **2021**, *22*, 2458–2477. [CrossRef]
27. Asim, M.; Khan, R.; Ahmed, A.; Ali, Q. Numerical modeling of nonlinear behavior of asphalt concrete. *Development* **2018**, *5*.
28. Ban, H.; Im, S.; Kim, Y.-R. Nonlinear viscoelastic approach to model damage-associated performance behavior of asphaltic mixture and pavement structure. *Can. J. Civ. Eng.* **2013**, *40*, 313–323. [CrossRef]
29. Luo, R.; Liu, H. Improving the Accuracy of Dynamic Modulus Master Curves of Asphalt Mixtures Constructed Using Uniaxial Compressive Creep Tests. *J. Mater. Civ. Eng.* **2017**, *29*, 04017032. [CrossRef]
30. Aigner, E.; Lackner, R.; Eberhardsteiner, J. Multiscale viscoelastic–viscoplastic model for the prediction of permanent deformation in flexible pavements. *Int. J. Mult. Comp. Eng.* **2012**, *10*, 615–634. [CrossRef]
31. Horvat, B.; Ducman, V. Influence of Particle Size on Compressive Strength of Alkali Activated Refractory Materials. *Materials* **2020**, *13*, 2227. [CrossRef]
32. Quarzwerke GmbH. Quarzmehl 6.400, (Sicherheitsdatenblatt (gemäß Verordnung (EG) 1907/2006 und Verordnung (EG) 1272/2008)). 2014.
33. Scherf GmbH. Basaltsand 0,2-1,9 mm. 2021.
34. Strobel Quarzsand GmbH. Kristall Quarzsand feuergetrocknet o. haldenfeucht Feinstquarzsande. (Sicherheitsdatenblatt (gemäß Verordnung (EG) 1907/2006, Verordnung (EG) 1272/2008, und Verordnung (EG) 830/2015)). 2017.
35. Aurangzeb, Q.; Ozer, H.; Al-Qadi, I.L.; Hilton, H.H. Viscoelastic and Poisson's ratio characterization of asphalt materials: Critical review and numerical simulations. *Mater. Struct.* **2017**, *50*, 49. [CrossRef]
36. González, J.M.; Canet, J.M.; Oller, S.; Miró, R. A viscoplastic constitutive model with strain rate variables for asphalt mixtures—Numerical simulation. *Comput. Mater. Sci.* **2007**, *38*, 543–560. [CrossRef]
37. Hofko, B. *Hot Mix Asphalt under Cyclic Compressive Loading: Towards an Enhanced Characterization of Hot Mix Asphalt under Cyclic Compressive Loading (Zugl.: Wien, TU, Diss., 2011)*; Südwestdeutscher Verlag für Hochschulschriften: Saarbrücken, Germany, 2012.

38. Graziani, A.; Bocci, M.; Canestrari, F. Complex Poisson's ratio of bituminous mixtures: Measurement and modeling. *Mater. Struct.* **2014**, *47*, 1131–1148. [CrossRef]
39. Liu, J.; Zheng, Y.; Zhao, Z.; Yuan, M.; Tsige, M.; Wang, S.-Q. Investigating nature of stresses in extension and compression of glassy polymers via stress relaxation. *Polymer* **2020**, *202*, 122517. [CrossRef]

Disclaimer/Publisher's Note: The statements, opinions and data contained in all publications are solely those of the individual author(s) and contributor(s) and not of MDPI and/or the editor(s). MDPI and/or the editor(s) disclaim responsibility for any injury to people or property resulting from any ideas, methods, instructions or products referred to in the content.

Article

Probing the Effect of Linear and Crosslinked POE-*g*-GMA on the Properties of Asphalt

Yujuan Zhang [1], Pei Qian [1], Peng Xiao [1,2,*], Aihong Kang [1,2], Chenguang Jiang [3], Changjiang Kou [1], Zhifeng Wang [4] and Yuqing Li [4]

[1] College of Civil Science and Engineering, Yangzhou University, Yangzhou 225127, China; 008211@yzu.edu.cn (Y.Z.); mz120210993@stu.yzu.edu.cn (P.Q.); ahkang@yzu.edu.cn (A.K.); changjiang.kou@yzu.edu.cn (C.K.)

[2] Research Center for Basalt Fiber Composite Construction Materials, Yangzhou 225127, China

[3] College of Chemistry and Chemical Engineering, Yangzhou University, Yangzhou 225002, China; dx120200067@yzu.edu.cn

[4] Testing Center, Yangzhou University, Yangzhou 225002, China; zfwang@yzu.edu.cn (Z.W.); liyuqing@yzu.edu.cn (Y.L.)

* Correspondence: pengxiao@yzu.edu.cn

Abstract: The copolymer ethylene–octene (POE) has good aging resistance and is an inexpensive asphalt additive compared to the styrene–butadiene–styrene copolymer (SBS). However, POE is easy to segregate in asphalt during storage at high temperatures. Grafting glycidyl methacrylate (GMA) onto the molecular backbone of POE (i.e., POE-*g*-GMA) may solve this problem, for the epoxy groups in GMA can react with the active groups in asphalt. Asphalt modified with linear and crosslinked POE-*g*-GMA were prepared, and the hot storage stability, physical properties and thermal oxidation aging properties were discussed in detail. The results show that linear and low-degree crosslinked POE-*g*-GMA-modified asphalts are storage-stable at high temperatures via measurements of the difference in softening points and small-angle X-ray scattering (SAXS) characterizations from macro and micro perspectives. The difference in softening points (ΔSP) between the upper and lower ends is no more than 3.5 °C for modified asphalts after 48 h of being in an oven at 163 °C. More importantly, the crosslinking modification of POE-*g*-GMA can further increase the softening point and reduce the penetration as well as rheological properties via conventional physical property, dynamic shear rheometer (DSR) and multiple-stress creep recovery (MSCR) tests. Furthermore, asphalt modified with crosslinked POE-*g*-GMA reveals better aging resistance via measurements of the performance retention rate and electron paramagnetic resonance (EPR) characterizations after a rolling thin film oven test (RTFOT). This work may provide further guidelines for the application of polymers in asphalt.

Keywords: crosslinking modification of POE-*g*-GMA; asphalt; storage stability; physical properties; thermal oxidation aging resistance

1. Introduction

Asphalt modified with polymers has a long history because polymers can improve the physical and rheological performance of asphalt. The triblock copolymer styrene–butadiene–styrene (SBS) can remarkably enhance the high-temperature rutting resistance and low-temperature cracking resistance of a matrix asphalt simultaneously, and is the most widely used modifier for asphalt [1]. However, SBS-modified asphalt faces several challenges, including poor aging resistance caused by the presence of double bonds in the backbone and high costs [2,3]. The copolymer ethylene–octene (POE) is one kind of polyolefin elastomer, and its phase structure is similar to that of SBS [4]. According to the chemical formula (Figure 1a), as a kind of thermoplastic elastomer with a saturated main chain, POE has excellent heat resistance and oxygen aging performance [5]. Moreover, it is much cheaper than SBS is. Therefore, asphalt modified with POE is necessary to be regarded as the research object.

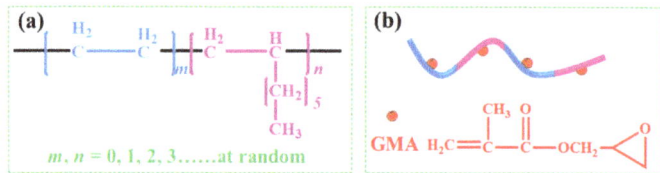

Figure 1. (a) The chemical formula of POE and (b) structural model of POE-g-GMA.

It is difficult for polymer-modified asphalt to achieve compatibility and it easily tends toward phase separation, which mainly arises from differences in molecular structure and weight, density and viscosity [6]. In view of the kinetic point, the system of polymer-modified asphalt (PMA) tends to segregate at high temperatures [7], which restricts the application of PMA in large-scale settings. In general, it contains a high probability to form a homogenous mixture between materials with similar polarities after physical blending [8]. However, the polarity of POE is weak, resulting in the poor compatibility of the two phases and low thermal storage stability. Therefore, it is difficult to prepare POE-modified asphalt with good compatibility only via mechanical blending [9].

Many efforts have been devoted to improving the compatibility between the polymer and asphalts phase, and chemical modification is certified as an effective approach via a chemical reaction of both phases [10]. It is known that some functional groups (e.g., epoxy groups) can react with the several active groups (e.g., hydroxyl groups) that exist in asphalt. Therefore, introducing functional groups into the molecular chain of a polymer, specifically a functional polymer, is one of the most popular methods to alleviate and overcome the problem of the poor compatibility of PMA [6]. Thermoplastic elastomers grafted with maleic anhydride (MAH) or glycidyl methacrylate (GMA) and the copolymers of ethylene-containing epoxy groups, which can be referred to as functionalized polymers, have been widely studied as asphalt modifiers [11,12]. Therefore, grafting GMA onto the molecular backbone of POE (Figure 1b) may solve the problem of POE segregation in asphalt. The production technology of POE grafted using GMA (POE-g-GMA) has become more and more advanced. Moreover, the crosslinking modification of a polymer can further improve its mechanical properties, thermal stability, and aging resistance, and the designability of the crosslinked network structure leads to the adjustability of material properties.

In this work, different crosslinked POE-g-GMAs were synthesized by changing the content of dicumyl peroxide (DCP) using the melt blending strategy, and then asphalt was modified via linear and crosslinked POE-g-GMA, respectively. The storage stability, physical and rheological properties were evaluated, and the resistance of thermal oxidation aging was also discussed in detail. POE-g-GMA improves the properties of asphalt, and compared with that modified with POE, POE-g-GMA-modified asphalt shows superior storage stability. Other than that, low-degree crosslinked POE-g-GMA has a small effect on storage stability, and can further enhance the properties of asphalt, which may satisfy the multi-functional needs of modern transportation.

2. Materials and Methods

2.1. Materials

Penetration 70 asphalt was used as the research object in this work, and the physical properties are listed in Table 1. POE and POE-g-GMA were provided by Xiamen Coace Chemical Co., Ltd. (Xiamen, Fujian, China). The melting index of POE-g-GMA is 6.0 g/min (load of 2.16 kg, at 190 °C), and the grafted rate is 1.2–2.0%. Dicumyl peroxide (DCP) was obtained from Sinopharm Chemical Reagent Co., Ltd. (Shanghai, China).

Table 1. The basic properties of Pen70 asphalt.

Properties	Value	Standard
Penetration (25 °C, 100 g, 5 s) (0.1 mm)	63.2	ASTM D5
Softening Point (°C)	47.7	ASTM D36
Ductility (5 cm/min, 10 °C) (cm)	62.0	ASTM D113
Viscosity (60 °C), Pa·s	213	ASTM D2171

2.2. Preparation Methods

2.2.1. Preparation of Crosslinked POE-g-GMA

Before processing, POE-g-GMA was dried at 80 °C for 6 h in a vacuum-dried oven. A Haake internal mixer (Thermo Scientific Co., Waltham, MA, USA) was employed to prepare crosslinked POE-g-GMA via melt blending with different DCP contents at 190 °C and 40 rpm for 5 min. The as-prepared samples were cut into small pieces with scissors. Hereafter, the crosslinked POE-g-GMA samples are marked as PG-x, where x is the concentration (wt%) of DCP in blends.

2.2.2. Preparation of the Modified Asphalts

First, matrix asphalt was melted at 170 °C until it could completely flow in the container, and different modifiers with predetermined ratios were added into the flowing asphalt, which was left to swell at 170 °C for 30 min after operating manual stirring. Then, the asphalt blends were sheared at about 170 °C for 30 min with a speed of 5000–6000 rpm using a high-speed shear mixer. At last, the modified asphalt was incubated at 170 °C for 30 min. The asphalt modified with y wt% POE-g-GMA is coded as A-yPG, while the asphalt modified with PG-x is referred to as A-3PG-x. The content of PG-x was fixed at 3 wt%. A detailed composition of the blends is shown in Table 2. As a comparison, asphalt modified with 3 wt%POE (A-3P) was chosen as the control sample.

Table 2. Fabrication of asphalts modified with PG-x.

Code	x	Content of PG-x (wt%)
A-1PG	0	1
A-2PG	0	2
A-3PG	0	3
A-4PG	0	4
A-3PG-0.05	0.05	3
A-3PG-0.1	0.1	3
A-3PG-0.2	0.2	3

2.2.3. Laboratory Aging

A rolling thin film oven test (RTFOT) was carried out to simulate the short-term thermal oxygen aging that happens to different asphalts (for 85 min, at 163 °C with rotation and the blowing of air at 4 L/min) according to ASTM D2872.

2.3. Characterizations

2.3.1. Attenuated Total Reflectance Fourier Transform Infrared Spectroscopy (ATR-FTIR)

The Cary 610/670 micro-infrared spectrometer produced by Varian Company (Palo Alto, CA, USA) was employed to carry out ATR-FTIR for analyzing changes in chemical structure. The scanning range of the instrument is 4000–500 cm^{-1} with a resolution of 4 cm^{-1}, and it has an accumulation of 32 scans continuously.

2.3.2. Gel Contents Measurement

The gel content analysis of crosslinked POE-g-GMA was carried out via extraction with toluene in a Soxhlet apparatus at 130 °C for about 72 h, which can be used to qualitatively analyze the degree of crosslinking of POE-g-GMA. The insoluble products obtained via filtration were dried at 130 °C in an oven to a constant weight. The gel contents (%) were measured using Equation (1):

$$\text{Gel content} = w_2/w_1 \times 100\% \quad (1)$$

where w_1 is the weight of the as-prepared samples without extraction, while w_2 is the weight of the insoluble products.

2.3.3. Differential Scanning Calorimetry (DSC)

A DSC (8500, PerkinElmer Co., Wilmington, DE, USA) test was performed to detect the melting behaviors of PG-x (~10 mg). The heat program was as follows: heating from −65 to 100 °C with a heating rate of 5 K/min. The measurements were conducted in a nitrogen atmosphere.

2.3.4. Storage Stability and Conventional Physical Properties

The test on storage stability under high- temperatures was carried out in accordance with the standard T 0661-2011 [13], while the ΔSP between the top and bottom of the aluminum tube was used to evaluate the storage stability of modified asphalts at high temperatures. Modified asphalt with a ΔSP of less than 3.5 °C can be supposed to have good storage stability. The tests on conventional physical properties, including penetration, softening point and ductility, were carried out in accordance with standards T 0604-2011, T 0606-2011 and T 0605-2011, respectively [13]. Three replications of each test were performed to obtained averages for each test project.

2.3.5. Small-Angle X-ray Scattering (SAXS)

The compatibility of modified asphalts in view of microscopic perspective was characterized using a Nano STAR small-angle X-ray scatter (SAXS) meter produced by Bruker, Saarbrucken, SL, Germany. The asphalt was wrapped in tinfoil and pressed into sheets of a thickness of less than 1 mm. The test was performed at room temperature, and the incident X-rays of CuKα radiation (1.54 A) were monochromated using a cross-coupled Göbel mirror and passed through the sheet sample. The distance between the sample and detector was calibrated using silver behenate, giving a scattering vector q range of 0.07 to 0.25 nm^{-1}.

2.3.6. Dynamic Shear Rheometer (DSR)

Temperature Sweep Test

A DHR-2 rotational rheometer equipped with a pair of 25 mm parallel plates was employed to conduct the rheological experiments, and was produced by TA Co., New Castle, DE, USA. The measurements were conducted at 10 rad/s with a 1.0 mm gap and 0.2% strain, and the range of the temperature sweep test was from 40 to 80 °C.

Multiple-Stress Creep and Recovery Test

The MSCR test was conducted at 64 °C to evaluate the resistance to permanent deformation of different asphalts under shear stress values of 0.1 kPa and 3.2 kPa.

2.3.7. Morphological Characterization

A LSM700-3D laser microscope (CARL ZEISS, Co., Oberkochen, Germany), which was equipped with a blue filter system with a wavelength of excitation ranging from 390 to 490 nm, was employed to investigate the morphology of the modified asphalts at a magnification of 200. The heated, liquid modified asphalts were poured into a square mold, and then left to cool down to room temperature to obtain a flat surface.

2.3.8. Electron Paramagnetic Resonance (EPR) Test

The free radicals produced due to aging were detected via A300 EPR spectroscopy (Bruker, Saarbrucken, Germany). The test conditions were as follows: the sweep time was 167.772 s, the time constant was 163.840 ms, and the sweep width was 500 G. After the RTOFT test, the sample was dropped into trichloroethylene immediately to stop free radicals from quickly quenching. The solution was extracted to be measured with a capillary, and then the capillaries containing the samples were placed in a standard 4 mm quartz sample tube.

3. Results and Discussion

3.1. Formation of Crosslinked POE-g-GMA

The crosslinking modification of POE-g-GMA was performed via the melt blending of POE-g-GMA with DCP at 190 °C. Figure 2a gives the torque curves of crosslinked POE-g-GMA with different contents of DCP. As the mixing time increased, the torque gradually increased at the beginning of mixing, which indicates the occurrence of a crosslinking reaction. The torque also increased with the increasing content of DCP, indicating that the density and degree of crosslinking increase with the loading of DCP [14]. The gel contents of the PG-x were determined by dissolving it in toluene and via weighting. Figure 2b shows that with the increased loading of DCP from 0.05 to 0.5%, the gel contents of PG-x increased. The gel content increased from 39.65% to 71.12% while the content of DCP was 0.05 wt% and 0.5 wt%, respectively. The glass transition temperature (Tg) of PG-xs was determined via a DSC trace. As shown in Figure 2c, the Tg gradually shifts to a higher temperature with the growth content of DCP. The Tg of PG-0.5 is 65.8 °C, which compared to that of POE-g-GMA represents an increase of about 10 °C. The growth of Tg indicates the restriction of the chain mobility of the polymer [15]. The results indicate that it is possible to successfully conduct the crosslinking modification of POE-g-GMA with DCP, and that it is easy to adjust the degree of crosslinking by varying the content of DCP.

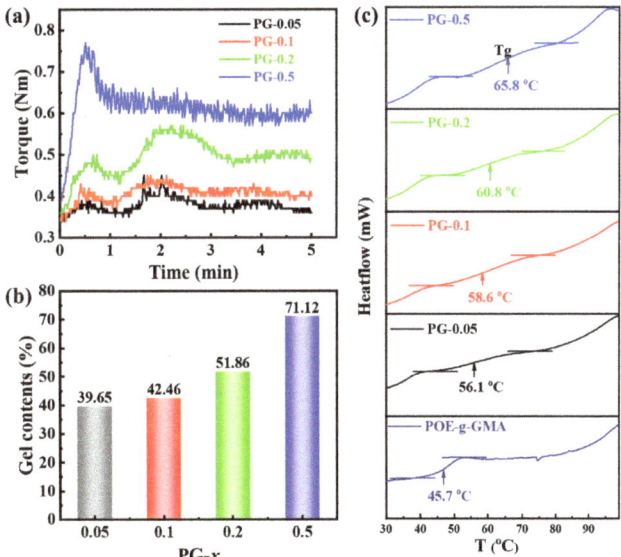

Figure 2. (a) Torque evolution curves, (b) gel contents and (c) DSC heating curves of PG-x.

The characterization of FT-IR is displayed in Figure 3. As shown in Figure 3a, the crosslinking modification of POE-g-GMA has a small effect on the functional groups. A peak around 1735 cm^{-1} refers to the stretching vibration of C=O in GMA. The peak located at about 2916 cm^{-1} and 2850 cm^{-1} refers to stretching vibration of C-H$_2$, and the peak located at 1471 cm^{-1} and 1378 cm^{-1} refers to the bending vibration of C-H and C-H$_3$. The epoxy group located at about 910 cm^{-1} is the active group of glycidyl methacrylate (GMA), which can react with the carboxyl, carbonyl and other active groups in the matrix asphalt to enhance high-temperature stability. As shown in Figure 3b, the characteristic peak of the epoxy groups still existed in the crosslinked POE-g-GMA with various DCP contents. The results demonstrate that the olefin chain of POE-g-GMA was attacked by the free radicals generated by DCP without the consumption of the epoxy groups during crosslinking modification. In the other words, the crosslinked POE-g-GMA prepared in this paper still exhibited reactivity, and could further react with the matrix asphalt in the following melt blending procedure.

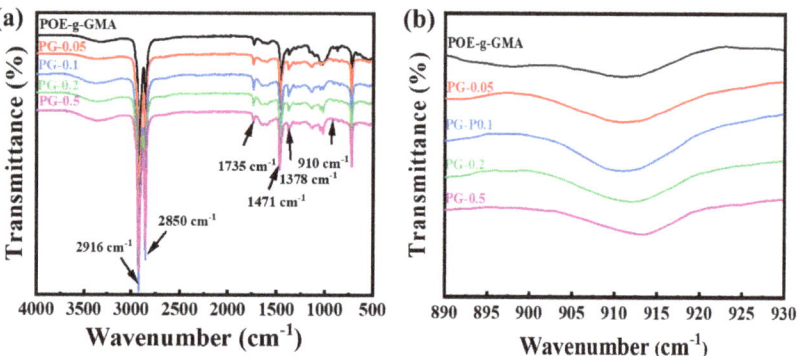

Figure 3. (a) The range of 500–4000 cm^{-1} and (b) 890–930 cm^{-1} in the FT-IR spectra of POE-g-GMA with various DCP contents.

3.2. Storage Stability of Modified Asphalts

The storage stability of modified asphalts at high temperatures is commonly evaluated from a macro perspective via a calculation of the value of the ΔSP, and the smaller the ΔSP value, the more stable it is. According to the literature, asphalt modified with polyolefin usually has low high-temperature storage stability [16–18], which has serious adverse impacts on pavement performance. Values of the ΔSP for different modified asphalts and microstructure characteristics are studied in this paper. As shown in Figure 4a, the ΔSP value of A-3P is 18.2 °C, indicating obvious phase separation between POE and asphalt in the process of hot storage. However, POE-g-GMA is stably and uniformly dispersed in asphalt, which is confirmed by the value of the ΔSP (Figure 4a). Although the ΔSP value of modified asphalt slightly increases with the rise in the POE-g-GMA content, all values are less than 1.0 °C, which is indicative of satisfactory compatibility. This may be a result of the reaction that happens between the epoxy groups in POE-g-GMA and the hydroxyl, carboxyl, and other active groups in the asphalt during processing, which improved its storage stability [6,19,20]. The results of the ΔSP of the asphalt modified with crosslinked POE-g-GMA are given in Figure 4b. With the rise in x, the value of the ΔSP increases, and asphalt modified with weakly crosslinked POE-g-GMA ($x \leq 0.2$) shows good storage stability with a ΔSP value of less than 3.5 °C, while the ΔSP of A-3PG-0.5 is more than 3.5 °C, reaching 11.7 °C. The results indicate that the crosslinking modification of POE-g-GMA has an adverse impact on the compatibility of the modified asphalts. This is because the migration of molecular chains is restricted by the crosslinking network structure, resulting in a decrease in the reactivity of epoxy groups in PG-x [14]. Hence, the thermal storage stability of asphalt modified with highly crosslinking PG-x is poor. Figure 4c describes the schematics of the modification and storage stability of different modified asphalts.

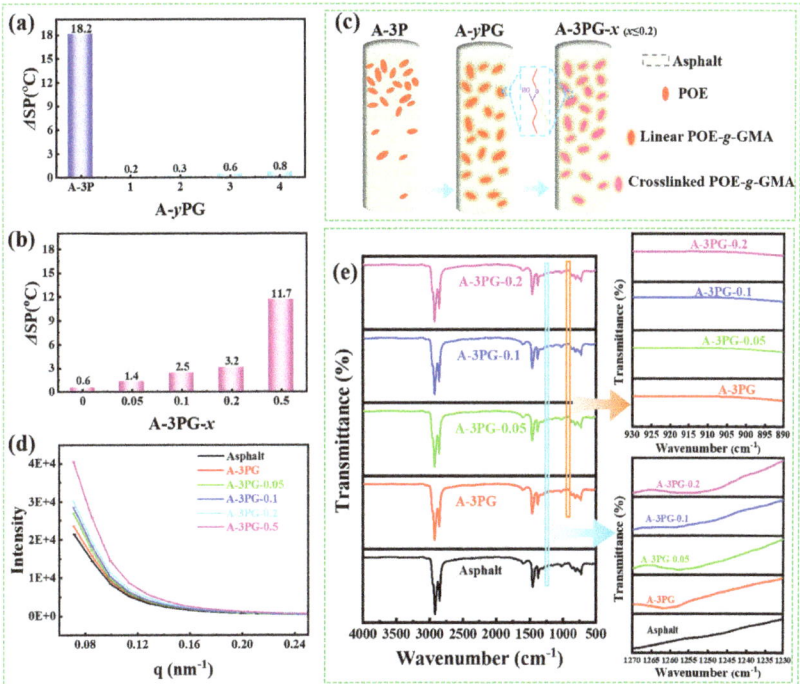

Figure 4. The storage stability of (**a**) A-yPG and (**b**) A-3PG-x; (**c**) the schematics for storage stability, (**d**) the SAXS spectra and (**e**) the FT-IR spectra of different modified asphalts.

Figure 4d displays the SAXS spectra of different asphalts. It can be seen that all samples have a high scattering intensity, and that the scattering intensity of A-3PG-x increases with the increase in x, which is especially the case for A-3PG-0.5. The high scattering intensity in the SAXS spectra arises from the fluctuations in molecular density caused by phase separation [21]; thus, the results indicate that the composition of asphalt itself is very complex, and the crosslinking modification of POE-g-GMA exacerbates the phase separation of modified asphalt at a micro level. The result is consistent with that of the macroscopic compatibility test.

Figure 4e depicts the FT-IR spectra of different modified asphalts. The results show the disappearance of the epoxy group (~910 cm^{-1}) and the emergence of new characteristic peaks (~1255 cm^{-1}), which may be attributed to C-O in the aromatic ether. FT-IR spectra confirm the existence of the reaction between POE-g-GMA and asphalt. Therefore, asphalt modified with POE-g-GMA shows good storage stability.

3.3. Penetration, Softening Point and Ductility

The test of penetration was conducted to evaluate the consistency and hardness of modified asphalt, and the results of penetration are given in Table 3. With the growth content of POE-g-GMA, the penetration of modified asphalts gradually decreases, and the penetration of asphalt decreases from 63.2 to 40.2 after loading with 4 wt% POE-g-GMA, which indicates that the hardness and consistency of asphalt can be enhanced via modification with POE-g-GMA. The penetration of asphalt modified with PG-x slightly decreases, indicating that weakly crosslinked POE-g-GMA has a small impact on the consistency and hardness of asphalt [22]. The softening point was determined to evaluate the properties under high temperatures, and properties under low temperatures were evaluated via the ductility under 10 °C. All the test data are given in Table 3, and the table shows that along with the rise in the loading of POE-g-GMA, the softening point gradually increases. The softening point of asphalt increases from 47.7 °C to 54.9 °C after being modified with 4 wt% POE-g-GMA. The softening point of A-3PG-x continuously increases with the increase in x. When x = 0.2, the softening point continues to increase from 52.6 °C to 59.2 °C, indicating that the weak-crosslinking modification of POE-g-GMA can further enhance the properties under high temperatures. A ductility test was performed at 10 °C. The results in Table 3 show that ductility reduces from 62.0 cm to 15.5 cm after modification with 1 wt% POE-g-GMA, reflecting that POE-g-GMA has a visible harmful influence on ductility. However, ductility slightly increases as the content of POE-g-GMA continues to increase. The ductility of asphalt modified with PG-x decreases with an increase in x, indicating that crosslinking modification is not conducive to its properties under low temperatures.

Table 3. The physical properties of modified asphalts before thermal oxidation aging.

Sample	Penetration (0.1 mm) (25 °C, 100 g, 5 s)	Softening Point (°C)	Ductility (cm) (5 cm/min, 10 °C)
Asphalt	63.2 ± 0.1	47.7 ± 0.3	62.0 ± 0.4
A-1PG	52.4 ± 0.3	50.9 ± 0.2	15.5 ± 0.6
A-2PG	47.0 ± 0.2	51.3 ± 0.1	15.7 ± 0.6
A-3PG	44.0 ± 0.3	52.6 ± 0.6	17.6 ± 0.8
A-4PG	40.2 ± 0.6	54.9 ± 0.3	18.8 ± 0.5
A-3PG	44.0 ± 0.1	52.6 ± 0.3	17.6 ± 0.9
A-3PG-0.05	44.1 ± 0.4	56.6 ± 0.1	16.5 ± 0.3
A-3PG-0.1	44.4 ± 0.1	57.6 ± 0.2	14.1 ± 0.6
A-3PG-0.2	45.4 ± 0.2	59.2 ± 0.7	13.1 ± 0.8

3.4. Dynamic Rheological Properties

The difference in chemical composition or structure between modifier and asphalt has an impact on its rheological performance, which is strongly associated with the processing and paving of the asphalt mixture and the properties of the pavement [23,24]. Therefore, it is important to study the rheological performance of modified asphalts, including their complex shear modulus (G^*), phase angle (δ) and rutting factor ($G^*/\sin\delta$). G^* and δ were directly obtained via a DSR test. G^* represents the stiffness and the ability of resistance to shear deformation. δ is the ratio of the elastic to the viscous component of asphalt, and a smaller δ indicates that there are more elastic components in the asphalt, reflecting that it is easier to be deformed and that there are more unrecoverable parts in the deformation of asphalt. The ability to resist high-temperature rutting can be evaluated by calculating the value of $G^*/\sin\delta$, which is an important index and defined in the Superpave specification [25]. The G^*, δ and $G^*/\sin\delta$ curves with the temperature for the matrix asphalt and different modified asphalts are shown in Figure 5. As the temperature increases, G^* and $G^*/\sin\delta$ become smaller and smaller, while δ becomes larger and larger.

In Figure 5a, G^* increases along with the increasing content of POE-g-GMA, and the G^* of A-3PG-x increases with the rise in x, reflecting that the weak crosslinking modification of POE-g-GMA further enhances the stiffness and resistance to shear deformation of asphalt [26]. The δ of the modified asphalts is smaller than that of the matrix asphalt (Figure 5b), and the change rule is opposite to that of G^*, which is indicative of stronger elastic properties [27]. The rutting factor is displayed in Figure 5c, and the change rule is consistent with that of G^*. The results of $G^*/\sin\delta$ show that the anti-rutting factor of asphalt is improved via modification with linear POE-g-GMA, and crosslinked POE-g-GMA can further strengthen the rutting resistance of modified asphalts [28]. Overall, it is beneficial to the improvement of the properties of asphalt under high temperatures via modification with linear and crosslinked POE-g-GMA.

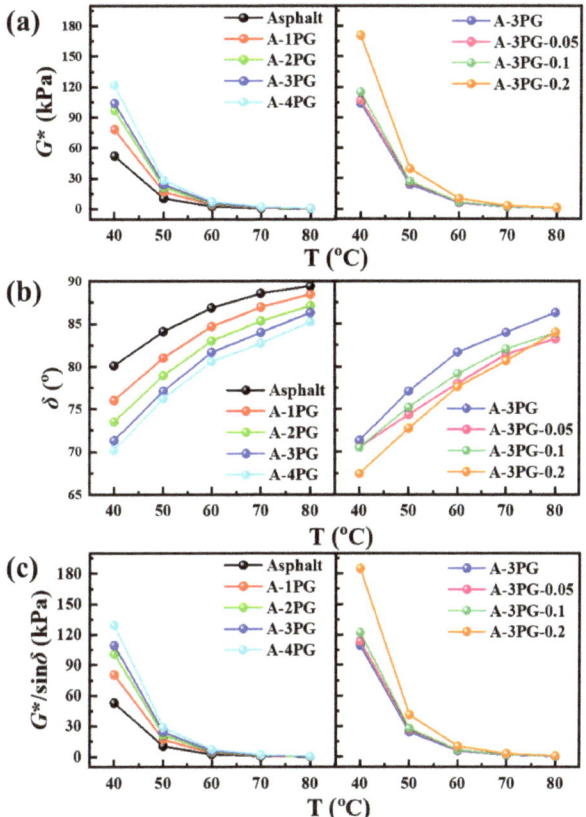

Figure 5. (**a**) The complex modulus, (**b**) the phase angle and (**c**) the rutting factor of different modified asphalts.

The evaluation parameters, non-recoverable creep compliance (Jnr) and percent recovery (R), were calculated and are shown in Figure 6; these parameters are more representative and convincing than are rutting parameter for evaluating the resistance of asphalt to rutting deformation under high temperatures. Figure 6a shows the Jnr of different modified asphalts at 0.1 and 3.2 kPa (Jnr01 and Jnr3.2), while the R values (R01 and R3.2) are given in Figure 6b. Jnr01 and Jnr3.2 of POE-*g*-GMA-modified asphalts are lower than that of unmodified asphalt, and decrease with the increasing content of POE-*g*-GMA, which indicates that POE-*g*-GMA improves rutting resistance. Moreover, crosslinked POE-*g*-GMA can further enhance the high-temperature performance. The Jnr value represents the unrecoverable strain after removing the preset load, while R is employed to characterize the ability of asphalt to restore its original state. As shown in Figure 6b, the R of POE-*g*-GMA-modified asphalts is higher than that of unmodified asphalt, and crosslinked POE-*g*-GMA significantly increases the value of R, especially in the case of A-3PG-0.2. Overall, it is beneficial to the improvement of the rheological properties of asphalt to modify it with linear and crosslinked POE-*g*-GMA. This is because that the functional groups in POE-*g*-GMA have the ability to react with the acidic compound in asphalt and therefore the network structure. The movement of the modified asphalt molecule was restricted by the network structure. As a result, the flow and deformation of asphalt were restricted under high temperatures.

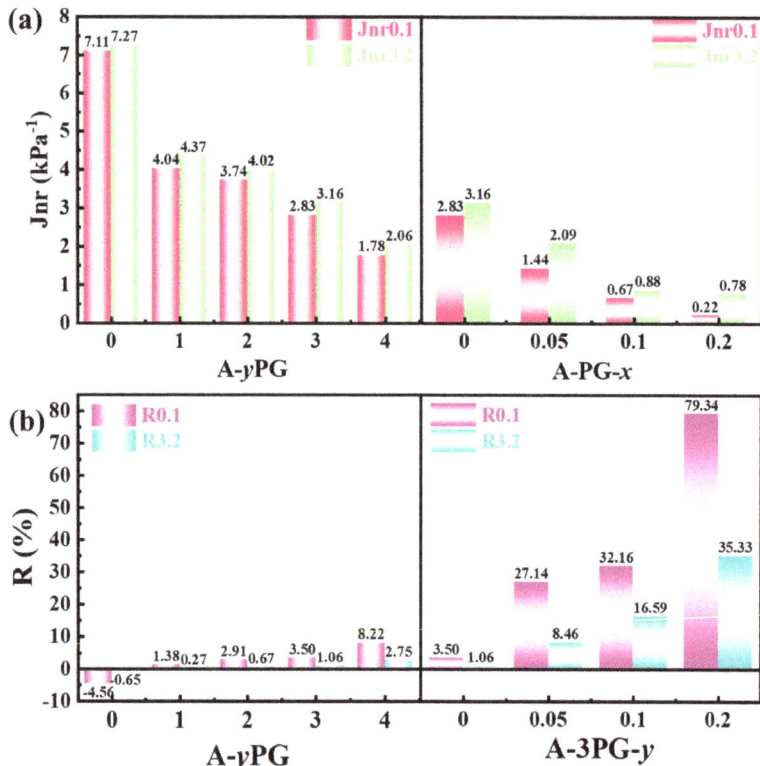

Figure 6. (a) Non-recoverable creep compliance and (b) Percent recovery of different modified asphalts.

3.5. Morphology of Modified Asphalts

The dispersibility of a polymer in asphalt is extremely important when it comes to the properties of asphalt [24,29]. The fluorescence microscope test is considered to be the simplest but most valuable method to analyze the morphology of asphalt modified with a polymer [30,31]. The distribution and phase structure of linear and crosslinked POE-*g*-GMA in the asphalt was characterized using a 3D laser microscope, and the fluorescent images with a 200× magnification are shown in Figure 7. The POE-*g*-GMA particles with different degrees of crosslinking lit up with a greenish-yellow glow; they show a dot structure and good dispersion in the asphalt. As the figure inserted in Figure 7 shows, the grain size distribution curve gradually shifted toward a larger size, indicating that the particle size of PG-*x* increases with the increase in the degree of crosslinking, and the average sizes are 2.95 μm, 3.72 μm, 4.02 μm and 4.85 μm. This may be because crosslinking modification increased the viscosity of the polymer, and made it more and more difficult for the POE-*g*-GMA phase to break into smaller sizes [14]. Weak crosslinking had a small impact on the dispersibility of POE-*g*-GMA.

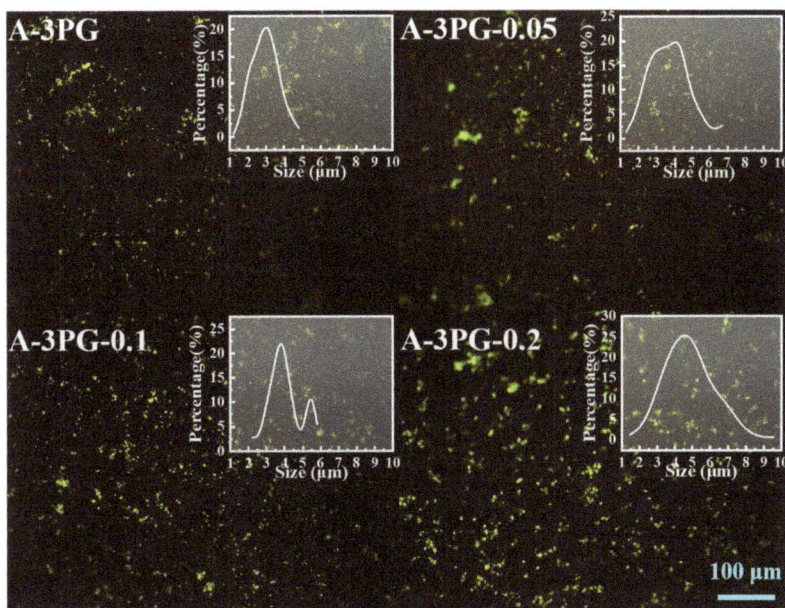

Figure 7. The fluorescence micrographs of A-3PG-*x* before thermal oxidation aging.

3.6. Thermal Oxidation Aging Resistance of Modified Asphalts

The comparison of physical parameters, including the penetration, softening point, and ductility of asphalts, obtained before and after RTFOT experiment can be applied to evaluate the aging resistance of asphalt, and the results are given in Table 4. It can be seem that the performance change trend of PG-*x* modified-asphalt is free of thermal oxidation aging. Furthermore, the physical parameters of modified asphalt after aging change in the same way as do those of unmodified asphalt; the penetration and ductility of all samples decrease, but the softening point is higher than that before aging.

Table 4. The physical properties of modified asphalts after thermal oxidation aging.

Sample	Penetration (0.1 mm) (25 °C, 100 g, 5 s)	Softening Point (°C)	Ductility (cm) (5 cm/min, 10 °C)
Asphalt	41.4 ± 0.3	54.4 ± 0.2	25.3 ± 0.6
A-3PG	30.4 ± 0.7	58.8 ± 0.6	8.0 ± 0.9
A-3PG-0.05	30.5 ± 0.2	61.5 ± 0.3	7.7 ± 0.7
A-3PG-0.1	30.9 ± 0.7	61.9 ± 0.2	7.5 ± 0.3
A-3PG-0.2	32.0 ± 0.3	62.5 ± 0.7	7.2 ± 0.6

Figure 8 displays the change in both the morphology of modified asphalts and the particle size of the polymer after aging. Because of degradation [32], the average size of polymer particles decreases to 2.65 μm, 3.02 μm, 3.49 μm and 4.3 μm. Moreover, the results of fluorescence microscopy show the difference between the residual dosages of PG-*x* in asphalt after aging, and the greater the *x*, the more residual dosages. This is because that crosslinking modification can inhibit the degradation of polymers during aging [33].

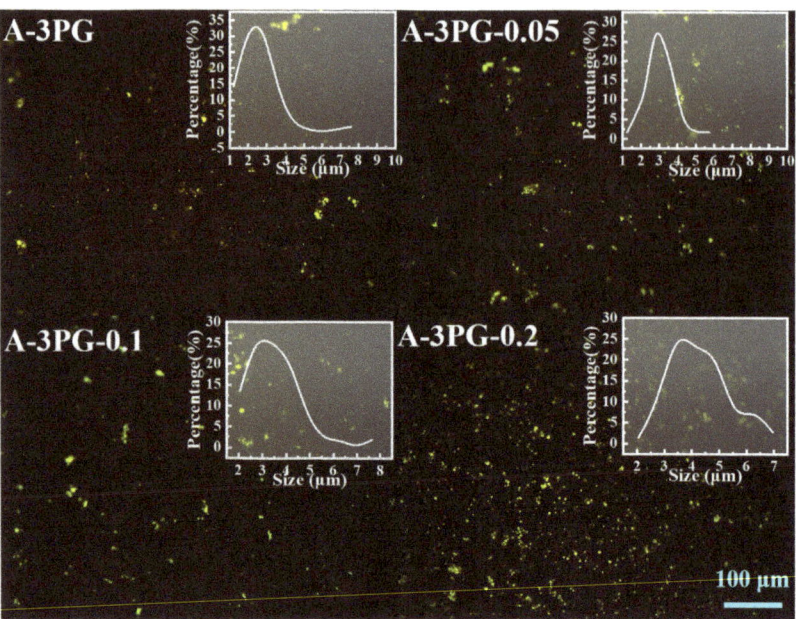

Figure 8. The fluorescence micrographs of A-3PG-*x* after thermal oxidation aging.

The property retention ratio (PRR) of penetration and ductility and the softening point increment (SPI) can be defined as aging indexes and they are useful for evaluating the extent of aging in asphalt; a large PRR and small SPI correspond to a low degree of aging [34–36]. Both the PRRs and SPI were conducted to study the aging resistance of the matrix and modified asphalts in this paper. The PRR and SPI and were calculated according to Equations (2) and (3), and the results of PRR and SPI are shown in Figure 9a. For all asphalts, the retention ratio of ductility is lower than that of penetration, indicating that aging has a great impact on ductility. The PRRs of all modified asphalts are greater than those of the matrix asphalt, while the SPIs are smaller than those of the matrix asphalt, indicating that both linear and crosslinked POE-*g*-GMA can effectively enhance the thermal oxidation aging resistance of asphalt. Modified asphalts have good anti-aging performance, mainly because POE-*g*-GMA is a kind of polymer with a saturated molecular backbone [37]. It is noteworthy that with an increase in x, the change trends of PRR and SPI are consistent for all asphalts, while PRRs gradually increase and SPIs decrease, reflecting a reduction in the aging degree of asphalt. The crosslinking modification of POE-*g*-GMA can further enhance the aging resistance of modified asphalt.

$$\mathrm{PRR} = \frac{\text{Property after aging}}{\text{Property before aging}} \times 100\% \qquad (2)$$

$$\mathrm{SPI} = \text{Softening point after aging} - \text{Softening point before aging} \qquad (3)$$

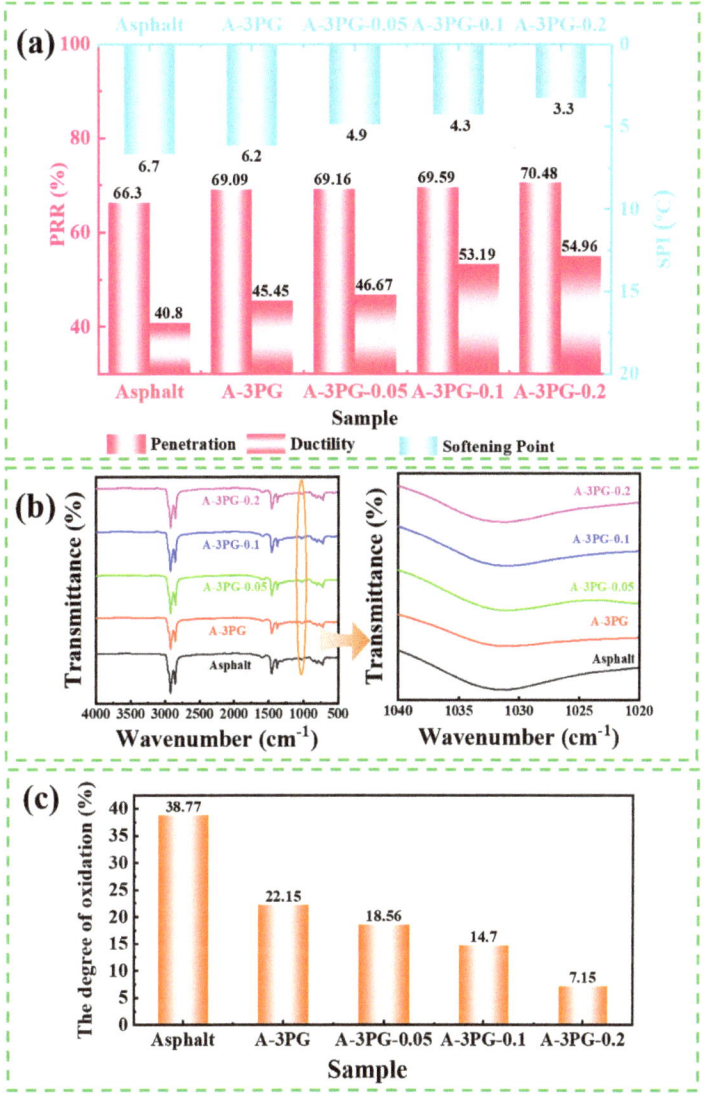

Figure 9. (a) The aging indexes of different modified asphalts; (b) the FT-IR spectra and (c) degree of oxidation of different modified asphalts after thermal oxidation aging.

ATR-FTIR was employed to detect the change in characteristic functional groups in asphalt during thermal oxidation aging. Oxygen-containing functional groups (e.g., sulfoxide and carbonyl) are often used to reflect the aging characteristics and indicators of asphalt. In this work, the sulfoxide index ($I_{S=O}$) was applied to reflect the changes in S=O in the asphalt. The vibration of the sulfoxide S=O functional group is located at about 1031 cm^{-1} [38], and the related FTIR spectra are shown in Figure 9b. The $I_{S=O}$ can be determined as follows [39]:

$$I_{S=O} = \frac{A_{1031}}{\sum A} \quad (4)$$

where A_{1031} is the peak area of S=O, and $\sum A$ represents the total peak area sum from 4000 cm^{-1} to 500 cm^{-1}. The degree of oxidation (%) was used to study the impact of PG-x on the anti-aging performance of asphalt, and was calculated via the following equation:

$$\text{The degree of oxidation} = \frac{I^*_{S=O} - I_{S=O}}{I_{S=O}} \times 100\% \tag{5}$$

where $I^*_{S=O}$ is the sulfoxide index after aging. As shown in Figure 9c, the result of $I_{S=O}$ of the matrix asphalt is much greater than that of modified asphalts, and decreases with the increased degree of the crosslinking of POE-g-GMA. This indicates that PG-x inhibits the oxidation reaction occurring in asphalt during thermal -oxidation aging. Moreover, the greater the crosslinking of POE-g-GMA, the better the effect.

Free radical theory can be applied to explain the mechanism of the thermal oxidation aging of asphalt [40,41]. The EPR technique has provided a unique and powerful tool to elucidate radical mechanisms [42,43]. The free radicals produced via thermal oxidation aging in matrix and modified asphalts were detected via EPR measurements. Via the double integration of experimental data, the overall yield of radicals was semi-quantitatively determined. As shown in Figure 10, the signal of the intensity of asphalt is the highest, while the signal intensity dramatically decreases after modification, and reduces with the increased degree of the crosslinking of POE-g-GMA. The results indicate that the yield of radicals decrease. This is because crosslinking places a restriction on the mobility of molecular chain segments, and the movement ability of macromolecular free radicals decreases, slowing down the reaction rate of free radical chain growth [44]. As a result, the yield of radicals decreases, which is indicative of good characteristics of resistance to thermal oxidation aging. Therefore, it is effective to enhance the thermal aging resistance of modified asphalt by regulating the mobility of polymer molecular chains through the crosslinking modification of a polymer modifier.

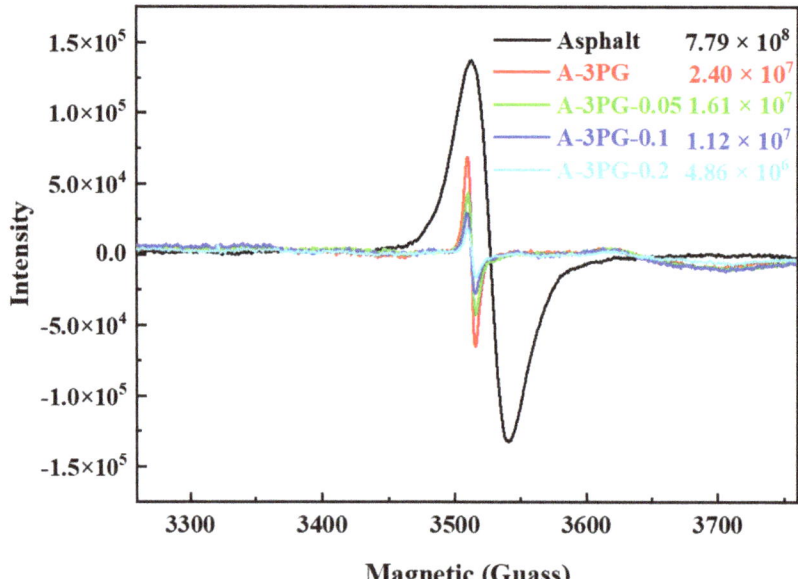

Figure 10. EPR spectra of the different modified asphalts after thermal oxidation aging.

4. Conclusions

POE-g-GMA was taken as the research object of this paper, and it was crosslinked via melt blending with DCP. Subsequently, samples of asphalt modified with linear and crosslinked POE-g-GMA were prepared. Thereout, the effect of the polymer molecular network structure on the properties of asphalt was studied. The conclusions are as follows:

1. The compatibility between linear or crosslinked POE-g-GMA and asphalt can be evaluated from macroscopic and microscopic perspectives via measuring the difference in softening points and SAXS characterizations. It is found that asphalt modified with linear or low-degree-crosslinked POE-g-GMA shows excellent hot storage stability compared to POE-modified asphalt. However, high crosslinking may restrain the reactivity of epoxy groups in POE-g-GMA, which has an adverse effect on its compatibility with asphalt.
2. With the modification of POE-g-GMA, the penetration reduces and the rheological properties increase as well as the softening point, which endows asphalt with a good ability to resist high-temperature rutting, while the crosslinking modification of POE-g-GMA further enhances the modification effect.
3. Moreover, the crosslinking modification of POE-g-GMA has a positive impact on the thermal oxidation aging resistance of modified asphalt for oxidation reactions are inhibited during the process of aging, and the movement ability of macromolecular free radicals are restricted, thereby slowing down the reaction rate of free radical chain growth. The EPR technique provided a unique and powerful tool to elucidate the radical mechanisms.

Because of its excellent performance and low cost, POE-g-GMA may be chosen as a good candidate to modify asphalt with good engineering benefits.

Author Contributions: Y.Z.: conceptualization, methodology, writing—original draft preparation, investigation, formal analysis, and visualization; P.Q.: visualization, methodology, investigation, and formal analysis; P.X.: validation, supervision, project administration, and funding acquisition; A.K.: validation, supervision, and project administration; C.J.: conceptualization, writing—reviewing and editing, and methodology; C.K.: investigation, writing—reviewing and editing, and methodology; Z.W.: investigation, visualization, and methodology; Y.L.: conceptualization, visualization, and investigation. All authors have read and agreed to the published version of the manuscript.

Funding: This work was financially supported by Jiangsu Funding Program for Excellent Postdoctoral Talent, the National Natural Science Foundation of China (grant number 52178439), the Jiangsu Science and Technology Association Promotion Project for the Youths (grant number 2021-090) and the Qing Lan Project of Jiangsu Province.

Institutional Review Board Statement: Not applicable.

Informed Consent Statement: Not applicable.

Data Availability Statement: Not applicable.

Conflicts of Interest: The authors declare no conflict of interest.

References

1. Liang, B.; Shi, K.; Niu, Y.F.; Liu, Z.C.; Zheng, J.L. Probing the modification mechanism of and customized processing design for SBS-modified asphalts mediated by potentiometric titration. *Constr. Build. Mater.* **2020**, *234*, 117385. [CrossRef]
2. Behnood, A.; Olek, J. Rheological properties of asphalt binders modified with styrene-butadiene-styrene (SBS), ground tire rubber (GTR), or polyphosphoric acid (PPA). *Constr. Build. Mater.* **2017**, *151*, 464–478. [CrossRef]
3. Xiao, F.P.; Amirkhanian, S.; Wang, H.N.; Hao, P.W. Rheological property investigations for polymer and polyphosphoric acid modified asphalt binders at high temperatures. *Constr. Build. Mater.* **2014**, *64*, 316–323. [CrossRef]
4. Bensason, S.; Stepanov, E.V.; Chum, S.P.; Hiltner, A.; Baer, E. Deformation of elastomeric ethylene-octene copolymers. *Macromolecules* **1997**, *30*, 2436–2444. [CrossRef]
5. Bensason, S.; Minick, J.; Moet, A.A.; Chum, S.P.; Hiltner, A.; Baer, E. Classification of homogeneous ethylene-octene copolymers based on comonomer content. *J. Polym. Sci. Polym. Phys.* **1996**, *34*, 1301–1315. [CrossRef]
6. Behnood, A.; Gharehveran, M.M. Morphology, rheology, and physical properties of polymer-modified asphalt binders. *Eur. Polym. J.* **2019**, *112*, 766–791. [CrossRef]
7. Li, M.R.; Chen, X.; Cong, P.L.; Luo, C.J.; Zhu, L.Y.; Li, H.Y.; Zhang, Y.M.; Chao, M.; Yan, L.K. Facile synthesis of polyethylene-modified asphalt by chain end-functionalization. *Compos. Commun.* **2022**, *30*, 101088. [CrossRef]

8. Polacco, G.; Stastna, J.; Biondi, D.; Zanzotto, L. Relation between polymer architecture and nonlinear viscoelastic behavior of modified asphalts. *Curr. Opin. Colloid Interface Sci.* **2006**, *11*, 230–245. [CrossRef]
9. Behnood, A.; Olek, J. Stress-dependent behavior and rutting resistance of modified asphalt binders: An MSCR approach. *Constr. Build. Mater.* **2017**, *157*, 635–646. [CrossRef]
10. Nien, Y.H.; Yeh, P.H.; Chen, W.C.; Liu, W.T.; Chen, J.H. Investigation of flow properties of asphalt binders containing polymer modifiers. *Polym. Compos.* **2008**, *29*, 518–524. [CrossRef]
11. Padhan, R.K.; Leng, Z.; Sreeram, A.; Xu, X. Compound modification of asphalt with styrene-butadiene-styrene and waste polyethylene terephthalate functionalized additives. *J. Clean. Prod.* **2020**, *277*, 124286. [CrossRef]
12. Navarro, F.J.; Partal, P.; García-Morales, M.; Martínez-Boza, F.J.; Gallegos, C. Bitumen modification with a low-molecular-weight reactive isocyanate-terminated polymer. *Fuel* **2007**, *86*, 2291–2299. [CrossRef]
13. Research Institute of Highway Ministry of Transport. *JTG E20-2011*; Standard Test Methods of Bitumen and Bituminous Mixtures for Highway Engineering. Ministry of Transportation: Beijing, China, 2011.
14. Qu, Y.D.; Chen, Y.H.; Ling, X.Y.; Wu, J.L.; Hong, J.T.; Wang, H.T.; Li, Y.J. Reactive micro-crosslinked elastomer for supertoughened polylactide. *Macromolecules* **2022**, *55*, 7711–7723. [CrossRef]
15. Zhang, Y.J.; Liu, X.Y.; Li, Y.Q.; Wu, D.F.; Zhang, M. Understanding the fracture toughness of gadolinium- and lead-containing plexiglass. *Polym. Eng. Sci.* **2023**, *63*, 441–453. [CrossRef]
16. Ouyang, C.F.; Gao, Q.; Shi, Y.T.; Shan, X.Q. Compatibilizer in waste tire powder and low-density polyethylene blends and the blends modified asphalt. *J. Appl. Polym. Sci.* **2011**, *123*, 485–492. [CrossRef]
17. Padhan, R.K.; Sreeram, A. Enhancement of storage stability and rheological properties of polyethylene (PE) modified asphalt using cross linking and reactive polymer based additives. *Constr. Build. Mater.* **2018**, *188*, 772–780. [CrossRef]
18. Yu, C.H.; Hu, K.; Yang, Q.L.; Wang, D.D.; Zhang, W.G.; Chen, G.X.; Kapyelata, C. Analysis of the storage stability property of carbon nanotube/recycled polyethylene-modified asphalt using molecular dynamics simulations. *Polymers* **2021**, *13*, 1658. [CrossRef]
19. Li, J.; Zhang, Y.X.; Zhang, Y.Z. The research of GMA-g-LDPE modified Qinhuangdao bitumen. *Constr. Build. Mater.* **2008**, *22*, 1067–1073. [CrossRef]
20. Redelius, P.; Soenen, H. Relation between bitumen chemistry and performance. *Fuel* **2015**, *140*, 34–43. [CrossRef]
21. Zhang, Y.J.; Wang, C.H.; Wu, D.F.; Guo, X.T.; Yu, L.; Zhang, M. Probing the effect of straight chain fatty acids on the properties of lead-containing plexiglass. *React. Chem. Eng.* **2021**, *6*, 1628–1634. [CrossRef]
22. Ge, D.D.; Yan, K.Z.; You, L.Y.; Wang, Z.X. Modification mechanism of asphalt modified with Sasobit and Polyphosphoric acid (PPA). *Constr. Build. Mater.* **2017**, *143*, 419–428. [CrossRef]
23. Zhang, M.Y.; Hao, P.W.; Dong, S.; Li, Y.; Yuan, G.A. Asphalt binder micro-characterization and testing approaches: A review. *Measurement* **2020**, *151*, 107255. [CrossRef]
24. Li, M.R.; Luo, C.J.; Zhu, L.Y.; Li, H.Y.; Cong, P.L.; Feng, Y.Y.; Yan, L.K. A novel epoxy-terminated polyethylene modified asphalt with low-viscosity and high storage stability. *Constr. Build. Mater.* **2022**, *335*, 127473. [CrossRef]
25. Azarhoosh, A.; Koohmishi, M. Investigation of the rutting potential of asphalt binder and mixture modified by styrene-ethylene/propylene-styrene nanocomposite. *Constr. Build. Mater.* **2020**, *255*, 119363. [CrossRef]
26. Xiao, Y.; Chang, X.W.; Yan, B.X.; Zhang, X.S.; Yunusa, M.; Yu, R.E.; Chen, Z.W. SBS morphology characteristics in asphalt binder and their relation to viscoelastic properties. *Constr. Build. Mater.* **2021**, *301*, 124292. [CrossRef]
27. Gong, Y.F.; Pang, Y.Z.; Li, F.Y.; Jin, W.D.; Bi, H.P.; Ma, Y.L. Analysis of the influence of SBS content and structure on the performance of SBS/CR composite modified asphalt. *Adv. Mater. Sci. Eng.* **2021**, *2021*, 5585891. [CrossRef]
28. Lv, S.T.; Xia, C.D.; Yang, Q.; Guo, S.C.; You, L.Y.; Guo, Y.P.; Zheng, J.L. Improvements on high-temperature stability, rheology, and stiffness of asphalt binder modified with waste crayfish shell powder. *J. Clean. Prod.* **2020**, *264*, 121745. [CrossRef]
29. Vargas, M.A.; Vargas, M.A.; Sánchez-Sólis, A.; Manero, O. Asphalt/polyethylene blends: Rheological properties, microstructure and viscosity modeling. *Constr. Build. Mater.* **2013**, *45*, 243–250. [CrossRef]
30. Sengoz, B.; Topal, A.; Isikyakar, G. Morphology and image analysis of polymer modified bitumens. *Constr. Build. Mater.* **2009**, *23*, 1986–1992. [CrossRef]
31. Dong, F.Q.; Zhao, W.Z.; Zhang, Y.Z.; Wei, J.M.; Fan, W.Y.; Yu, Y.J.; Wang, Z. Influence of SBS and asphalt on SBS dispersion and the performance of modified asphalt. *Constr. Build. Mater.* **2014**, *62*, 1–7. [CrossRef]
32. Yan, C.Q.; Huang, W.D.; Xiao, F.P.; Lv, Q. Influence of polymer and sulphur dosages on attenuated total reflection Fourier transform infrared upon Styrene-Butadiene-Styrenemodified asphalt. *Road Mater. Pavement Des.* **2017**, *20*, 1586–1600. [CrossRef]
33. Luo, Y.H.; Li, G.L.; Chen, L.; Hong, F.F. Preparation and evaluation of bacterial nanocellulose/hyaluronic acid composite artificial cornea for application of corneal transplantation. *Biomacromolecules* **2023**, *24*, 201–212. [CrossRef] [PubMed]
34. Geng, J.G.; Meng, H.H.; Xia, C.Y.; Chen, M.Y.; Lu, T.Y.; Zhou, H. Effect of dry-wet cycle aging on physical properties and chemical composition of SBS-modified asphalt binder. *Mater. Struct.* **2021**, *54*, 120. [CrossRef]
35. Peng, C.; Guo, C.; You, Z.P.; Xu, F.; Ma, W.B.; You, L.Y.; Li, T.J.; Zhou, L.Z.; Huang, S.F.; Ma, H.C.; et al. The effect of waste engine oil and waste polyethylene on UV aging resistance of asphalt. *Polymers* **2020**, *12*, 602. [CrossRef] [PubMed]
36. Li, L.M.; Guo, Z.Y.; Ran, L.F.; Zhang, J.W. Study on low-temperature cracking performance of asphalt under heat and light together conditions. *Materials* **2020**, *13*, 1541. [CrossRef]

37. Wang, S.F.; Xie, Y.G. Crumb tire rubber polyolefin elastomer modified asphalt with hot storage stability. *Prog. Rubber Plast. Recycl. Technol.* **2016**, *32*, 25–38. [CrossRef]
38. Liu, S.J.; Zhou, S.B.; Xu, Y.S. Evaluation of cracking properties of SBS-modified binders containing organic montmorillonite. *Constr. Build. Mater.* **2018**, *175*, 196–205. [CrossRef]
39. Ouyang, C.F.; Wang, S.F.; Zhang, Y.; Zhang, Y.X. Improving the aging resistance of styrene-butadienestyrene tri-block copolymer modified asphalt by addition of antioxidants. *Polym. Degrad. Stab.* **2006**, *91*, 795–804. [CrossRef]
40. Liu, S.; Shan, L.Y.; Li, G.N.; Underwood, B.S.; Qi, C. Molecular-based asphalt oxidation reaction mechanism and aging resistance optimization strategies based on quantum chemistry. *Mater. Des.* **2022**, *223*, 111225. [CrossRef]
41. Smith, L.M.; Aitken, H.M.; Coote, M.L. The fate of the peroxyl radical in autoxidation: How does polymer degradation really occur? *Acc. Chem. Res.* **2018**, *51*, 2006–2013. [CrossRef]
42. Li, J.W.; Ye, X.; Zhao, Y.K.; Yang, D.; Li, D.D.; Han, C.C.; Li, X. Structure-performance evolution on thermal-oxidative aging of CeO_2/SBS co-modified asphalt. *Int. J. Pavement Eng.* **2022**, *23*, 1–9. [CrossRef]
43. Pipintakos, G.; Soenen, H.; Ching, H.Y.V.; Velde, C.V.; Van Doorslaer, S.; Lemière, F.; Varveri, A.; Van den Bergh, W. Exploring the oxidative mechanisms of bitumen after laboratory short- and long-term ageing. *Constr. Build. Mater.* **2021**, *289*, 123182. [CrossRef]
44. Zhang, Y.J.; Chen, Z.Y.; Zhao, R.; Wang, K.; Wu, D.F.; Wang, C.H.; Zhang, M. Insight into the role of free volume in irradiation resistance to discoloration of lead-containing plexiglass. *J. Appl. Polym. Sci.* **2022**, *139*, 51545. [CrossRef]

Disclaimer/Publisher's Note: The statements, opinions and data contained in all publications are solely those of the individual author(s) and contributor(s) and not of MDPI and/or the editor(s). MDPI and/or the editor(s) disclaim responsibility for any injury to people or property resulting from any ideas, methods, instructions or products referred to in the content.

Article

Performance of High-Dose Reclaimed Asphalt Mixtures (RAPs) in Hot In-Place Recycling Based on Balanced Design

Lei Jiang [1], Junan Shen [2,*] and Wei Wang [3]

1. Jiangsu Technology Industrialization and Research Center of Ecological Road Engineering, Suzhou University of Science and Technology, Suzhou 215011, China; jl951329414@163.com
2. Department of Civil Engineering and Construction, Georgia Southern University, Statesboro, GA 30458, USA
3. School of Civil Engineering, Chongqing Jiaotong University, Chongqing 400074, China; wwangcqjtu@outlook.com
* Correspondence: jshen@georgiasouthern.edu

Abstract: This study endeavors to employ a balanced design methodology, aiming to equilibrate the resistance to rutting and cracking exhibited by hot in-place recycling asphalt mixtures containing a high dose of reclaimed asphalt pavement (RAP). The primary goal is to ascertain the optimal amount of new binder necessary for practical engineering applications, ensuring a balanced rutting and crack resistance performance of recycled asphalt mixtures. The investigation mainly employed wheel-tracking tests and semi-circular bending tests to assess the rutting and cracking performance of recycled asphalt mixtures with a different dose of RAP (in China, it is common to use RAP with 80% and 90% content as additives for preparing hot in-place recycling asphalt mixtures), and varying quantities of new binders (10%, 20%, and 30% of the binder content in the total RAP added). The results indicated that the addition of new binder reduced the resistance to rutting of the recycling asphalt mixtures but improved their resistance to cracking. Furthermore, for the recycling asphalt mixture with 80% RAP content aged for 5 days, the optimal new binder content is 1.52%, while the mixture with 90% RAP content requires 1.23% of new binder. After 10 days of aging, the optimal new binder content for the recycling asphalt mixture with 80% RAP content is 1.55%, while the mixture with 90% RAP content requires 1.28% of new binder.

Keywords: high RAP content; laboratory-accelerated aging test; hot in-place recycling asphalt mixtures; road performance

Citation: Jiang, L.; Shen, J.; Wang, W. Performance of High-Dose Reclaimed Asphalt Mixtures (RAPs) in Hot In-Place Recycling Based on Balanced Design. *Materials* **2024**, *17*, 2096. https://doi.org/10.3390/ma17092096

Academic Editor: F. Pacheco Torgal

Received: 12 March 2024
Revised: 25 April 2024
Accepted: 26 April 2024
Published: 29 April 2024

Copyright: © 2024 by the authors. Licensee MDPI, Basel, Switzerland. This article is an open access article distributed under the terms and conditions of the Creative Commons Attribution (CC BY) license (https://creativecommons.org/licenses/by/4.0/).

1. Introduction

Flexible pavements often experience various types of on-site damage due to factors such as prolonged and increased vehicle loading and severe climate fluctuations. Rutting and cracking are two of the common damages on flexible pavements as a result, which not only reduce the smoothness of the road surface but also affect driving comfort and safety. The performance of pavement materials is a critical factor contributing to these damages. To improve pavement material performance, adjustments to the composition or the development of new performance evaluation standards and design methods can be implemented.

Recycled/reclaimed asphalt mixture (RAP) has gradually gained widespread adoption in the highway industry, with an increased dose of RAP in recycling mixtures, enabling the 100% recycling of reclaimed asphalt pavement road surfaces, yielding multiple benefits. Firstly, it contributes to the conservation of natural aggregates and binder, among other natural resources. Secondly, it effectively reduces the emission of industrial waste, thereby mitigating environmental pollution. Additionally, recycling asphalt mixture helps lower energy consumption [1,2]. The recycled material, HMA, exhibited superior mechanical and resilient modulus performances. Furthermore, higher tensile strength ratios of the recycled material mixtures indicated a greater resistance to water damage [3]. However, due to durability issues associated with a high dose of RAP in recycling asphalt mixtures, and considering factors such as fatigue and water damage, many countries impose restrictions on the maximum allowable RAP content [4,5]. Currently, in hot in-place recycling technologies, the percentage of RAP in the asphalt mixture is relatively high but generally does not exceed 70–80%. During the mixture design phase, the primary consideration is that the aged binder should fully dissolve and mix with rejuvenating agents or new binder during the mechanical blending process, aiming for the complete regeneration of the aged binder. However, studies indicate that achieving this premise is challenging in practical engineering applications [6–10]. Therefore, the recycling degree of RAP plays a crucial role in the preparation of hot recycled mixtures, and current recycling methods lack such an essential characterization in this aspect [11]. Furthermore, the rejuvenating degree of RAP depends on its degree of aging, presenting another crucial factor, i.e., the fusion recycling level between the aged RAP binder and the newly introduced virgin binder. Recycling of aged RAP binder is considered an effective form of binder that forms a new binder mortar, bonding the aggregates together to create recycled asphalt mixtures [12]. Researchers categorize the mixing scenario between RAP and asphalt mixtures into three possible situations: (a) all RAP binder is mobilized and mixed with the original binder; (b) no RAP binder is activated, and RAP acts like black aggregates; and (c) some RAP binder is activated and mixed with the virgin binders [13–15].

Therefore, in designing recycling asphalt mixtures, it is necessary to better understand the mixing phenomenon and adjust the design procedures accordingly. Some studies have introduced an intrinsic property called the Degree of binder Activity (DoA) and redefined two well-known concepts, namely the Degree of Blending (DoB) and the Degree of binder Availability (DoAv). Practical recommendations have been proposed to incorporate them into the mixture design process [16].

Traditional design methods such as Superpave and Marshall methods typically determine the optimal binder content for mixture design based on volumetric parameters and estimate mixture performance. However, they do not guarantee the performance of asphalt mixtures. These methods have high requirements for material performance, especially for the stability, deformation resistance, and durability of binders. However, the situation of binder in high-dose RAP is complex, and the intrinsic properties of aged binder (DoA) as well as the DoB and DoAv of aged binder with rejuvenating agents are not yet fully known at this point. The cracking of asphalt pavement during use is often one of the primary factors affecting the service life of asphalt pavement designed by conventional methods [17].

Nowadays, there is an increasing emphasis on utilizing performance-based evaluation for asphalt mixture design and tests. The main aim of these tests is to employ established performance specifications to identify substandard asphalt mixtures, such as those prone to rutting or cracking, at various stages of the asphalt mixture design or production. Several studies have proved the ability of such tests to assess the performance of asphalt mixtures [18–35].

In 2006, Zhou et al. [36] introduced the Balanced Mix Design (BMD), a method aimed at finding a balance between rutting and cracking resistance to determine the optimum binder content of asphalt mixtures. Feng et al. [37], in 2011, applied rutting tests and low-temperature flexural beam tests for the balanced design of asphalt mixtures regarding cracking and rutting resistance. In 2012, Walubita and Hu et al. [38] conducted comparative experiments using the Accelerated Loading Facility (ALF) system, comparing asphalt mixtures designed by the Texas Department of Transportation (TxDOT) method with those designed by the Balanced Mix Design. The results demonstrated that asphalt mixtures designed by the Balanced Mix Design method exhibited superior performance. In the same year, Zhou et al. [39] first applied the Balanced Mix Design to RAP mixtures to determine an optimal binder content.

This study endeavors to investigate high-percentage hot in-place recycling asphalt mixtures, specifically focusing on augmenting the proportion of RAP content. Through the application of performance-balanced design, a specific quantity of supplementary new binder is incorporated to offset the unactivated fraction of aged binder present in RAP, thereby guaranteeing superior resistance to rutting and cracking in the recycling asphalt mixture.

To attain this goal, this study employed dense-graded asphalt mixture AC-13 and prepared different levels of aged RAP mixtures through laboratory short-term and long-term aging processes. The research entails the formulation of recycling asphalt mixtures incorporating RAP content levels of 80% and 90% with different dosages of virgin binders. Subsequently, this study assesses the rutting and cracking resistance performance subsequent to the incorporation of diverse quantities of new binder.

2. Laboratory Experimental Program

Figure 1 illustrates the experimental plan of this study. Firstly, raw materials (i.e., binder, aggregates, and rejuvenating agents) were collected from China, Suzhou, Suzhou Sanchuang Road Engineering Co., Ltd. Then, laboratory-aged asphalt mixtures were prepared using the accelerated aging process at different aging levels (5 days and 10 days). Next, the preparation of hot in-place recycling asphalt mixtures with different lab-prepared RAP contents (80% and 90%) was carried out. Subsequently, representative tests for evaluating the cracking and rutting performance were selected. Finally, based on the concept of BMD, the optimal additional binder content was determined. The following sections discuss the cracking and rutting performance tests selection, properties of raw materials, sample preparation, and testing procedure.

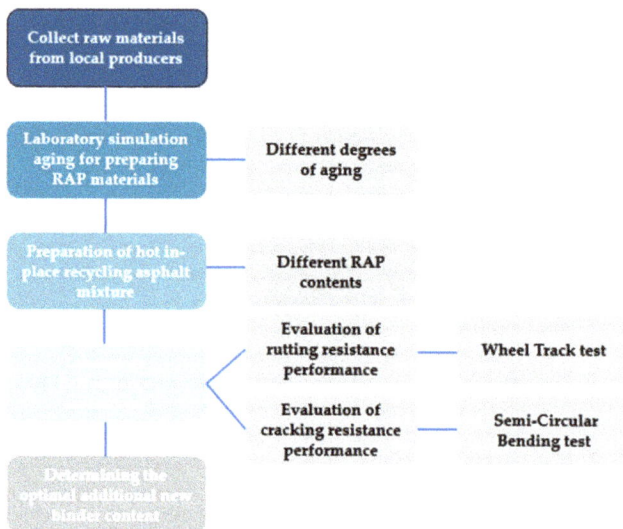

Figure 1. The experimental plan.

2.1. Properties of Raw Materials

2.1.1. Binder

Utilizing styrene–butadiene–styrene (SBS)-modified binder, the key technical specifications are presented in Table 1. These properties are determined by the supplier. Testing methods were all conducted in accordance with the specifications [40].

Table 1. Main properties of binder.

Item	25 °C Penetration Degree/(0.1 mm)	Softening Point/°C	5 °C Ductility/cm	Residue after RTFOT		
				Mass Change/%	Penetration Ratio/%	5 °C Residual Ductility/cm
Specification	30–70	>65	>20	≤±1.0	≥65	≥15
Test results	54	78	36	0.22	73	25
Test methods	T0604	T0606	T0605	T0609	T0604	T0605

2.1.2. Mineral Material

The basalt aggregate used in this study is detailed in Table 2, which provides the particle size distribution for each aggregate gradation and mineral powder. The properties for both coarse and fine aggregates, as well as mineral powder, meet the specifications as required.

Table 2. The particle size distribution of the aggregates used in the mix design.

Sieve Size/mm	Percentage Passing/%.				
	10–15 mm	5–10 mm	3–5 mm	0–3 mm	Mineral Powder
16	100.0	100.0	100.0	100.0	100.0
13.2	62.8	100.0	100.0	100.0	100.0
9.5	11.5	96.6	100.0	100.0	100.0
4.75	0.2	7.1	93.6	100.0	100.0
2.36	0.2	0.2	5.5	83.5	100.0
1.18	0.2	0.2	1.6	62.1	100.0
0.6	0.2	0.2	1.1	41.7	100.0
0.3	0.2	0.2	1.1	27.8	100.0
0.15	0.2	0.2	1.1	20.0	100.0
0.075	0.2	0.2	1.1	13.8	98.3

2.1.3. Rejuvenator

The primary component of the rejuvenating agent is low-viscosity mineral oil derived from petroleum. Its main function is to replenish the saturated and aromatic fractions lost during the aging process of binder, aiming to restore some of the pavement performance of modified binder and secondarily recycled mixtures. Therefore, essential characteristics of a high-quality rejuvenating agent include appropriate viscosity, favorable rheological properties, sufficient aromatic content, and a lower ratio of thin film oven test viscosity.

The technical specifications of the rejuvenating agent used in this study are presented in Table 3. These properties are determined by the supplier. It must be noted that the dose of rejuvenator is very efficient to soften the aged binders. The dose should be determined according to the target (i.e., PG grade or penetration) of the recycled binders. Based on our previous laboratory research results, the dose of this project incorporated into the recycling asphalt mixture is 5% of the binder content in the RAP [41].

Table 3. The technical specifications of the rejuvenating agent.

Test Item	Test Results	Technical Requirement
60 °C viscosity/(mm^2·s^{-1})	59.2	50~175
Flash point/°C	242	≥220
Saturated fraction/%	17.31	≤30
25 °C density/(g·cm^{-3})	1.017	Actual measurement
Film oven viscosity ratio	1.22	≤3
Quality change in film oven/%	−1.174	≤4, ≥−4
Appearance	Brown-black viscous liquid	

2.2. Lab Preparation of RAP

Accelerated aging was conducted according to JTG E20-2011 [40]. The aging steps are as shown in Figure 2. The aggregates were placed into the mixing pot heated to 185 °C, after being heated to 170 °C in an oven. After mixing aggregates with SBS-modified binder for 90 s, the mixing machine was paused to add the preheated mineral powder. Mixing continued for another 90 s to obtain the SBS-modified asphalt mixtures. The specimens must undergo short-term aging before long-term aging. The mixtures were first spread and placed in an oven at a temperature of 135 °C for 4 h under forced ventilation conditions as short-term aging. Then, the asphalt mixture was compacted into Marshall specimens and placed in an oven at a temperature of 85 °C. It underwent continuous heating for 5 days (assumed 3–5 years' aging in the field) and 10 days (assumed 5–10 years' aging) under forced ventilation conditions as long-term aging [40].

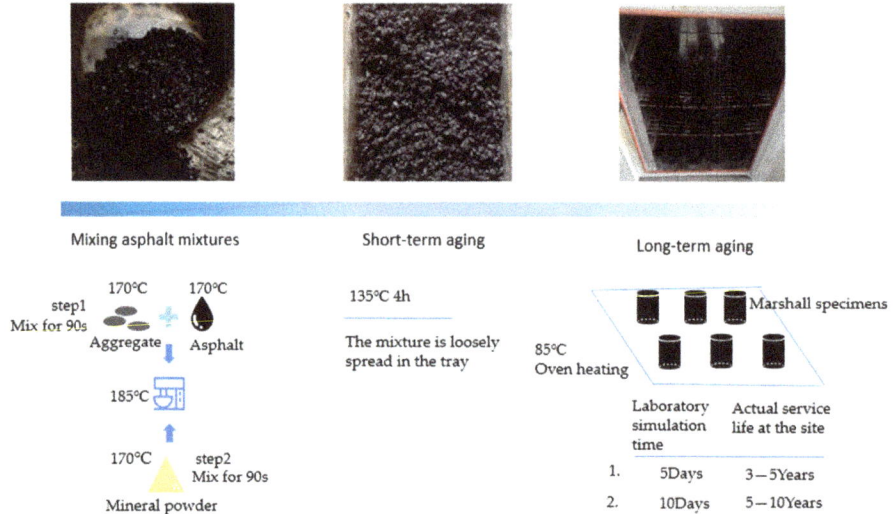

Figure 2. The aging steps (delete one of the layers to save space).

2.3. Lab-Prepared Hot In-Place Recycling Asphalt Mixture

The preparation process is illustrated in Figure 3. In the hot in-place recycling asphalt mixture, due to the presence of RAP, the RAP should be preheated at 170 °C for 2 h before mixing the recycling asphalt mixture. For optimal results, the rejuvenating agent should be mixed first with the old material during the mixing process. This helps soften and disperse the old material while promoting uniform mixing between the new and old materials. According to research findings, extending the mixing time can improve the performance of recycling asphalt mixtures in actual engineering applications. The laboratory mixing pot is simple in structure, but based on previous research experience, it is necessary to mix RAP with the rejuvenating agent for 120 s to achieve uniform mixing of the hot recycling asphalt mixture. Then, the preheated new aggregates are added to the mixing pot and mixed for 90 s. Subsequently, the preheated new binder, heated to 170 °C, is added and mixed for another 90 s. Finally, the preheated mineral powder is added and mixed for 90 s.

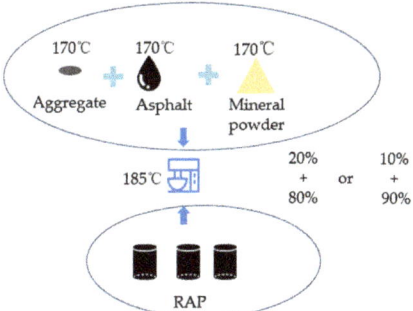

Figure 3. The process of preparing hot in-place recycling asphalt mixture.

This study selected two different RAP blending percentages (80% and 90%) with the aim of increasing the utilization rate of RAP in actual engineering applications. The issue arising from high percentages of RAP utilization is the impact on the availability of aged binder during the recycling process. This study addresses this issue by incorporating additional amounts of new binder (i.e., 10%, 20%, and 30% of the binder content in the RAP).

The samples are grouped and labeled according to different variables, as shown in the Table 4:

Table 4. Grouping of recycling asphalt mixtures.

Degree of Aging	RAP Content 80%	90%
5 Days	A	C
10 Days	B	D

Group A represents the recycling asphalt mixture (RAP) prepared with 80% RAP content after five days of aging. Groups B, C, D, and so on follow in sequence.

2.3.1. Mix Proportion

As the RAP used in this study underwent simulated aging in the laboratory, the damage and loss of aggregates during its aging process are controllable. Under both RAP content levels, the gradation curves remain consistent. The gradation curves of the recycling asphalt mixture are shown in Figure 4.

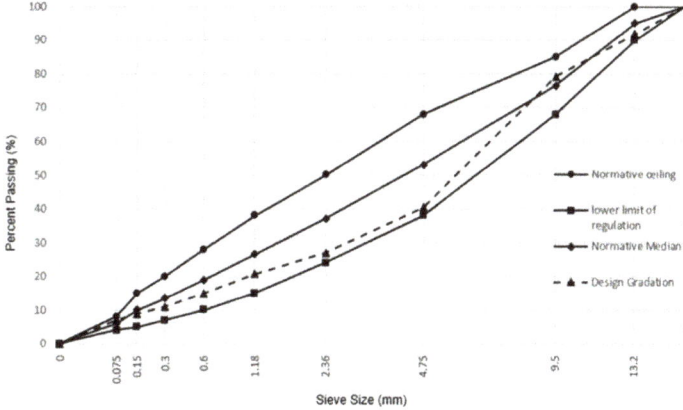

Figure 4. The gradation curves of recycling asphalt mixtures.

2.3.2. The Binder Content of Recycling Asphalt Mixture

The binder content was experimentally obtained, and the binder content of the AC-13 conventional hot mix asphalt mixture was around 4.7%. This study took 4.7% as the initial estimated binder content value and calculated the initial estimated new binder content value (Pnb) based on the RAP content. Formed specimens were prepared at three different levels of new binder content: 10%, 20%, and 30% of the aged binder content in the RAP (control group at 0% RAP). The optimal new binder content value was determined using the balanced design method. The aged binder content extracted from the RAP was determined to be 4.6%. According to Equation (1), the estimated new binder content values for 80% and 90% RAP contents were calculated, as shown in Table 5.

$$P_{nb} = P_b - P_{ob} \times \frac{n}{100} \quad (1)$$

where P_{nb} is the estimated new binder content (%), P_b is the total binder content of the hot recycling mixture (%), P_{ob} is the aged binder content in the RAP material (%), n is the RAP material mixing rate (%).

Table 5. Estimated new binder content values.

Estimated Total Binder Content/%	RAP Content/%	Aged Binder Content/%	Estimated New Binder Content/%	New Binder Addition Content/%
4.7	80	4.6	1.02	1.02 1.388 1.756 2.124
4.7	90	4.6	0.56	0.56 0.974 1.388 1.802

2.4. Selected Performance Tests and Indicators

2.4.1. Performance Assessment Tests

In this study, the authors selected the two most representative tests for cracking and rutting resistance: the semi-circular bending (SCB) test and wheel track test.

The wheel-track test, conducted according to JTG E20-2011, forms rutting test specimens with dimensions of 300 mm × 300 mm × 50 mm using a rolling method. After forming, the specimens were placed at 60 °C for at least 5 h but not exceeding 12 h. The specimens, along with the test mold, were placed into the wheel-track testing machine. The contact pressure between the test wheel and the specimen was set at 0.7 ± 0.05 Mpa, the test temperature was maintained at 60 °C, and the rolling duration was 60 min with a rolling speed (N) of 42 times per minute. The shaping apparatus and the rutting tester are depicted in Figure 5. Each sample group was tested with four identical samples, totaling 64 samples.

The semi-circular bending test was conducted according to the reference [42]. The recycling mixtures were molded into Marshall specimens. Circular specimens with a thickness of 50 mm were sawed from the middle of the Marshall specimens, and then symmetrically split open from the middle to obtain semi-circular specimens. Finally, a notch was cut at the bottom of the semi-circular specimens with a depth of 10 mm and a width of 2 mm, thus obtaining the semi-circular bending specimens [43–46]. Each sample group was tested with four identical samples, totaling 64 samples.

 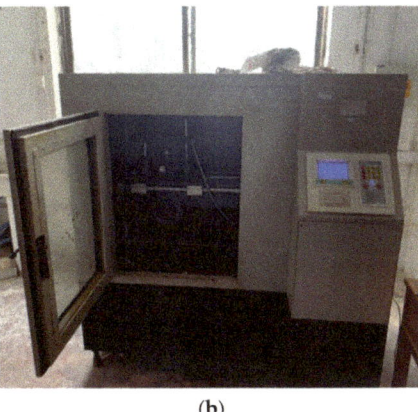

(a)   (b)

Figure 5. The rutting test equipment: (**a**) rutting sample forming machine, (**b**) wheel-track test machine.

Before the test, the specimens were placed in a constant temperature chamber and kept at 15 °C for 4 h. Subsequently, the specimens were symmetrically placed on supports with a spacing of 8 cm, and the test was conducted under the set temperature and loading rate conditions. The test was terminated when the specimen failed. The semi-circular bending strength test was conducted in the UTM130 multifunctional material testing system. To avoid impact failure, it was necessary to ensure contact between the upper loading head and the specimen before loading. Once the contact force stabilized at around 0.1 KN, loading of the specimen began until failure. The displacement change data at the mid-span loading head were automatically recorded using a linear variable displacement transducer (LVDT), while the change data of the vertical pressure and vertical displacement were recorded using an automatic data acquisition instrument. Figure 6 shows the semi-circular specimen during the test.

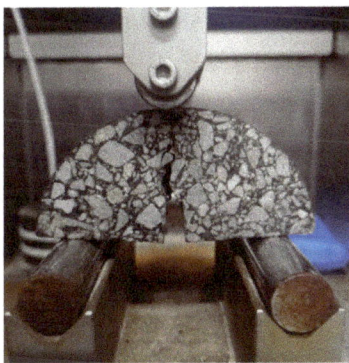

Figure 6. Semi-circular specimen test set.

2.4.2. Performance Assessment Indicators

The performance evaluation index for rutting resistance is the dynamic stability, calculated according to the Formula (2) specified in JTG E20-2011 T0719-2011. It reflects the deformation degree of the asphalt mixture under the action of external forces during high-speed driving. On highways, vehicles exert significant dynamic loads on the road surface. If the dynamic stability of the asphalt mixture is insufficient, deformation, looseness, or damage may occur, thereby affecting the service life of the pavement and driving safety. The formula is as follows:

$$DS = \frac{(t_2 - t_1) \times N}{d_2 - d_1} \times C_1 \times C_2 \quad (2)$$

where DS is the dynamic stability (times/mm), t_1 is 45 min, t_2 is 60 min, d_1, d_2 is the deformation amount (mm) corresponding to time t_1, t_2, C_1 is the coefficient for the type of testing machine, which operates in a reciprocating manner driven by a crank-link mechanism and is 1.0, C_2 is the coefficient for the specimen, which is prepared in the laboratory with a width of 300 mm and is 1.0, N is the reciprocating rolling speed of the test wheel, which is usually 42 times per minute.

According to the "Technical Specification for Highway Asphalt Pavement Construction: JTGF40-2004" [47], the dynamic stability of the modified asphalt mixture should be greater than 3000 times/mm.

The performance evaluation index for crack resistance is the flexibility index (FI). "FI" reflects the flexibility and crack resistance of the asphalt mixture. In cold climates, the road surface is susceptible to temperature fluctuations. If the asphalt mixture's crack resistance is insufficient, cracks and damage are more likely to occur, thereby affecting the pavement's service life and driving safety. The calculation Formulas (3)–(6) are as follows:

$$G_f = \frac{W_f}{A_{lig}} \quad (3)$$

where G_f represents the fracture energy, J/m^2; W_f represents the work of fracture, J, A_{lig} represents the ligament area, m^2.

The calculation formulas of W_f and A_{lig} are as follows: (4), (5)

$$W_f = \int_a^b P \, du \quad (4)$$

where P represents the load, N; u represents the deformation, m

$$A_{lig} = (r - a) \times t \quad (5)$$

where r represents the SCB specimen radius, m; a represents the depth of the pre-cut, m; t represents the thickness of SCB specimen

$$FI = \frac{G_f}{|m|} A \quad (6)$$

where $|m|$ represents the absolute value of the post-peak slope, m; A represents the unit scaling factor, which is recommended to be 0.01 in the IL-SCB test specification.

According to the research by Imad L and Al-Qadi et al. [27], it was found that when FI is greater than 4, the cracking rate of the mixture significantly decreases. Additionally, based on field trial results, it was observed that the FI values for AC-type mixtures with poor crack resistance are concentrated in the range of 1.3 to 3.9.

2.4.3. Designing Balanced Asphalt Mixture (BMD)

Balanced Asphalt Mixture (BMD) is a method for designing asphalt mixtures, which involves conducting performance tests on specimens subjected to various distress conditions under appropriate conditions, taking into account traffic, climate, mixture aging, and specimen position in the pavement structure. In essence, BMD comprises two or more mechanical tests, such as rutting and cracking tests, to assess the performance of mixtures against common pavement distresses [48].

The information obtained from material characterization and performance testing is then utilized in the optimization phase of BMD. This phase identifies the range of binder content that yields the desired performance characteristics [49]. FHWA has developed and improved four types of balanced design methods: (A) volumetric design with performance verification, (B) volumetric design with performance optimization, (C) performance-modified volumetric mix design, and (D) performance design.

This study adopted the "D" method. Initially, the synthetic gradation of the aggregate was determined, and an estimated value for the new binder content was obtained. Performance tests were conducted using the new binder addition content. The results of the rutting and cracking tests were plotted on the same graph, identifying the new binder content point that simultaneously satisfies both the rutting and cracking requirements.

3. Results and Discussions

3.1. Evaluation of Rutting Resistance Performance

The dynamic stability of the hot in-place recycling asphalt mixtures is illustrated in Figure 7.

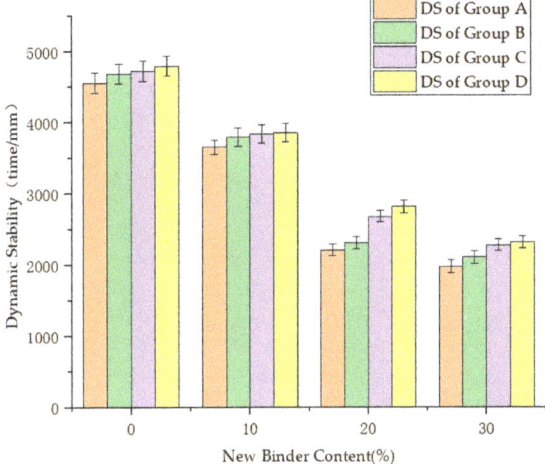

Figure 7. Dynamic stability of the hot in-place recycling asphalt mixtures.

According to the results shown in Figure 2, as the additional new binder content increases from 0% to 30%, the dynamic stability decreases. This variation can be explained by the increasing viscosity of the binder with the addition of new binder, leading to a weakening of the adhesion performance between the binder and aggregates. As a result, the mixture becomes more prone to shear under stress, reducing the dynamic stability. Simultaneously, the filling effect between the binder and aggregates intensifies, but it also reduces the contact area between the aggregates, increasing the gaps between the binder and aggregates, resulting in a looser mixture structure that is unfavorable for maintaining the dynamic stability [50–52].

Comparing the data between Group A and Group C, it can be observed that the dynamic stability has improved by 2.85% to 6.63%. Similarly, comparing the data between Group B and Group D, an improvement in the dynamic stability by 0.36% to 5.34% is shown. This indicates that with the deepening of aging, it is advantageous for the rutting resistance performance. The reason for this is that with the longer aging time, the binder undergoes oxidation and deterioration, altering its properties and enhancing the deformation resistance of the mixture [53].

Comparing the data between groups A and B, it can be observed that the dynamic stability improved by 3.67% to 21.40%. Similarly, comparing the data between groups C and D, the dynamic stability showed an improvement of 1.50% to 22.23%. This indicates that with the increase in the RAP content, the resistance to rutting of the recycling mixture improves [54,55]. The reason for this is that the higher content of aged binder in the RAP leads to greater viscosity, resulting in reduced plastic deformation at high temperatures.

In this experimental study, the dynamic stability of hot in-place recycling asphalt mixtures is influenced by factors such as the amount of new binder, the degree of RAP aging, and the content of RAP. These factors do not affect the dynamic stability of hot in-place recycling asphalt mixtures in exactly the same way, and further research is needed to investigate whether there are interactions among these factors. This paper employs multiple regression analysis to examine the impact of various factors on the dynamic stability of hot in-place recycling asphalt mixtures, and to predict their patterns of change. The aim is to understand the interactions among multiple factors and to provide feasible methods for improving the resistance to rutting of hot in-place recycling asphalt mixtures.

This paper conducts a multivariate regression analysis of the dynamic stability of hot in-place recycling asphalt mixtures under the influence of multiple factors using IBM SPSS Statistics 27.0.1 software. Prior to the regression analysis, it is necessary to address the issue of multicollinearity among the various factors in the data analysis, in order to avoid situations where multiple independent variables exhibit high correlation or linear correlation.

From Table 6, it is evident that the VIF values for each factor are all 1, much less than 5, indicating the absence of multicollinearity among variables. This suggests that a regression model for the dynamic stability of hot in-place recycling asphalt mixtures can be established and utilized for multiple linear regression calculations.

Table 6. Results of the linear regression analysis on dynamic stability.

Model	B	p-Value (P)	Variance Inflation Factor (VIF)	R^2
Constant	2329.425	/	/	
The amount of new binder	−88.395	0.001	1.000	
The degree of RAP aging	19.350	0.460	1.000	0.954
The content of RAP	2515.000	0.071	1.000	

Table 6 reflects the linear correlation between various factors and the dependent variable. The correlation coefficient R^2 represents the degree of fit of the model. Three factors—the amount of new binder, the degree of RAP aging, and the content of RAP—account for 95.4% of the variation in the dynamic stability of hot in-place recycling asphalt mixtures.

In terms of significance P, the significance of the amount of new binder P is less than 0.05, while the significance P of the aging degree and RAP content is greater than 0.05. Moreover, the significance value of the RAP aging degree, 0.460, is greater than that of the RAP content, 0.071, indicating that the effect of the RAP aging degree on the dynamic stability is minimal (insignificant and can be ignored), while the effect of the additional new binder content is maximal. Based on the above analysis, the regression Equation (7) can be derived as follows:

$$Y = 2329.425 - 88.395 \times X1 + 2515.000 \times X2 \tag{7}$$

In the equation, Y represents the dynamic stability, time/mm; X1 denotes the amount of new binder, %; and X2 signifies the content of RAP, %.

The Percent–Percent Plot (P-P Plot) is commonly used to visually inspect whether data follow a normal distribution. Its principle lies in the fact that if the data are normally distributed, the cumulative proportion of the data should closely match the cumulative proportion of a normal distribution. This is achieved by plotting the actual cumulative proportion of the data on the X-axis against the corresponding cumulative proportion of a normal distribution on the Y-axis, creating a scatter plot. If the scatter plot approximately forms a diagonal line, it indicates that the data follow a normal distribution. Conversely, if it deviates from a straight line, it suggests that the data are non-normal. From Figure 8, it can be observed that the scatter points are mostly along the diagonal line, suggesting that the residuals of this regression model follow a normal distribution.

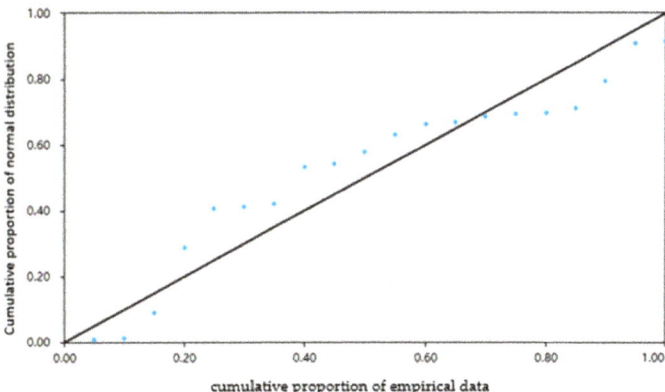

Figure 8. Percent–Percent Plot of dynamic stability.

3.2. Evaluation of Cracking Resistance Performamce

The flexibility index of the hot in-place recycling asphalt mixtures with the addition of different amounts of new binder are illustrated in Figure 9.

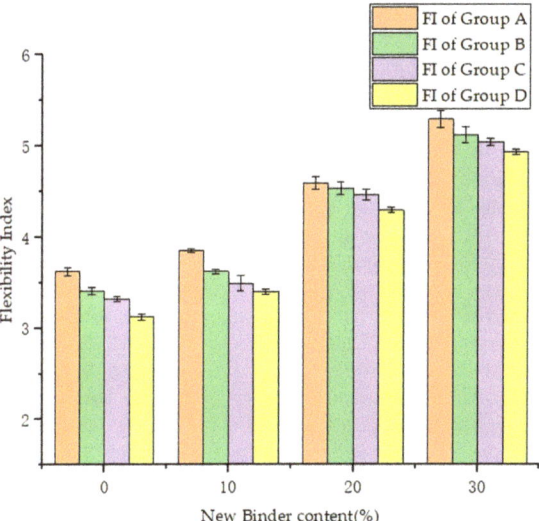

Figure 9. The flexibility index of the hot in-place recycling asphalt mixtures.

Based on Figure 9, it can be observed that with the increase in the binder content, the recycling asphalt mixture exhibits several positive effects, leading to a gradual increase in the flexibility index. The increased binder content may result in the formation of more binder films on the surface of the aggregates, improving the bonding performance between the binder and aggregates and helping to slow down the cracking of the mixture. Lastly, the higher binder content increases the flowability and deformability of the binder, leading to an increase in the elastic modulus of the mixture [27,56,57]. This makes the mixture more adaptable to external stresses, aiding in better shape retention and slowing down crack development. The combined effect of these factors results in positive improvements in low-temperature performance and resistance to cracking for the recycling asphalt mixture.

The phenomenon of embrittlement occurring after binder aging, accompanied by a decrease in the elastic modulus, makes the mixture more prone to cracking, leading to a continuous deterioration in the cracking resistance as reflected by the flexibility index. Comparing the data of Group A with Group C, it can be observed that the flexibility index decreases by 1.31% to 5.97%. Similarly, comparing the data of Group B with Group D, the flexibility index shows a decrease of 2.18% to 6.02%.

The addition of reclaimed materials can worsen the cracking resistance of the mixture. In particular, for recycling mixtures with a high proportion of reclaimed materials, their cracking resistance is significantly affected by the dosage. Comparing the data of Group A with Group B, it can be observed that the flexibility index decreases by 2.83% to 9.35%. Similarly, comparing the data of Group C with Group D, the flexibility index decreases by 3.71% to 8.50%.

According to the analysis method described in Section 3.1, similarly, how the flexibility index of the asphalt mixture is affected by the amount of new binder, the degree of RAP aging, and the content of RAP will be analyzed. The results are shown in Table 7.

Table 7. Results of the linear regression analysis on flexibility index.

Model	B	p-Value (P)	Variance Inflation Factor (VIF)	R^2
Constant	5.557	/	/	
The amount of new binder	0.061	0.001	1.000	
The degree of RAP aging	−0.031	0.094	1.000	0.956
The content of RAP	−2.475	0.013	1.000	

From Table 7, it is evident that the VIF values for each factor are all 1, much less than 5, indicating the absence of multicollinearity among variables. This suggests that a regression model for the flexibility index of hot in-place recycling asphalt mixtures can be established and utilized for multiple linear regression calculations.

Table 7 reflects the linear correlation between various factors and the dependent variable. The correlation coefficient R^2 represents the degree of fit of the model. Three factors—the amount of new binder, the degree of RAP aging, and the content of RAP—account for 95.6% of the variation in the flexibility index of hot in-place recycling asphalt mixtures.

From the perspective of significance levels P, the significance levels of the amount of new binder and RAP content are both less than 0.05. Additionally, the significance level of the RAP content (0.013) is greater than the significance level of the amount of new binder (0.001). However, the significance level of the RAP aging degree (0.094) is greater than 0.05, indicating that the effect of the RAP aging degree on the flexibility index is minimal, while the effect of the additional binder content is the most significant. Based on the above analysis, the regression Equation (8) is derived as follows:

$$Y = 5.557 + 0.061 \times X1 - 0.031 \times X2 - 2.475 \times X3 \tag{8}$$

In the equation, Y represents the flexibility index; X1 denotes the amount of new binder, %; X2 signifies the degree of RAP aging, days; and X3 signifies the content of RAP, %.

From Figure 10, it can be observed that the scatter points are mostly along the diagonal line, suggesting that the residuals of this regression model follow a normal distribution.

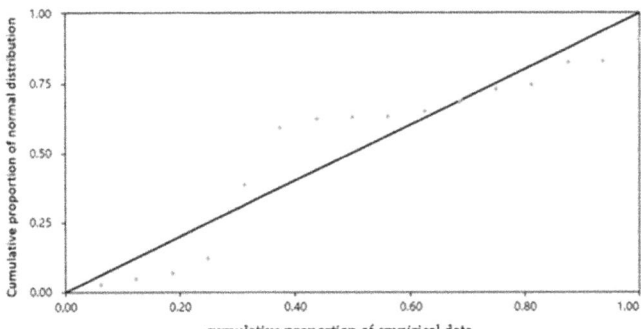

Figure 10. Percent–Percent Plot of flexibility index.

3.3. Determining the Optimal Binder Dosage

To visually present the experimental results, the test results of different hot in-place recycling asphalt mixtures are plotted as dual-axis line graphs, as shown in Figure 11. The grey area in the graph represents the range of new binder content values that meet the performance test criteria (dynamic stability > 3000 times/mm and FI > 4, according to Section 2.4.2).

The equilibrium-designed binder content graph for Group A is depicted in Figure 11a. The new binder content corresponding to a dynamic stability of 3000 times/mm is 1.54%. The new binder content corresponding to an FI value of 4 is 1.49%. Therefore, the optimal new binder content for the asphalt mixture obtained using the equilibrium design method is 1.52%. The summary of the experimental results for the different test groups is shown in Table 8.

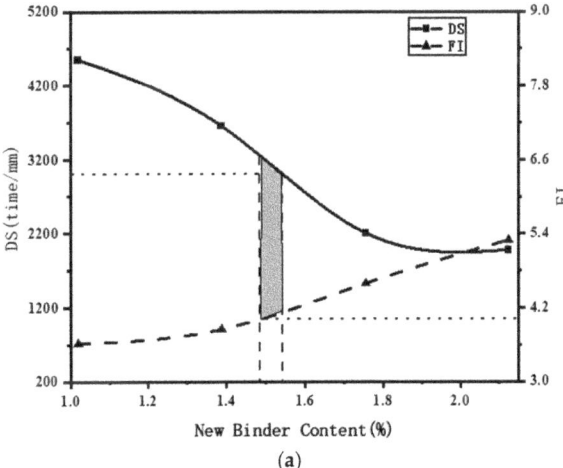

(a)

Figure 11. *Cont.*

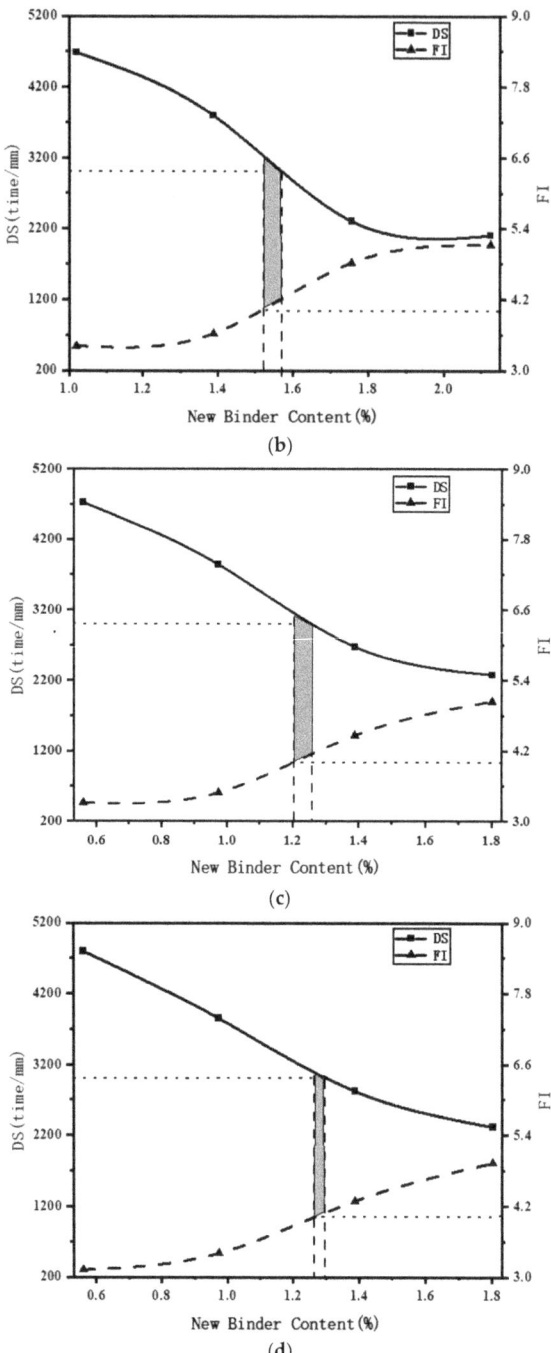

Figure 11. Binder content range chart for asphalt mixtures: (**a**) binder content of Group A, (**b**) binder content of Group B, (**c**) binder content of Group C, (**d**) binder content of Group D.

Table 8. Summary of experimental results.

RAP Content/%	Degree of Aging/Days	New Binder Content When FI = 4/%	New Binder Content When DS = 3000 times/mm/%	Optimal New Binder Content/%	Total Binder Content/%
80	5	1.49	1.54	1.52	5.20
80	10	1.52	1.57	1.55	5.23
90	5	1.20	1.26	1.23	5.37
90	10	1.26	1.30	1.28	5.42

From Table 8, it can be seen that the performance of the aged binder in the RAP is poor. As the RAP content increases, the balance design method requires more new binder to improve the cracking resistance performance of the recycling mixture, which leads to a continuous increase in the total binder content of the recycling asphalt mixture. As the aging time increases, the degree of binder aging deepens, leading to an increase in the total binder content and the addition of new binder in the recycling asphalt mixture. Prolonged aging makes the mixture more prone to embrittlement, reducing its viscosity and elasticity, thereby decreasing its resistance to cracking. To achieve a balance between resistance to rutting and resistance to cracking, it is necessary to increase the new binder content appropriately.

4. Conclusions and Recommendations

The main objective of this study is to utilize the concept of balanced design by balancing the rutting and cracking resistance of a recycling asphalt mixture, in order to determine the optimal new binder content for a high-dose RAP hot in-place recycling asphalt mixture in engineering applications. The conclusions are as follows:

(1) With the increase in the additional new binder content, the cracking resistance of the recycling asphalt mixture has improved. Significant differences exist between the recycling asphalt mixture with a high new binder content (such as 20% and 30%) and the control group (0%).

(2) As the additional binder content increases, the rutting resistance of recycling asphalt mixtures significantly decreases. With each 10% increase in new binder content, the dynamic stability of recycling asphalt mixtures decreases by nearly 19%. Additionally, the addition of new binder can enhance the crack resistance of high-dose RAP mixtures, but it reduces their rutting resistance, especially at high levels of new binder addition. For instance, when an extra 20% of new binder is added, the crack resistance of recycling asphalt mixtures increases by 27%, but their rutting resistance decreases by 51%.

(3) The dynamic stability and flexibility index of the mixture exhibit a strong linear relationship with the dosage of new binder, the aging degree of RAP, and the dosage of RAP (R2 = 0.954, 0.956). Among these factors, the dosage of new binder has a significant impact.

(4) Based on the balanced design of rutting and cracking resistance, the optimal new binder content for the Group A recycling asphalt mixture is 1.52%; for Group B, it is 1.55%; for Group C, it is 1.23%; and for Group D, it is 1.28%. As the RAP content increases, the rejuvenated aged binder positively influences the performance, reducing the amount of new binder added. However, the unactivated aged binder also increases, requiring more binder to meet the demand for crack resistance, leading to an overall increase in binder usage. With deeper binder aging and decreased viscosity, additional new binder is needed to restore the crack resistance, further contributing to the overall increase in binder consumption.

(5) Adding additional new binder to a high dose of hot in-place recycling asphalt mixtures effectively enhances their crack resistance. However, since the RAP used in this study was prepared in the laboratory, future work should consider linking laboratory design with field experiments to balance the design of actual construction site RAP. This will

advance the performance balance design of in-place recycling asphalt mixtures with a high RAP content in the highway industry.

Author Contributions: Methodology, L.J.; Software, L.J. and W.W.; Formal analysis, L.J.; Data curation L.J.; Writing—original draft, L.J.; Project administration, J.S.; Funding acquisition, J.S.; Writing—review & editing, J.S.; Investigation, W.W.; Supervision, W.W. All authors have read and agreed to the published version of the manuscript.

Funding: This research received no external funding.

Institutional Review Board Statement: Not applicable.

Informed Consent Statement: Not applicable.

Data Availability Statement: Data are contained within the article.

Acknowledgments: The author wishes to acknowledge the Suzhou University of Science and Technology for the experimental equipment.

Conflicts of Interest: The authors declare no conflicts of interest.

References

1. Chen, J.Y. Experimental Study on Regeneration Method of Asphalt Pavement. Master's Thesis, Dalian University of Technology, Dalian, China, 2011.
2. Copeland, A. *Reclaimed Asphalt Pavement in Asphalt Mixtures: State of the Practice*; No. FHWA-HRT-11-021; United States Federal Highway Administration, Office of Research, Development, and Technology: McLean, VA, USA, 2011.
3. Sapkota, K.; Yaghoubi, E.; Wasantha, P.L.P.; Van Staden, R.; Fragomeni, S. Mechanical Characteristics and Durability of HMA Made of Recycled Aggregates. *Sustainability* **2023**, *15*, 5594. [CrossRef]
4. Claine, P.J. Chemical composition of asphalt as related to asphalt durability: State of the Art. *Transp. Res. Rec.* **1984**, *999*, 13–30.
5. Qian, R.Y.; Ji, X.W.; Shen, J.N. *Experimental Study on Fatigue Life of In-Situ Thermally Recycled Asphalt Mixtures*; Shanxi Construction: Xi'an, China, 2021; pp. 103–106.
6. Qin, Y.C.; Huang, S.C.; Xu, J. Research on the integration of new and aged binder in plant-mixed warm recycled asphalt mixtures. *Highw. Transp. Sci. Technol.* **2015**, *32*, 24–28+52.
7. Cheng, P.F.; Xiang, Y.J.; Kao, H.Y.; Zhou, Y.Z. Effect of waste asphalt mixture blending on mixing and compaction temperature of hot recycled modified asphalt mixture. *Sci. Technol. Eng.* **2018**, *18*, 273–279.
8. Zhang, K.; Zheng, X.; Zhang, X. Experimental study on fatigue characteristics of hot recycled asphalt mixtures. *Highw. Eng.* **2017**, 228–232.
9. Li, L.; Yin, Y.X. Research on the integration of new and old materials and factory production parameters of thermally recycled SMA mixes with high proportion of RAP admixture. *Highw. Eng.* **2016**, *41*, 6.
10. Li, M.X.; Wang, X.C. Effect of newly added asphalt grade on the strength and fatigue performance of high dosage RAP recycled mix. *J. Chang. Univ.* **2017**, *37*, 9–15.
11. Zhang, W.D.; Dai, W.J.; Zhou, Y. Research on road performance of in-situ thermally regenerated mix. *East China Highw.* **2018**, *6*, 101–105.
12. Ma, Y.; Polaczyk, P.; Hu, W.; Zhang, M.; Huang, B. Quantifying the effective mobilized RAP content during hot in-place recycling techniques. *J. Clean. Prod.* **2021**, *314*, 127953. [CrossRef]
13. Huang, B.; Li, G.; Vukosavljevic, D.; Shu, X.; Egan, B.K. Laboratory investigation of mixing hot-mix asphalt with reclaimed asphalt pavement. *Transp. Res. Rec. J. Transp. Res. Board* **2005**, *1929*, 37–45. [CrossRef]
14. Zhao, S.; Huang, B.; Shu, X.; Woods, M.E. Quantitative evaluation of blending and diffusion in high RAP and RAS mixtures. *Mater. Des.* **2016**, *89*, 1161–1170. [CrossRef]
15. Ding, Y.; Huang, B.; Shu, X. Characterizing blending efficiency of plant produced asphalt paving mixtures containing high RAP. *Constr. Build. Mater.* **2016**, *126*, 172–178. [CrossRef]
16. Lo Presti, D.; Vasconcelos, K.; Orešković, M.; Pires, G.M.; Bressi, S. On the degree of binder activity of reclaimed asphalt and degree of blending with recycling agents. *Road Mater. Pavement Des.* **2020**, *21*, 2071–2090. [CrossRef]
17. Mccarthy, L.M.; Callans, J.; Quigley, R. *Performance Specifications for Asphalt Mixtures, NCHRP SYNTHESIS 492*; Transportation Research Board: Washington, DC, USA, 2016.
18. Bayomy, F.; Mull-Aglan, M.A.; Abdo, A.A.; Santi, M.J. Evaluation of hot mix asphalt(HMA) fracture resistance using the critical strain energy release rate, Jc. In Proceedings of the Transportation Research Board 85th Annual Meeting, Washington, DC, USA, 22–26 January 2006.
19. Buttlar, W.G.; Roque, R.; Kim, N. Accurate asphalt mixture tensile strength. In Proceedings of the Materials Engineering Conference, Washington, DC, USA, 10–14 November 1996.

20. *AASHTO322*; Standard Method of Test for Determining the Creep Compliance and Strength of Hot Mix Asphalt (HMA) Using the Indirect Tensile Test Device. American Association of State Highway and Transportation Officials: Washington, DC, USA, 2011.
21. Molenaar, A.A.A.; Scarpas, A.; Liu, X.; Erkens, S.M.J.G. Semi-circular bending test; simple but useful. *Asph. Paving Technol. Assoc. Asph. Paving Technol.* **2002**, *71*, 794–815.
22. Mitchell, M.R.; Link, R.E.; Huang, L.; Cao, K.; Zeng, M. Evaluation of semicircular bending test for determining tensile strength and stiffness modulus of asphalt mixtures. *J. Test. Eval.* **2009**, *37*, 122–128.
23. Hofman, R.; Oosterbaan, B.; Erkens, S.M.J.G.; Van der Kooij, J. Semi-circular bending test to assess the resistance against crack growth. In Proceedings of the 6th International Rilem Symposium, Zurich, Switzerland, 14–16 April 2003; pp. 257–263.
24. West, R.C.; Copeland, A. High RAP Asphalt Pavements: Japan practice-lessons learned. *Natl. Asph. Pavement Assoc.* **2015**, *139*, 62.
25. Kaseer, F.; Yin, F.; Arámbula-Mercado, E.; Martin, A.E.; Daniel, J.S.; Salari, S. Development of an index to evaluate the cracking potential of asphalt mixtures using the semi-circular bending test. *Constr. Build. Mater.* **2018**, *167*, 286–298. [CrossRef]
26. West, R.C.; Winkle, C.V.; Maghsoodloo, S.; Dixon, S. Relationships between simple asphalt mixture cracking tests using ndesign specimens and fatigue cracking at fhwa's accelerated loading facility. *Road Mater. Pavement Des.* **2017**, *86*, 579–602. [CrossRef]
27. Al-Qadi, I.L.; Ozer, H.; Lambros, J.; Khatib, A.E.; Singhvi, P.; Khan, T.; Rivera-Perez, J.; Doll, B. *Testing Protocols to Ensure Performance of High Asphalt Binder Replacement Mixes Using RAP and RAS*; Illinois Center for Transportation Series No. 15-017; Illinois Center for Transportation: Urbana, IL, USA, 2015.
28. Majidifard, H.; Jahangiri, B.; Rath, P.; Buttlar, W.G. Development of a balanced cracking index for asphalt mixtures tested in semi-circular bending with load-LLD measurements. *Measurement* **2021**, *173*, 108658. [CrossRef]
29. Zhou, F.; Im, S.; Sun, L.; Scullion, T. Development of an IDEAL cracking test for asphalt mix design and QC/QA. *Road Mater. Pavement Des.* **2017**, *86*, 549–577. [CrossRef]
30. Alkuime, H.; Tousif, F.; Kassem, E.; Bayomy, F.M. Review and evaluation of intermediate temperature monotonic cracking performance assessment testing standards and indicators for asphalt mixes. *Constr. Build. Mater.* **2020**, *263*, 120121. [CrossRef]
31. Alkuime, H.; Kassem, E.; Al-Rousan, T.; Mujalli, R.O.; Alshraiedeh, K.A. Accounting for the effect of air voids on asphalt mix monotonic cracking testing results. *J. Test. Eval.* **2023**, *51*, 3662–3681. [CrossRef]
32. Alkuime, H. Impact of testing and specimen configurations on monotonic high-temperature indirect tensile(high-IDT) rutting assessment test. *Int. J. Pavement Res. Technol.* **2023**, 1–18. [CrossRef]
33. Alkuime, H.; Kassem, E.; Bayomy, F.M.; Nielsen, R.J. Development of a multi-stage semi-circle bending cyclic test to evaluate the cracking resistance of asphalt mixtures. *Road Mater. Pavement Des.* **2022**, *23*, 1–21. [CrossRef]
34. Alkuime, H.; Kassem, E. Comprehensive evaluation of wheel-tracking rutting performance assessment tests. *Int. J. Pavement Res. Technol.* **2020**, *13*, 334–347. [CrossRef]
35. Alkuime, H.; Kassem, E.; Bayomy, F.M.S.; Nielsen, R.J. Evaluation and development of performance-engineered specifications for monotonic loading cracking performance assessment tests and indicators. *J. Test. Eval.* **2021**, *49*, 4151–4169. [CrossRef]
36. Zhou, F.J.; Hu, S.; Scullion, T. *Integrated Asphalt (Overlay) Mixture Design, Balancing Rutting and Cracking Requirements*; FHWA/TX-06/0-5123-1; Texas Transportation Institute: Bryan, TX, USA, 2006.
37. Feng, X.J.; Cha, X.D.; Wang, W.Y. Balanced design of asphalt mixtures based on rutting and cracking resistance. *J. Wuhan Univ. Technol.* **2011**, *33*, 5.
38. Walubita, L.; Hu, X.D.; Scullion, T. Laboratory evaluation of two different HMA MixDesign Methods and Test Procedures. *Asfaltosy Pavimentos Magazine*, 4–16 June 2012; Edition 24.
39. Zhou, F.J.; Hu, S.; Das, G.; Lee, R.; Scullion, T.; Claros, G. Successful high RAP mixes designed with balanced rutting and cracking requirements. *Asph. Paving Technol. Assoc. Asph. Paving Technol.-Proc. Tech. Sess.* **2012**, *81*, 477–505.
40. *JTG E20-2011*; Ministry of Transport. Standard Test Methods of Bitumen and Bituminous Mixtures for Highway Engineering. People's Communications Press: Beijing, China, 2011.
41. Qian, R.Y.; Shen, J.N. Long-Term Performance Study of In-Place Hot Recycled Asphalt Mixtures. Master's Thesis, Suzhou University of Science and Technology, Suzhou, China, 2021.
42. *Q/SJKG YH B2-001-2020*; Jiangsu Provincial Department of Transportation. Standardization Guidelines for Highway Construction in Jiangsu Province. Pavement Engineering. People's Communications Press: Beijing, China, 2012.
43. Yu, D. Research on Balanced Design Method of Recycled Asphalt Mixture Based on Uniaxial Penetration Test. Master's Thesis, Shandong Jianzhu University, Jinan, China, 2018.
44. Luo, P. Research on Fracture Test Method and Evaluation Index of Asphalt Mixture Based on Semi-Circular Bending Test. Master's Thesis, Chang'an University, Xi'an, China, 2017.
45. Feng, D.; Cui, S.; Yi, J.; Chen, Z.; Qin, W. Research on Evaluation Index of Low-temperature Performance of Asphalt Mixture Based on SCB Test. *China J. Highw. Transp.* **2020**, *33*, 50–57.
46. Chen, Z.; Zhu, Y.; Zhang, H.; Song, B. Medium Temperature Cracking Resistance of Recycled Asphalt Mixture Based on Fracture Test. *J. Hefei Univ. Technol.* **2017**, *40*, 1260–1263+1288.
47. *JTGF40-2004*; Ministry of Transport of the People's Republic of China. Technical Specification for Highway Asphalt Pavement Construction. People's Communications Press: Beijing, China, 2005.
48. NCHRP. *Development of a Framework for Balanced Mix Design*; National Center for Asphalt Technology at Auburn University: Auburn, AL, USA, 2018.

49. Alkuime, H.; Kassem, E.; Alshraiedeh, K.A.; Bustanji, M.; Aleih, A.; Abukhamseh, F. Performance Assessment of Waste Cooking Oil-Modified Asphalt Mixtures. *Appl. Sci.* **2024**, *14*, 1228. [CrossRef]
50. Cao, K.; Xu, W.; Chen, D. Study on High Temperature Performance of Silica/SBS Composite Modified Asphalt Mixture. *New Build. Mater.* **2018**, *45*, 55–58.
51. Zhang, C.; Li, T.; Ding, W. Study on the Performance of High-viscosity and High-elastic Asphalt Mixture for Ultra-thin Wear-resistant Layer. *Sci. Technol. Eng.* **2023**, *23*, 12241–12249.
52. Luo, L.; Ye, Q. Study on the Influence of Temperature and Asphalt Content Ratio on the Performance of Asphalt Mixtures for Ultra-thin Wear-resistant Layers. *J. China Foreign Highw.* **2013**, *33*, 242–244.
53. Liu, J.X.; Tong, S.J. Study on the Factors Influencing the High-Temperature Performance of Asphalt Mixtures under Ultraviolet Aging. *Highway* **2016**, *61*, 201–205.
54. Ma, T.; Wang, H.; Huang, X.; Wang, Z.; Xiao, F. Laboratory performance characteristics of high modulus asphalt mixture with high-content RAP. *Constr. Build. Mater.* **2015**, *101*, 975–982. [CrossRef]
55. Busang, S.; Maina, J. Influence of aggregates properties on microstructural properties and mechanical performance of asphalt mixtures. *Constr. Build. Mater.* **2022**, *318*, 126002. [CrossRef]
56. Sharma, A.; Naga, G.R.R.; Kumar, P.; Rai, P. Mix design, development, production and policies of recycled hot mix asphalt: A review. *J. Traffic Transp. Eng.* **2022**, *9*, 765–794. [CrossRef]
57. Stimilli, A.; Canestrari, F.; Teymourpour, P.; Bahia, H.U. Low-temperature mechanics of hot recycled mixtures through Asphalt Thermal Cracking Analyzer (ATCA). *Constr. Build. Mater.* **2015**, *84*, 54–65. [CrossRef]

Disclaimer/Publisher's Note: The statements, opinions and data contained in all publications are solely those of the individual author(s) and contributor(s) and not of MDPI and/or the editor(s). MDPI and/or the editor(s) disclaim responsibility for any injury to people or property resulting from any ideas, methods, instructions or products referred to in the content.

Article

Investigation of Phenolic Resin-Modified Asphalt and Its Mixtures

Lieguang Wang [1], Lei Wang [1], Junxian Huang [2,*], Mingfei Wu [1], Kezhen Yan [2,*] and Zirui Zhang [1]

1 Zhejiang East China Engineering Consulting Co., Ltd., Hangzhou 310030, China
2 School of Civil Engineering, Hunan University, Changsha 410082, China
* Correspondence: huangjunxian@hnu.edu.cn (J.H.); yankz@hnu.edu.cn (K.Y.)

Abstract: This study comprehensively examines the influence of phenol-formaldehyde resin (PF) on the performance of base asphalt and its mixtures for road applications, emphasizing its innovative use in enhancing pavement quality. Optimal PF content was determined through the evaluation of standard indicators and rotational viscosity. In-depth analyses of PF-modified asphalt's high- and low-temperature rheological properties and viscoelastic behavior were conducted using dynamic shear rheometers and bending beam rheometers. Aging resistance was assessed through short-term aging and performance grade (PG) grading. Moreover, Marshall and water stability tests were performed on PF-modified asphalt mixtures. Findings indicate that the uniform dispersion of PF particles effectively inhibits asphalt flow at high temperatures, impedes oxygen penetration, and delays the transition from elasticity to viscosity. These unique properties enhance the high-temperature stability, rutting resistance, and aging resistance of PF-modified asphalt. However, under extremely low temperatures, PF's brittleness may impact asphalt flexibility. Nonetheless, the structural advantages of PF-modified asphalt, such as improved mixture density and stability, contribute to enhanced high-temperature performance, water stability, adhesion, and freeze–thaw cycle stability. This research demonstrates the feasibility and effectiveness of using PF to enhance the overall performance of base asphalt and asphalt mixtures for road construction.

Keywords: PF-modified asphalt; viscoelastic properties; rheological characteristics; mixture; high-temperature stability

Citation: Wang, L.; Wang, L.; Huang, J.; Wu, M.; Yan, K.; Zhang, Z. Investigation of Phenolic Resin-Modified Asphalt and Its Mixtures. *Materials* **2024**, *17*, 436. https://doi.org/10.3390/ma17020436

Academic Editor: Simon Hesp

Received: 18 December 2023
Revised: 7 January 2024
Accepted: 12 January 2024
Published: 16 January 2024

Copyright: © 2024 by the authors. Licensee MDPI, Basel, Switzerland. This article is an open access article distributed under the terms and conditions of the Creative Commons Attribution (CC BY) license (https:// creativecommons.org/licenses/by/ 4.0/).

1. Introduction

Petroleum asphalt, known for its high viscosity, is widely employed as a binder in road constructions due to its economic advantages, noise reduction capabilities, facilitation of a comfortable driving experience, and ease of use in construction compared with cement road surfaces [1,2]. However, as the global economy rapidly evolves, conventional asphalt pavements are continually plagued by issues such as cracking, ruts, potholes, and surface deformation, rendering them inadequate for meeting the demands of today's road traffic loads and the harsh and ever-changing natural environment [3–5]. Consequently, there is an urgent need to further enhance the performance of asphalt in highway construction.

The use of polymers to enhance the performance of asphalt binders and thereby improve asphalt pavement quality is a popular direction for current research [6–8]. For instance, the styrene-butadiene-styrene (SBS) block copolymer, with its strength and elasticity that are derived from a three-dimensional network of crosslinked molecules, has shown significant potential. The polystyrene end blocks in the SBS copolymer impart strength to the polymer, while the polybutadiene rubber matrix provides exceptional viscosity [9]. Researchers like Danial Mirzaiyan et al. [10] have explored the impact of SBS on the fatigue response of asphalt mixtures through indirect tensile fatigue tests (ITFTs), discovering that SBS significantly enhances fatigue resistance by swelling crosslinking in the saturated and aromatic fractions of asphalt. Faheem Sadiq Bhat et al. [11] have confirmed through linear amplitude sweep tests that SBS additions notably improve the high-temperature stability of asphalt binders, reducing the risk of pavement cracking. Simultaneously, a growing number of researchers are exploring rubber modifiers, such as crumb rubber, to enhance pavement performance [12,13]. Peerapong Jitsangiam et al. [14] have found that rubber from recycled tires effectively improves the rutting resistance of base asphalt at high temperatures. Unlike elastomers, Aboelkasim Diab et al. [15] have shown that using ethylene-vinyl acetate (EVA), which is a thermoplastic polymer, to modify asphalt significantly boosts its flexibility and crack resistance at low temperatures. This improvement is closely related to the content and molecular weight of vinyl acetate in EVA. The increase in vinyl acetate content lowers the glass transition temperature (Tg) of EVA, maintaining its flexibility even at lower temperatures.

Despite these advancements, these polymers face challenges in terms of high costs, environmental concerns, and performance limitations under specific conditions [16–18]. For instance, Ali Behnood et al. [19] found that polymer-modified asphalt is 1.5 times costlier than traditional asphalt, leading to a 20% increase in the cost of the modified mixtures. SK Sohel Islam et al. [20] have observed that the degradation of SBS at high temperatures significantly reduces the Marshall stability, indirect tensile strength, fatigue, and rutting resistance of asphalt mixtures regardless of the grade and source of the base asphalt. Xiong et al. [21] noted that achieving a high performance in SBS-modified asphalt not only requires chemical adjustments to restore its properties but also the reconstruction of the polymer network through the chemical structure of products from SBS degradation products. Nonde Lushinga et al. [22] have found that enhancing the compatibility of rubber particles with asphalt requires microwave radiation to break the sulfur network on the rubber particle surface for depolymerization and desulfurization. These polymers' applications are limited by costs, aging, and compatibility issues [23,24]. Simultaneously, Mukul Rathore et al. [25] pointed out that while common polymer modifiers enhance asphalt's rut resistance, they also significantly increase its viscosity. To reduce production temperatures and promote energy saving and emission reduction, warm mix asphalt additives are often incorporated. Therefore, from both engineering and economic perspectives, selecting appropriate modifiers that balance desired performance with practical application values becomes crucial.

Phenol-formaldehyde resin (PF), as a binding agent derived from the condensation reaction of phenol (C_6H_5OH) and formaldehyde (HCHO), is characterized by readily available raw materials, cost-effective production, and a straightforward manufacturing process [26,27]. This compound's chemical structure exhibits a remarkable stability that confers resistance to oxidation and decomposition, thereby bestowing PF with robust anti-aging capabilities [28]. The three-dimensional polymeric network formed through condensation reactions is particularly noteworthy; it generates a multitude of stable chemical bonds, thereby contributing to the high hardness and strength of phenol-formaldehyde resin [29,30]. As such, PF exhibits superior mechanical properties along with heat and cold resistance, dimensional stability, water stability, and flame resistance, thus making it a versatile material widely applied across various fields [31]. Li et al. [32] discovered excellent compatibility in PF when incorporated into styrene-butadiene rubber (SBR)-modified asphalt. These polar functional groups can form interactions with the polar molecules present in asphalt, facilitating an effective and uniform dispersion of PF within the asphalt mixtures [33]. The ideal internal structure of PF-modified asphalt should consist of a network structure formed through the cross-linking of high binding strength PF and asphalt molecules, leveraging the high-temperature resistance and load-buffering attributes of PF [34]. Despite these promising features, the academic community is yet to exhaustively explore the influence of PF on the mechanical properties of base asphalt and mixed materials. In particular, the current literature provides limited insight into the effects of varying PF contents on these properties.

Therefore, this study systematically evaluated the impact of different PF levels on the conventional mechanical properties, high- and low-temperature rheological properties, and aging resistance of the base asphalt. We also investigated the microstructure and mode of action of PF within the asphalt binder. It uncovered the governing rules of PF on the mechanical properties of base asphalt and, both innovatively and for the first time, correlated the influence mechanism of PF-modified asphalt binders on the stability and resistance to water damage of their mixtures. This research aims to explore the potential of integrating PF into asphalt to enhance pavement performance and extend pavement life, thereby providing novel solutions for sustainable road construction.

2. Materials and Preparation

2.1. Asphalt

In this study, we selected the domestically produced Baoli 70# road petroleum asphalt. The various performance parameters were tested in accordance with the "ASTM specifications for asphalt binder properties", as detailed in Table 1.

Table 1. Parameter index of 70# matrix asphalt used in the experiment.

Index	Test Result	Specification Requirements	Test Method
Penetration (100 g 5 s 25 °C) (0.1 mm)	67.2	60–80	ASTM D5 [35]
Ductility (5 cm/min 15 °C) (cm)	150	>100	ASTM D113 [36]
Softening point (°C)	47.3	>46	ASTM D36 [37]
Kinematic viscosity (60 °C) (Pa·S)	110	----	ASTM D4402 [38]

2.2. PF

The phenol-formaldehyde resin (PF) used in this study was sourced from Hebei Zetian Chemical Co., Ltd. (Hengshui, China). Relevant parameters are presented in Table 2.

Table 2. PF parameter indicators.

Technical Index	Unit	Test Result
Specific surface area	m^2/g	140
Fineness	mesh	>200
Moisture content	%	<4.0
Flow length	mm	25~40
Melting point	°C	85~92
Free phenol content	%	3.0~4.0

2.3. Aggregates

In accordance with the aforementioned standard, limestone aggregate sourced from Yunzhong Technology Zijing Factory in Changsha, China, was tested. Tables 3–5 present the indices for coarse and fine aggregates obtained from the experiment.

Table 3. Coarse aggregate performance index.

Technical Index	Unit	Test Result	Specification Requirements	Test Method
Los Angeles abrasion loss	%	16.7	≤30	ASTM C131 [39]
Crushing value	%	20.5	≤28	ASTM D5821 [40]
Soundness	%	3.72	≤12	ASTM C88 [41]
Soft stone content	%	1.33	≤3.0	-
Needle and flake content	%	10.13	≤15	ASTM D4792 [42]
Particle size > 9.5 mm	%	9.54	≤12	ASTM C136 [43]
Particle size < 9.5 mm	%	12.66	≤18	ASTM C136 [43]

Table 4. Fine aggregate performance index.

Technical Index	Unit	Test Result	Specification Requirements	Test Method
Mud content (<0.075 mm)	%	2.37	≤3	ASTM D1140 [44]
Apparent specific gravity	-	2.64	≥2.50	ASTM C128 [45]
Sand equivalent	%	77.81	≥30	ASTM D2419 [46]
Soundness	%	4.82	≤12	ASTM C88

Table 5. Density index of different grades of aggregates.

Technical Index	Unit	10–20 Gears	5–10 Gears	Aggregate Chips
Apparent density	t/m^3	2.71	2.786	2.635
Water content	%	0.32	0.69	0.48
Gross volume relative density	t/m^3	2.69	2.734	2.602

2.4. Preparation of PF-Modified Asphalt

The preparation of PF composite-modified asphalt was performed according to the method proposed by the previous scholars from [47]. Firstly, the base asphalt was heated to a molten state at 145 °C. Subsequently, varying proportions of PF (representing 1%, 2%, 3%, 4%, and 5% of the total weight of the asphalt) were gradually introduced into a low-speed mixer at 145 °C. The mixture was then subjected to high-speed shearing at a rate of 4000 rpm for 45 min at 165 °C followed by a final stirring at 2000 rpm for 30 min. After this process, the mixture was allowed to cool, yielding the PF-modified asphalt. The preparation procedure for the 70# base asphalt is consistent with the points provided above, with a PF dosage of 0%.

3. Mixture Design and Test Method

3.1. Design and Preparation of PF-Modified Asphalt Mixtures

3.1.1. Design of Aggregate Gradation

The dense-graded AC-13C was selected for the preparation of the asphalt mixture. The gradation curve of the aggregates is shown in Figure 1, and the median curve of the gradation was determined as the design for the aggregate gradation.

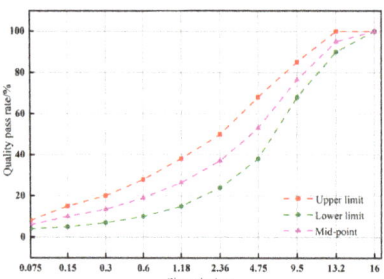

Figure 1. Asphalt mixture grading curve.

3.1.2. Determination of Mixing and Compaction Temperature

In accordance with the requirements of the "D6926/D6926M—standard practice for preparation of bituminous specimens using Marshall apparatus", the mixing and compaction temperatures were determined based on the corresponding temperatures of 0.17 Pa·s ± 0.02 Pa·s and 0.28 Pa·s ± 0.03 Pa·s, respectively (Figure 2). Therefore, the mixing and compaction temperatures for the matrix asphalt were 160 °C and 150 °C, respectively. When the mixing and compaction temperatures of modified asphalt with a PF content range of 2–5% at 175 °C; the highest temperature of 175 °C was selected.

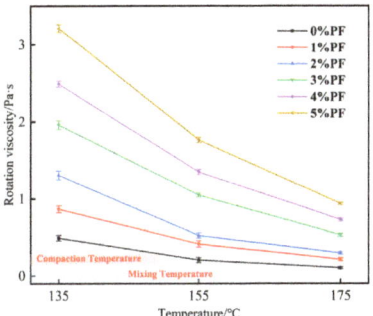

Figure 2. Rotational viscosity diagram for different asphalts.

3.1.3. Determination of the Optimal Asphalt–Aggregate Ratio and Volumetric Parameters

According to the ASTM D2726 and AASHTO T166-07 standards [48], Marshall specimens were prepared, selecting five different mass percentages of 4%, 4.5%, 5%, 5.5%, and 6%. The Marshall stability test was used to measure the bulk relative density (γf), stability (MS), flow value (FL), void ratio (VV), aggregate void ratio (VMA), and effective asphalt saturation (VFA). As shown in Table 6, the optimum asphalt content (OAC) for different asphalt mixtures was calculated through Formulas (1) and (2).

$$OAC_1 = (a_1 + a_2 + a_3 + a_4)/4 \tag{1}$$

$$OAC_2 = (OAC_{max} + OAC_{min})/2 \tag{2}$$

Table 6. Marshall test index at optimal asphalt dosage.

Index	70#	2% PF	3% PF	Specification
OAC/%	4.77	5.03	4.97	----
γ_f	2.14	2.38	2.38	----
VV/%	5.52	5.22	4.78	4~6
MS/kN	8.28	11.35	12.26	≥8
FL/mm	3.78	3.65	3.48	1.5~4
VMA/%	17.32	15.79	15.32	≥14
VFA/%	66.32	67.35	68.28	65~75

In the formulas, a_1 represents the asphalt content corresponding to the maximum value of γf, a_2 corresponds to the asphalt content at which maximum MS is achieved, a_3 is the asphalt content at the median value of VV, a_4 indicates the asphalt content at the median value of VFA, OAC_{min} is the minimum range of optimum asphalt content as specified by the standard criteria, and OAC_{max} is the maximum range as specified by the standard criteria, as referred to in Table 6.

3.2. Test Method

3.2.1. Mechanical Properties of Asphalt Binders

Firstly, the durability, stability, and resistance to deformation of different asphalt materials are assessed using the three major indicators. The optimal dosage range of PF was determined through the Brookfield viscosity test. The dynamic shear rheometer (DSR) is utilized to measure the phase angle (δ) and complex modulus (G^*) of asphalt at varying strains and frequencies, thereby evaluating its high-temperature rheological performance. By subjecting the asphalt to loading at different stress levels and frequencies, the multiple stress creep recovery test (MSCR) simulates complex traffic loads and climate variations experienced on roads. The bitumen bending beam rheometer (BBR) method is employed to impose constant shear stresses on various types of asphalt at temperatures of −12 °C and −18 °C, thereby generating the bending beam rheometer stiffness (S) and creep rate (m) and further evaluating the asphalt's deformation behavior and brittle characteristics under low-temperature conditions. By utilizing the performance grade (PG) grading system for different asphalt types, a more holistic and scientifically informed prediction of the fundamental properties and application scenarios of PF-modified asphalt can be achieved. Lastly, the ratio of the rutting factor ($G^*/\sin\delta$) for different asphalts before and after short-term aging is calculated. The impact of PF on the chemical composition of asphalt binders and its micro-morphology within the asphalt binder were characterized using fourier transform infrared spectroscopy (FT-IR) and fluorescence microscopy (FM).

3.2.2. Mechanical Properties of Asphalt Mixture

The Marshall test is utilized to quantify both the compressive strength and deformation behavior of the mixture, thereby determining the optimal asphalt–aggregate ratio and mixture composition. Moreover, it serves to simulate and assess the stability and endurance of asphalt mixtures under varying traffic loads. The immersion Marshall test evaluates the wet stability and resistance to erosion of the water in the mixture. Concurrently, the freeze–thaw splitting test identifies the mixture's vulnerability to cracking and its ability to withstand freeze–thaw cycles, thus guaranteeing the robustness and dependability of roads exposed to adverse climatic conditions. The above-mentioned tests all adhere to the ASTM standards for asphalt and asphalt mixture testing. For each type of sample, three parallel experiments were conducted, and the standard error was calculated. Error bars have been annotated in the figure to indicate this variability.

4. Results and Discussion

4.1. Empirical Indicator Testing

As can be seen from Figure 3, with the addition of PF, the penetration of asphalt gradually decreases as the amount of PF increases. The ductility of asphalt shows the same trend, and the rate of decrease accelerates. The softening point of PF-modified asphalt initially rises and then falls as the PF content increases, indicating that compatibility issues between PF and asphalt become prominent when the amount of PF is excessive. Overall, the incorporation of PF results in decreased penetration and ductility and an increased softening point of the asphalt. Compared with the base asphalt, PF-modified asphalt has improved elasticity and high-temperature performance but reduced plasticity. These findings are consistent with other research results on PF-modified asphalt [49].

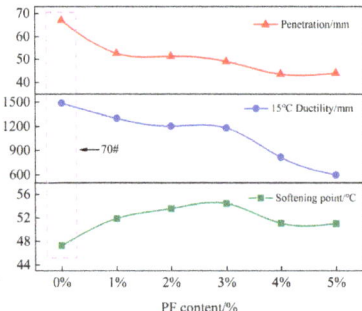

Figure 3. The three major indicators of different types of PF doping.

4.2. Rotational Viscosity Testing

As can be seen from Figure 4, as the temperature rises the intermolecular binding strength of the asphalt decreases and its fluidity increases. Meanwhile, the viscosity values of PF-modified asphalt are all higher than that of the 70# base asphalt, and they increase with the addition of PF. When the PF content is between 2 and 3%, the increase in asphalt's viscosity is the most significant result. However, the rotational viscosity of 5% PF-modified asphalt at 135 °C exceeds 3.0 Pa·s, suggesting that the use of 5% PF-modified asphalt to prepare asphalt mixtures may lead to uneven mixing and an unstable performance due to excessively high viscosity. The results indicate that the appropriate addition of PF improves the adhesion and high-temperature flow deformation resistance of the asphalt.

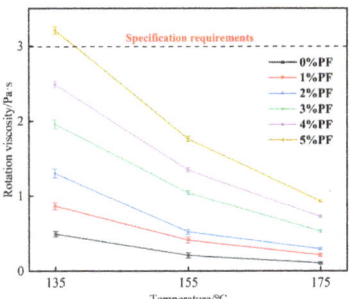

Figure 4. PF-modified asphalt rotational viscosity data graph.

By combining the results from the three main indicators and the rotational viscosity test, it can be observed that 2% PF- and 3% PF-modified asphalts have numerous advantages. They exhibit an improved high-temperature performance, a lesser degree of decrease in ductility, and an increase in rotational viscosity while also meeting the specification requirements for mixing viscosity.

4.3. Temperature Sweep Testing

The rheological properties of the asphalt were evaluated using the DSR test in accordance with the ASTM standard D7175. As can be seen from Figure 5, the G^* of different types of asphalt decreases with the increase in temperature, and the rate of decrease slows down at 60 °C. This suggests that the deformation resistance of asphalt decreases under the continuous action of temperature and repeated loads, and all types of asphalt essentially lose their deformation resistance upon reaching 90 °C. The G^* curve of PF-modified asphalt is higher than that of the base asphalt, indicating that PF has a positive effect on the high-temperature shear deformation resistance of asphalt.

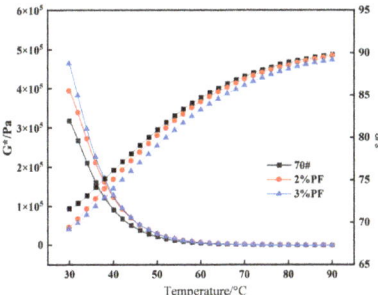

Figure 5. Different asphalt temperature scanning test results.

Further evaluation of the viscoelastic boundary state of asphalt specimens during the experiment through the δ revealed that the δ value increases with a rising temperature, indicating that the asphalt in question gradually tends toward transitioning to a viscous state. The addition of PF alleviates the degree to which asphalt transitions to a viscous state in a gradually heating environment. The δ value of the base asphalt approaches 90°

at 90 °C, indicating that the elastic properties of the base asphalt are essentially lost and irreversible deformation is gradually accumulated.

Figure 6 shows the trend of $G^*/\sin\delta$ with temperature changes. It can be observed that the value of $G^*/\sin\delta$ decreases with the increase in temperature and increases with the addition of PF, indicating that the addition of PF can significantly improve the rutting resistance of asphalt. The temperature corresponding to $G^*/\sin\delta = 1.0$ kPa is the failure temperature of asphalt (details can be seen in Table 7). This suggests that phase separation has occurred in the polymer–asphalt blend system and the high-temperature rutting resistance of the modified asphalt is essentially lost. Through calculations, it was found that the failure temperature of 3% PF-modified asphalt is 5.6 °C higher than that of the base asphalt.

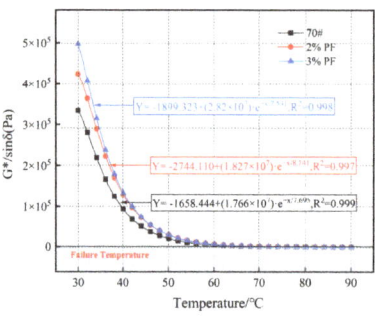

Figure 6. Rutting factors of different asphalts.

Table 7. Critical temperature of different asphalts.

Asphalt Type	Critical Temperature/°C
70#	69.5
2% PF	74.2
3% PF	75.1

4.4. MSCR Testing

According to the ASTM D6648 standard, the temperature corresponding to $G^*/\sin\delta = 1.0$ kPa is defined as the failure temperature of modified asphalt. Figure 7 shows the creep recovery rate (R) of different asphalts under loads of 0.1 kPa and 3.2 kPa, respectively. When the stress load is constant, the R value of asphalt decreases with the rise in temperature. However, the R value of PF-modified asphalt achieves a significant increase compared with the base asphalt. This suggests that the addition of PF can effectively alleviate the loss of asphalt elasticity caused by the temperature rise, thereby enhancing the overall creep recovery ability. Interestingly, the nonlinear fitting of the above data reveals that the rate of reduction in the R-value of PF-modified asphalt slows down with increasing temperature when the stress is at 0.1 KPa, further confirming the contribution of PF to the asphalt's elasticity and high-temperature stability. Meanwhile, when the applied stress increases to 3.2 KPa, the R-value of the base asphalt approaches zero, indicating near-complete failure.

As shown in Figure 8, the Jnr values of all asphalts exhibit nonlinear growth with the increase in temperature. This suggests that the fluidity of the asphalt gradually decreases and its elasticity gradually diminishes, especially at high temperatures where the viscosity of the asphalt dominates, thereby resulting in larger irreversible deformation and a higher loss of creep recovery ability [50]. The addition of PF makes the creep compliance of modified asphalt much smaller than that of base asphalt and reduces the rate of creep compliance increases when faced with temperature rises and stress increases. This is because phenol-formaldehyde resin has excellent adhesion and heat resistance, which can effectively fill the microscopic voids in the asphalt and form a strong chemical bond with them. This can enhance the shear resistance and tensile resistance of asphalt pavements, thereby reducing the deformation of pavements under repetitive loads and extending the pavements' service lives. For them, the high-temperature deformation resistance is optimal when the PF dosage is 3%.

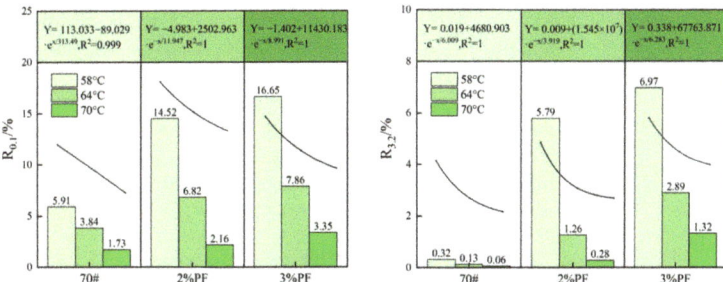

Figure 7. Variation of R-values with temperatures for different asphalts under two stresses.

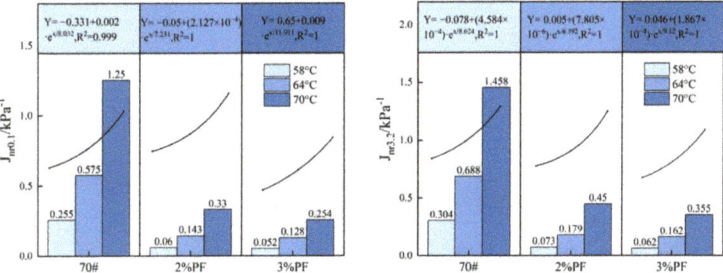

Figure 8. Variation of J_{nr} values with temperatures for different asphalts under two stresses.

4.5. Analysis of Low-Temperature Rheological Properties

The asphalt's low-temperature performance and susceptibility to cracking were assessed using the BBR test, following the guidelines of the ASTM standard D6648. Figure 9 shows the S values of three types of asphalt at two test temperatures. It is evident that as the temperature drops from $-12\ °C$ to $-18\ °C$ the S value of the asphalt sharply increases, thereby indicating that the decrease in temperature causes the asphalt to become harder and more brittle. At $-12\ °C$, the S values of PF-modified asphalt (142 MPa and 163 MPa) suggest that PF-modified asphalt meets the specification requirements (300 MPa). It is noteworthy that at temperatures as low as $-18\ °C$, all three types of asphalt lose flexibility due to inherent material properties. Additionally, the brittleness of PF at low temperatures may further diminish the asphalt pavement's flexibility and toughness.

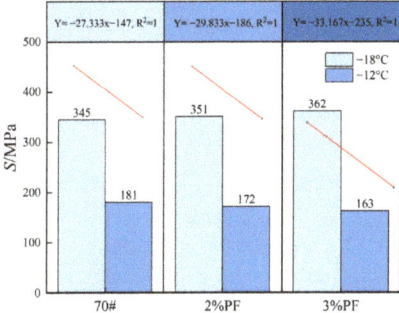

Figure 9. S-values of different asphalts.

Figure 10 presents the m values of different asphalts at two test temperatures. Consistent with the change in S values, the improvement effect of PF on low-temperature performance at −18 °C is not significant, with the m values of both 2% PF-modified asphalt and 3% PF-modified asphalt being less than that of base asphalt. This suggests that at −18 °C, the brittleness of PF impacts the creep capability of asphalt, thereby resulting in a decrease of the m value. However, at −12 °C, the addition of PF can increase the m value of asphalt to a certain extent. This is because low temperatures reduce the viscosity of asphalt, worsening the dispersion of PF in asphalt. The aggregated material that forms becomes more rigid at low temperatures, making the asphalt more brittle overall. Therefore, when the temperature drops to −18 degrees, PF molecules in the asphalt are more likely to cluster together, thereby forming larger aggregates. However, when the temperature drops to −12 degrees, the degree of PF molecule aggregation in the asphalt is lower, so the brittleness of the asphalt is relatively less.

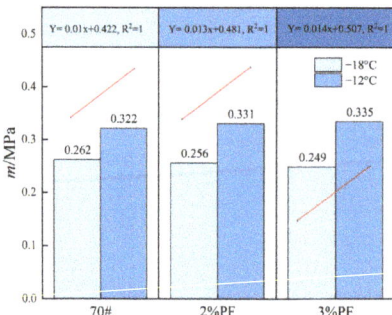

Figure 10. m-values of different asphalts.

4.6. PG Grading

4.6.1. High-Temperature Classification

The Superpave performance grading (PG) system follows the ASTM standard D6373. To simulate the effects of short-term aging on asphalt properties, asphalt was subjected to the rolling thin film oven test (RTFOT) in accordance with the ASTM standard D2872. The PG grades are divided into six levels as follows: 46 °C, 52 °C, 58 °C, 64 °C, 70 °C, 76 °C, and 82 °C. Different asphalts are classified through these PG grades according to the standard (see Table 8).

Table 8. Superpave's asphalt high temperature PG grading standards.

Asphalt State	Test Method	PG Grading Basis
Not aged	DSR	$G^*/\sin\delta \geq 1.0$ kPa
Short term aging	DSR	$G^*/\sin\delta \geq 2.2$ kPa

From the PG grading of various asphalts (Table 9), it can be seen that the high-temperature PG grade for base asphalt is 70 °C, while the high-temperature PG grade for PF-modified asphalt is consistently 76 °C. This suggests that PF significantly extends the high-temperature range of asphalt use.

4.6.2. Low-Temperature Classification

The low-temperature PG grades are set at −46 °C, −40 °C, −34 °C, −28 °C, −22 °C, −16 °C, and −10 °C. According to Table 10, different asphalts are classified under low-temperature PG grades, and the critical low temperatures are determined using interpolation methods. The statistical results can be found in Table 11.

Table 9. High temperature PG classification of different asphalts.

Asphalt Type	Temperature/°C	$G^*/\sin\delta$/kPa	Asphalt State	PG	High Temperature PG Level/°C
70#	64	2.92	Before aging	70	70
	70	1.31			
	70	2.89	After aging	76	
	76	1.20			
2% PF	70	2.10	Before aging	76	76
	76	1.15			
	70	1.81	After aging	76	
	76	3.85			
3% PF	70	2.08	Before aging	76	76
	76	1.09			
	76	1.59	After aging	76	
	82	0.75			

Table 10. Superpave's asphalt low temperature PG grading standards.

Asphalt State	Test Method	PG Grading Basis
After long-term aging	BBR	$S \leq 300$ MPa, $m \geq 0.3$

Table 11. Low temperature PG classification of different asphalts.

Asphalt Type	S-Value Critical Temperature/°C	m-Value Critical Temperature/°C	Low Temperature PG Level/°C	Low Temperature PG Continuous Grading/°C
70#	−25.3	−22.5	−22	−22.5
2% PF	−25.3	−22.6	−22	−22.6
3% PF	−24.6	−22.5	−22	−22.5

4.7. Aging Property

Figure 11 shows the aging index (AI) obtained by calculating the ratio of $G^*/\sin\delta$ before and after aging. It was found that the AI value decreased with an increasing temperature within a certain range. This suggests that the temperature in the temperature scan experiment is not a key factor affecting the aging index. Furthermore, the aging index (AI) values of the base asphalt exceed those of other asphalts, peaking at 2.1 at 52 °C. This highlights the base asphalt's heightened aging and diminished stability following short-term aging. The results show that the addition of PF can enhance the asphalt's resistance to aging and stability. This can be attributed to the excellent thermal and chemical stability of PF, which effectively protects the asphalt from environmental factors such as oxidation that lead to degradation. Moreover, the phenolic groups within PF possess antioxidative properties that are capable of reacting with free radicals, thus reducing the quantity of free radicals in the asphalt and consequently diminishing the degree of aging of the asphalt surface.

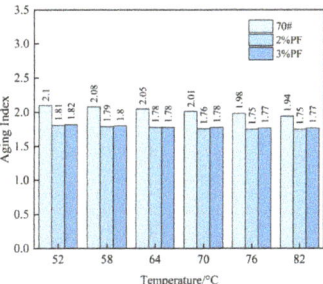

Figure 11. Variations of the aging index with temperatures for different asphalts.

4.8. Marshall Testing

As shown in Figure 12, all asphalt mixtures meet the specification requirement of a Marshall stability higher than 8 kN for hot regions (ASTM D6927). Among them, the stability of the base asphalt mixture is the lowest, indicating that the addition of PF has a positive effect on the high-temperature deformation resistance and high-temperature stability of the asphalt mixture. At the same time, the stability of the 3% PF-modified asphalt mixture is higher than that of the 2% PF-modified asphalt mixture, indicating that PF has a positive response to the high-temperature performance and adhesion of the asphalt mixture within the 0–3% PF content range. By observing the deformation quantity (flow value) when different asphalt mixtures encounter damage, it was found that the content of PF is negatively correlated with the flow value. This suggests that the addition of PF can reduce the deformation of the asphalt mixture. This may be due to PF providing sufficient adhesion and a more stable structure and toughness to the asphalt.

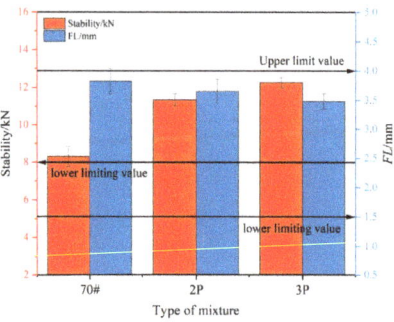

Figure 12. Marshall stability and flow value of different asphalt mixtures.

By calculating the ratio of stability to FL, the Marshall modulus (T) can be introduced to further explore the impact of PF on the asphalt mixture. The results are shown in Figure 13. With the addition of PF and the increase in its content, the value of T gradually increases and the standard deviation gradually decreases. This suggests that the modified asphalt mixture with added PF has a stronger load-bearing capacity and a high-temperature resistance under the same deformation and temperature conditions.

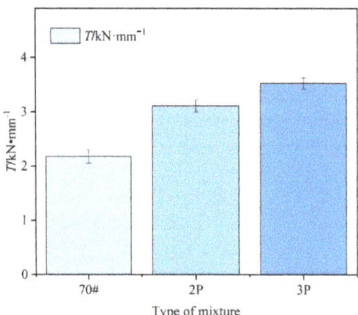

Figure 13. T-value of different asphalt mixtures.

4.9. Water Stability Analysis

4.9.1. Immersion Marshall Testing

The water stability of PF-modified asphalt mixtures can be evaluated by calculating the ratio of Marshall stability before and after water immersion to obtain immersion residual stability (IRS), as shown in Figure 14. It can be seen that the immersion residual stability of all kinds of asphalt mixtures meets the corresponding standards for rainy regions (the IRS of the matrix asphalt mixture is ≥80% and the IRS of the modified asphalt mixture is ≥85%). At the same time, the immersion residual stability of 2% PF- and 3% PF-modified asphalt mixtures increased by 7% and 9%, respectively, compared with the matrix asphalt. This is due to the strong water resistance of PF itself, which can effectively prevent the mixture from being eroded and damaged by water.

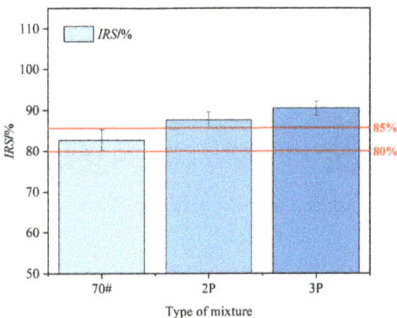

Figure 14. *IRS*-value of different asphalt mixtures.

4.9.2. Freeze–Thaw Splitting Testing

By conducting indirect tensile split tests on asphalt mixtures before and after freeze–thaw treatment, the freeze–thaw split strength ratio (*TSR*) can be calculated, as shown in Figure 15. It can be seen that all types of asphalt mixtures meet the corresponding standards for rainy regions (the *TSR* of the matrix asphalt mixture is ≥75%, and the *TSR* of the modified asphalt mixture is ≥80%). Meanwhile, the *TSR* value increases with the increase in PF content, and the addition of 2% PF and 3% PF increases the *TSR* value of asphalt mixtures by 17.7% and 20.7%, respectively. Through comparative analysis, it is found that PF-modified asphalt mixtures have better adhesion with regard to the asphalt–aggregate system, water permeability resistance, and stripping resistance. This is attributed to PF forming a hard, heat-resistant, water-resistant, and chemically stable polymer, which helps to enhance the internal bonding strength of the mixture, thus allowing the material to better resist breaking during the freeze–thaw process.

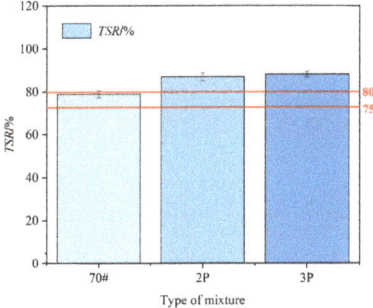

Figure 15. *TSR*-value of different asphalt mixtures.

4.10. FM Analysis

FM tests on base asphalt and 3% PF-modified asphalt were conducted to further investigate the impact and variations of PF's three-dimensional crosslinking within the asphalt. As depicted in Figure 16, the base asphalt appears black in Figure 16a due to its lack of fluorescence. In contrast, Figure 16b displays irregularly shaped and unevenly sized PF particles uniformly dispersed throughout the asphalt, with no signs of aggregation. This dispersion of rigid three-dimensional particles can further restrict the flow of asphalt at high temperatures.

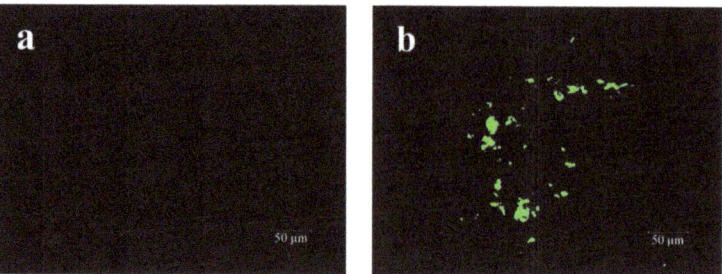

Figure 16. FM figures of asphalt: (**a**) base asphalt; (**b**) 3% PF-modified asphalt.

4.11. FT-IR Analysis

To better detect the impact of PF on the chemical structure and composition of asphalt, FT-IR testing was conducted on 3% PF-modified asphalt and the base 70# asphalt binder. As depicted in Figure 17, both samples exhibited absorption peaks at 2920 cm^{-1} and 2851 cm^{-1}, corresponding to the asymmetric and symmetric stretching vibrations of the C-H bonds. Additionally, peaks at 1456 cm^{-1} and 1376 cm^{-1} were observed, associated with the bending vibrations of the C-H bonds and the symmetric C-H bending vibrations of methyl (CH$_3$) groups, respectively. Of particular note is the peak at 1246 cm^{-1} (highlighted in the inset of the figure) observed for the 3% PF-modified asphalt, which is attributed to the characteristic structures of PF such as the stretching vibrations of the C-O-C bonds in aromatic rings. Overall, the FT-IR test results suggest that no significant chemical reactions occur between PF and the 70# base asphalt binder; PF predominantly exists in a physically dissolved state within the asphalt.

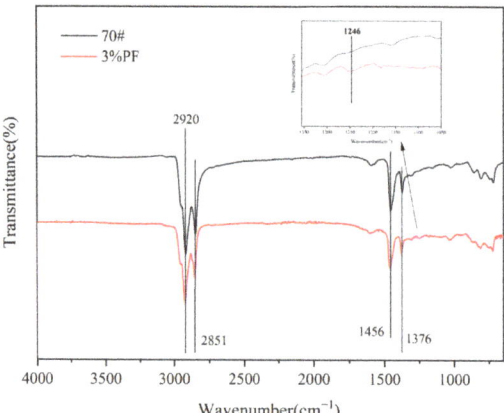

Figure 17. FT-IR figures of asphalts.

5. Conclusions

This study selected PF as a modifier for 70# base asphalt. Through comprehensive analysis including three major indicator tests and viscosity testing, the optimal PF content range was determined. The research investigated how varying PF contents influenced the high-temperature and low-temperature rheological properties as well as the anti-aging performance of the base asphalt. Furthermore, it delved into the microstructural morphology and interaction mechanisms of PF within the asphalt binder. An innovative aspect of this study involved a concurrent analysis of the impact mechanism of PF-modified asphalt binders on the stability and resistance to water damage of asphalt mixtures. Based on the results, the following conclusions were drawn:

(1) Adjusting PF content modulates the conventional physical properties of asphalt. Increased PF content decreases penetration and ductility while raising the softening point and viscosity. PF-modified asphalts with 2% and 3% PF showcased the most notable modification effects.
(2) PF contributes to delaying the transition of asphalt from an elastic state to a viscous state, with 3% PF raising the destruction temperature of the base asphalt by 5.6 °C. Owing to PF's exceptional mechanical properties, it greatly enhances the shear and tensile resistance of the base asphalt.
(3) The elevation of the high-temperature performance grade (PG) from 70 °C for the base asphalt to 76 °C for PF-modified asphalt highlights the potential of PF to extend the operational range of asphalt under high-temperature conditions.
(4) PF does not significantly enhance the low-temperature crack resistance of asphalt binders.
(5) The high viscosity and adhesive properties of PF-modified asphalt reduce porosity within the mixture, enhancing adhesion between different components and improving resistance to high-temperature deformation as well as stripping in the mixture.
(6) The uniform dispersion of PF particles without aggregation contributes to limiting the asphalt flow at high temperatures and inhibiting oxygen penetration.

In summary, this study reveals the potential mechanisms underlying PF-controlled rheological behaviors and the macroscopic mechanical properties of asphalt mixtures. It provides valuable insights into the construction of asphalt pavements tailored to withstand increased traffic loads and harsh environmental conditions. Future research will focus on understanding the interfacial mechanisms between PF and asphalt as well as elucidating the viscoelastic response to PF degradation and asphalt oxidation interactions.

Author Contributions: Conceptualization, L.W. (Lieguang Wang) and K.Y.; methodology, L.W. (Lieguang Wang) and K.Y.; formal analysis, J.H.; investigation, M.W.; data curation, L.W. (Lei Wang); writing—original draft preparation, Z.Z. and L.W. (Lei Wang); writing—review and editing, L.W. (Lieguang Wang); visualization, Z.Z. All authors have read and agreed to the published version of the manuscript.

Funding: This research received no external funding.

Institutional Review Board Statement: Not applicable.

Informed Consent Statement: Not applicable.

Data Availability Statement: The data presented in this study are available on request from the corresponding author.

Conflicts of Interest: Lieguang Wang, Lei Wang, Mingfei Wu, and Zirui Zhang are employed by the Zhejiang East China Engineering Consulting Co., Ltd. Other authors declare no conflicts of interest.

References

1. Mendonça, A.M.G.D.; Neto, O.D.M.M.; Rodrigues, J.K.G.; Silvani, C.; De Lima, R.K.B. Physicochemical and rheological effects of the incorporation of micronized polyethylene terephthalate in asphalt binder. *Pet. Sci. Technol.* **2022**, *40*, 822–838. [CrossRef]
2. Wu, S. Characterization of ductility of field-aged petroleum asphalt. *Pet. Sci. Technol.* **2018**, *36*, 696–703. [CrossRef]
3. Abdellatif, M.; Peel, H.; Cohn, A.G.; Fuentes, R. Pavement crack detection from hyperspectral images using a novel asphalt crack index. *Remote Sens.* **2020**, *12*, 3084. [CrossRef]
4. Huang, J.; Yan, K.; Wang, M.; Zhang, X. Enhancing rheological and aging performance of matrix asphalt through thermoplastic phenol-formaldehyde resin-based intercalated clay nanocomposites: Mechanisms and effects. *Constr. Build. Mater.* **2024**, *411*, 134351. [CrossRef]
5. Nalbandian, K.M.; Carpio, M.; González, Á. Analysis of the scientific evolution of self-healing asphalt pavements: Toward sustainable road materials. *J. Clean. Prod.* **2021**, *293*, 126107. [CrossRef]
6. BAl-Humeidawi, H.; Chafat, O.H.; Kadhim, H.A. Characterizing the properties of sustainable semi-flexible pavement produced with polymer modified bitumen. *Period. Eng. Nat. Sci.* **2021**, *9*, 1064–1072.

7. Habbouche, J.; Hajj, E.Y.; Sebaaly, P.E.; Piratheepan, M. A critical review of high polymer-modified asphalt binders and mixtures. *Int. J. Pavement Eng.* **2020**, *21*, 686–702. [CrossRef]
8. Polacco, G.; Filippi, S.; Merusi, F.; Stastna, G. A review of the fundamentals of polymer-modified asphalts: Asphalt/polymer interactions and principles of compatibility. *Adv. Colloid. Interface Sci.* **2015**, *224*, 72–112. [CrossRef]
9. Hassanpour-Kasanagh, S.; Ahmedzade, P.; Fainleib, A.M.; Behnood, A. Rheological properties of asphalt binders modified with recycled materials: A comparison with Styrene-Butadiene-Styrene (SBS). *Constr. Build. Mater.* **2020**, *230*, 117047. [CrossRef]
10. Mirzaiyan, D.; Ameri, M.; Amini, A.; Sabouri, M.; Norouzi, A. Evaluation of the performance and temperature susceptibility of gilsonite-and SBS-modified asphalt binders. *Constr. Build. Mater.* **2019**, *207*, 679–692. [CrossRef]
11. Bhat, F.S.; Mir, M.S. A study investigating the influence of nano Al2O3 on the performance of SBS modified asphalt binder. *Constr. Build. Mater.* **2021**, *271*, 121499. [CrossRef]
12. Nanjegowda, V.H.; Biligiri, K.P. Recyclability of rubber in asphalt roadway systems: A review of applied research and advancement in technology. *Resour. Conserv. Recycl.* **2020**, *155*, 104655. [CrossRef]
13. Badughaish, A.; Wang, J.; Hettiarachchi, C.; Xiao, F. A review on the crumb rubber-modified asphalt in the Middle East. *J. Mater. Cycles Waste Manag.* **2022**, *24*, 1679–1692. [CrossRef]
14. Jitsangiam, P.; Nusit, K.; Phenrat, T.; Kumlai, S.; Pra-ai, S. An examination of natural rubber modified asphalt: Effects of rubber latex contents based on macro-and micro-observation analyses. *Constr. Build. Mater.* **2021**, *289*, 123158. [CrossRef]
15. Diab, A.; You, Z.; Adhikari, S.; You, L.; Li, X.; El-Shafie, M. Investigating the mechanisms of rubber, styrene-butadiene-styrene and ethylene-vinyl acetate in asphalt binder based on rheological and distress-related tests. *Constr. Build. Mater.* **2020**, *262*, 120744. [CrossRef]
16. Ahmad, M.; Beddu, S.; Hussain, S.; Manan, A.; Itam, Z.B. Mechanical properties of hot-mix asphalt using waste crumber rubber and phenol formaldehyde polymer. *Aims Mater. Sci.* **2019**, *6*, 1164–1175. [CrossRef]
17. Archibong, F.N.; Orakwe, L.C.; Ogah, O.A.; Mbam, S.O.; Ajah, S.A.; Okechukwu, M.E.; Igberi, C.O.; Okafor, K.J.; Chima, M.O.; Ikelle, I.I. Emerging progress in montmorillonite rubber/polymer nanocomposites: A review. *J. Mater. Sci.* **2023**, *58*, 2396–2429. [CrossRef]
18. Joohari, I.B.; Maniam, S.; Giustozzi, F. Enhancing the Storage Stability of SBS-Plastic Waste Modified Bitumen Using Reactive Elastomeric Terpolymer. *Int. J. Pavement Res. Technol.* **2023**, *16*, 304–318. [CrossRef]
19. Behnood, A.; Gharehveran, M.M. Morphology, rheology, and physical properties of polymer-modified asphalt binders. *Eur. Polym. J.* **2019**, *112*, 766–791. [CrossRef]
20. Islam, S.S.; Singh, S.K.; RN, G.R.; Ravindranath, S.S. Performance deterioration of sbs-modified asphalt mix: Impact of elevated storage temperature and sbs concentration of modified binder. *J. Mater. Civ. Eng.* **2022**, *34*, 4021475. [CrossRef]
21. Xu, X.; Sreeram, A.; Leng, Z.; Yu, J.; Li, R.; Peng, C. Challenges and opportunities in the high-quality rejuvenation of unmodified and SBS modified asphalt mixtures: State of the art. *J. Clean. Prod.* **2022**, *378*, 134634. [CrossRef]
22. Lushinga, N.; Dong, Z.; Cao, L. Evaluating the High-Temperature Properties and Reaction Mechanism of Terminal Blend Rubber/Nano Silica Composite Modified Asphalt Using Activated Rubber. *Nanomaterials* **2022**, *12*, 4388. [CrossRef]
23. Diab, A.; Eneib, M.; Singh, D. Influence of aging on properties of polymer-modified asphalt. *Constr. Build. Mater.* **2019**, *196*, 54–65. [CrossRef]
24. Zhu, J.; Birgisson, B.; Kringos, N. Polymer modification of bitumen: Advances and challenges. *Eur. Polym. J.* **2014**, *54*, 18–38. [CrossRef]
25. Rathore, M.; Haritonovs, V.; Zaumanis, M. Performance evaluation of warm asphalt mixtures containing chemical additive and effect of incorporating high reclaimed asphalt content. *Materials* **2021**, *14*, 3793. [CrossRef]
26. Jiang, H.; Wang, J.; Wu, S.; Yuan, Z.; Hu, Z.; Wu, R.; Liu, Q. The pyrolysis mechanism of phenol formaldehyde resin. *Polym. Degrad. Stabil.* **2012**, *97*, 1527–1533. [CrossRef]
27. Li, C.; Wang, W.; Mu, Y.; Zhang, J.; Zhang, S.; Li, J.; Zhang, W. Structural properties and copolycondensation mechanism of valonea tannin-modified phenol-formaldehyde resin. *J. Polym. Environ.* **2018**, *26*, 1297–1309. [CrossRef]
28. Alonso, M.V.; Oliet, M.; Dominguez, J.C.; Rojo, E.; Rodriguez, F. Thermal degradation of lignin-phenol-formaldehyde and phenol-formaldehyde resol resins: Structural changes, thermal stability, and kinetics. *J. Therm. Anal. Calorim.* **2011**, *105*, 349–356. [CrossRef]
29. Chugh, B.; Thakur, S.; Pani, B.; Murmu, M.; Banerjee, P.; Al-Mohaimeed, A.M.; Ebenso, E.E.; Singh, M.; Singh, J.; Singh, A.K. Investigation of phenol-formaldehyde resins as corrosion impeding agent in acid solution. *J. Mol. Liq.* **2021**, *330*, 115649. [CrossRef]
30. Berdnikova, P.V.; Zhizhina, E.G.; Pai, Z.P. Phenol-Formaldehyde Resins: Properties, Fields of Application, and Methods of Synthesis. *Catal. Ind.* **2021**, *13*, 119–124. [CrossRef]
31. Ravindran, L.; MS, S.; Kumar, S.A.; Thomas, S. A comprehensive review on phenol-formaldehyde resin-based composites and foams. *Polym. Compos.* **2022**, *43*, 8602–8621. [CrossRef]
32. Ming, L.Y.; Feng, C.P.; Siddig, E.A. Effect of phenolic resin on the performance of the styrene-butadiene rubber modified asphalt. *Constr. Build. Mater.* **2018**, *181*, 465–473. [CrossRef]
33. Liu, J.; Wang, S.; Peng, Y.; Zhu, J.; Zhao, W.; Liu, X. Advances in sustainable thermosetting resins: From renewable feedstock to high performance and recyclability. *Prog. Polym. Sci.* **2021**, *113*, 101353. [CrossRef]

34. Yao, H.; Zhang, X.; Shen, L.; Bao, N. Tribological and anticorrosion properties of polyvinyl butyral (PVB) coating reinforced with phenol formaldehyde resin (PF). *Prog. Org. Coat.* **2021**, *158*, 106382. [CrossRef]
35. *ASTM D5-06*; Standard Test Method for Penetration of Bituminous Materials. ASTM International: West Conshohocken, PA, USA, 2006.
36. *ASTM D113-17*; Standard Test Method for Ductility of Asphalt Materials. ASTM International: West Conshohocken, PA, USA, 2017.
37. *ASTM D36/D36M-09*; Standard Test Method for Softening Point of Bitumen (Ring-and-Ball Apparatus). ASTM International: West Conshohocken, PA, USA, 2009.
38. *ASTM D4402*; Standard Test Method for Viscosity Determination of Asphalt at Elevated Temperatures Using a Rotational Viscometer. ASTM International: West Conshohocken, PA, USA, 2015.
39. *ASTM C131-06*; Standard Test Method for Resistance to Degradation of Small-Size Coarse Aggregate by Abrasion and Impact in the Los Angeles Machine. ASTM International: West Conshohocken, PA, USA, 2006.
40. *ASTM D5821-13*; Standard Test Method for Determining the Percentage of Fractured Particles in Coarse Aggregate. ASTM International: West Conshohocken, PA, USA, 2013.
41. *ASTM C88-05*; Standard Test Method for Soundness of Aggregates by Use of Sodium Sulfate or Magnesium Sulfate. ASTM International: West Conshohocken, PA, USA, 2005.
42. *ASTM D4792-00*; Standard Test Method for Potential Expansion of Aggregates from Hydration Reactions. ASTM International: West Conshohocken, PA, USA, 2000.
43. *ASTM C136-06*; Standard Test Method for Sieve Analysis of Fine and Coarse Aggregates. ASTM International: West Conshohocken, PA, USA, 2006.
44. *ASTM D1140*; Standard Test Methods for Determining the Amount of Material Finer than 75-μm (No. 200) Sieve in Soils by Washing. ASTM International: West Conshohocken, PA, USA, 2017.
45. *ASTM C127-15*; Standard Test Method for Relative Density (Specific Gravity) and Absorption of Coarse Aggregate. ASTM International: West Conshohocken, PA, USA, 2015.
46. *ASTM D2419-09*; Standard Test Method for Sand Equivalent Value of Soils and Fine Aggregate. ASTM International: West Conshohocken, PA, USA, 2009.
47. Huang, J.; Liu, Y.; Muhammad, Y.; Li, J.Q.; Ye, Y.; Li, J.; Li, Z.; Pei, R. Effect of glutaraldehyde-chitosan crosslinked graphene oxide on high temperature properties of SBS modified asphalt. *Constr. Build. Mater.* **2022**, *357*, 129387. [CrossRef]
48. *ASTM D2726/D2726M-17*; Standard Test Method for Bulk Specific Gravity and Density of Non-Absorptive Compacted Asphalt Mixtures. ASTM International: West Conshohocken, PA, USA, 2017.
49. Cheng, P.; Li, Y.; Zhang, Z. Effect of Phenolic Resin on the Rheological and Morphological Characteristics of Styrene-Butadiene Rubber-Modified Asphalt. *Materials* **2020**, *13*, 5836. [CrossRef] [PubMed]
50. Yuan, D.; Jiang, W.; Hou, Y.; Xiao, J.; Ling, X.; Xing, C. Fractional derivative viscoelastic response of high-viscosity modified asphalt. *Constr. Build. Mater.* **2022**, *350*, 128915. [CrossRef]

Disclaimer/Publisher's Note: The statements, opinions and data contained in all publications are solely those of the individual author(s) and contributor(s) and not of MDPI and/or the editor(s). MDPI and/or the editor(s) disclaim responsibility for any injury to people or property resulting from any ideas, methods, instructions or products referred to in the content.

Article

Experimental Investigation of the Size Effect on Roller-Compacted Hydraulic Asphalt Concrete under Different Strain Rates of Loading

Xiao Meng, Yunhe Liu *, Zhiyuan Ning, Jing Dong and Gang Liang

State Key Laboratory of Eco-hydraulics in Northwest Arid Region of China, Xi'an University of Technology, Xi'an 710048, China; mxlucky0928@163.com (X.M.)
* Correspondence: liuyhe@xaut.edu.cn

Highlights:

(1) Failure modes of RCHAC under the coupling effect of the strain rate effect and the size effect were analyzed.
(2) Dynamic mechanical properties under different size and strain rates were presented.
(3) The effect of strain energy on failure modes was analyzed.
(4) A modified dynamic size effect law was proposed.

Abstract: Asphalt concrete is widely used in hydraulic structure facilities as an impermeable structure in alpine cold regions, and its dynamic mechanical properties are influenced by the strain rate and specimen size. However, the specimen size has an important effect on mechanical properties; few systematic studies have investigated on the size effect of hydraulic asphalt concrete (HAC) under dynamic or static loading rates. In the present study, four sizes of cylindrical roller-compacted hydraulic asphalt concrete (RCHAC) specimens with heights of 50 mm, 100 mm, 150 mm, and 200 mm were prepared and tested under different loading rates ranging from 10^{-5} s^{-1} to 10^{-2} s^{-1} to investigate the size effects of mechanical properties and failure modes at the temperature of 5 °C. The effect of strain rate on the size effects of the compressive strength and the elastic modulus of RCHAC have also been explored. These tests indicate that when the specimen size increases, the compressive strength and failure degree decrease, while the elastic modulus increases. When the height increases from 50 mm to 200 mm, the compressive strength at different strain rates decreased by more than 50%. Furthermore, the elastic modulus increased by about 211.8% from 0.51 GPa to 1.59 GPa at a strain rate of 10^{-5} s^{-1}, and increased by 150% from 5.08 GPa to 12.71 GPa at a strain rate of 10^{-2} s^{-1}. As the strain rate increases, the variation trends with the size of the compressive strength, elastic modulus, and failure degree are distinctly intensified. A modified dynamic size effect law, which incorporates both the specimen size and strain rate, is proposed and verified to illustrate the dynamic size effect for the RCHAC under different loading rates.

Keywords: hydraulic asphalt concrete; failure modes; strain rate effect; size effect; dynamic size effect law

1. Introduction

Since the first roller-compacted hydraulic asphalt concrete (RCHAC) core embankment was built in Germany in 1962, this extremely competitive type of dam has been widely used in many countries due to its advantages of good seepage prevention, earthquake resistance, and deformation adaptability [1]. The damage to the asphalt core wall will greatly reduce its bearing capacity and impermeability, which will seriously threaten the safety performance of the whole dam. In many countries, lots of hydraulic structures are built in the alpine cold region; with low temperatures and frequent earthquakes, the dynamic mechanical safety performance of the RCHAC core wall has always been the concern of researchers. There have been many efforts to study the mechanical properties and failure mechanisms of RCHAC [2–6]; the majority of these, however, ignored the effect that size has on mechanical properties, while many studies showed that the size of the specimen has a significant effect on the mechanical properties of heterogeneous concrete-like materials [7–11]. Therefore, another noteworthy aspect is the influence of the specimen size on the mechanical properties and failure mechanisms in the mechanical research of HAC.

Considering the influence that size has on the mechanical properties of asphalt concrete, Wang [12], through bending tests of asphalt concrete of different sizes, concluded that the deformation of large-size specimens was larger, while the strength was slightly lower, initially confirming that the mechanical properties of HAC were affected by size. Kim [13] carried out static tensile tests and discrete element numerical simulations on disc specimens of asphalt concrete of differing sizes, further demonstrating that the dependence of the tensile strength of asphalt concrete on size is similar to that of ordinary cement concrete; with the increase in specimen size, the tensile strength decreases while the fracture energy increases. In addition, Liu [14] and Hagighat [15] also conducted numerical simulations and mechanical experimental studies on the size effect of asphalt concrete. In general, there are relatively few studies on the size effect of the mechanical properties of RCHAC. Moreover, due to the limitations of test conditions and other factors, there is a lack of data on the effect size has on the dynamic mechanical properties of asphalt concrete, which needs to be further improved.

RCHAC is also a rate-sensitive material [12,16]. The mechanical response under dynamic loads, such as explosions or earthquakes, is significantly different from that under static loads, which is the strain rate effect behavior. Nakumara [17] reported that the strain rate has a significant effect on the dynamic tensile strain failure of asphalt concrete. Wang [5] conducted shear tests on asphalt concrete and concluded that shear modulus, shear strength, and cohesion increased with an increase in strain rate. Ning [18] carried out compressive performance tests of RCHAC under dynamic loads with different strain rates, and their results showed that the strain rate significantly influenced the mechanical properties and failure modes of RCHAC. Therefore, it is necessary to attribute more attention to the effect of the strain rate when studying the dynamic mechanical properties of RCHAC.

In summary, both the size effect and the strain rate effect should be considered in research. However, there is relatively little research on the size effect of RCHAC under dynamic loads. Chen [19] and Albayati [20] reported that HAC has elastic–brittle mechanics and characteristics in low-temperature environments. Therefore, for the dynamic size effect of RCHAC in alpine cold regions, previous studies on the size effect of concrete-like materials carried out by other researchers can be used as references. Krauthammer [21] and Elfahal [22] performed experimental research on the cylindrical concrete specimen with different dimensions under loading rates ranging from 0.014 s^{-1} to 3.03 s^{-1}. They found that the dynamic compressive strength of the cylindrical concrete decreases with the increase in the specimen's size, presenting a significant size effect. Liang [23] tested the uniaxial compression of rock specimens with different ratios of height to diameter (i.e., L/D = 0.5, 0.75, 1, and 1.25) under the strain rates ranging from 10^{-5} s^{-1} to 10^{-2} s^{-1}. They found that with the increase in the specimen's height, the strength and peak strain decrease, the elastic modulus increases, the degree of failure declines, and the dynamic mechanical response is less sensitive towards the strain rate. Jin [24] established random aggregate models of concrete with different sizes under dynamic compression loading at different strain rates ranging from 10^{-5} s^{-1} to 100 s^{-1}. They found that the influence of the dimension on the compressive strength under static and dynamic loading was obviously different, and that the dynamic and static unified size effect of concrete compressive strength was established based on the influence mechanisms of the strain rate effect and the size effect.

Based on the above research method for the dynamic size effect, uniaxial compression experiments of RCHAC with different sizes under different strain rates were carried out to study the combined effects of the strain rate effect and the size effect on the mechanical properties and failure modes of RCHAC specimens with different sizes under dynamic loading rates from 10^{-5} s^{-1} to 10^{-2} s^{-1} (i.e., an earthquake) at a low temperature (5 °C) and establish a dynamic size effect law for mechanical properties based on the influencing mechanism of the strain rate effect on dynamic properties and the size effect. This research shows the need for a reasonable transition between scaled mechanical tests in the laboratory and the actual engineering situation, provides data and theoretical support for the study of the dynamic mechanical properties of hydraulic asphalt concrete, and promotes the development of asphalt concrete anti-seepage materials.

2. Materials and Methods

2.1. Specimen Preparation

In this experiment, the asphalt core specimen was prepared at a roller-compacted hydraulic asphalt concrete (RCHAC) core dam construction site in Northwestern China. The grading index of the asphalt mixture was 0.4, and the aggregate gradation curve is shown in Figure 1. The aggregate was crushed limestone. The added filler (<0.075 mm) was limestone powder and accounted for 13% of mineral mass. The bitumen was of grade B90, and the bitumen content was 7% of the mineral weight. The porosity was about 1.7% to 2.1%. The asphalt and aggregates were mixed in a mixing plant in accordance with the above proportions, and the discharging temperature was 165 °C. The asphalt mixture was paved and pre-rolled into a test section (Figure 2a) with a paver equipped with a vibrating iron compactor, and the pre-compaction coefficient was 86%. Then, when rolling the asphalt concrete, the initial rolling temperature was not less than 150 °C, and the final rolling temperature was not less than 110 °C. After curing, asphalt concrete cores with diameters of 100 mm were drilled and cut into four differently sized cylinders with heights of 50 mm, 100 mm, 150 mm, and 200 mm (Figure 2b). The ratios of height to diameter (L/D) of the specimens were 0.5, 1, 1.5, and 2. According to the Chinese standard DL/T 5362-2006 [25], the specimen with the height of 100 mm was used as the standard specimen for the uniaxial compression test of RCHAC.

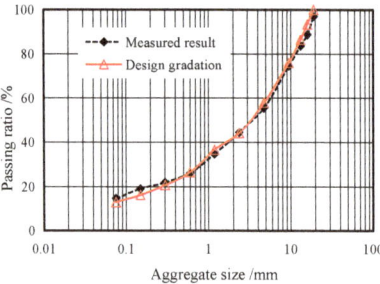

Figure 1. Aggregate gradation curve of RCHAC.

(a) paving, rolling and drilling

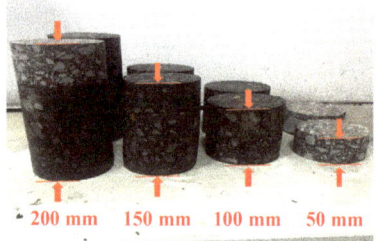

(b) cutting specimens into various height

Figure 2. Specimen preparation.

2.2. Experimental Methods

Dynamic compression tests were carried out on three specimens under different strain rates in each group using the MTS dynamic fatigue test system (Figure 3). The test temperature was set at 5 °C, which was determined according to the temperature in the asphalt core located in the central region of embankment dams in the alpine cold region [2]. Before loading, the specimens were kept at the constant temperature at 5 °C for at least 48 h. During the whole experiment, the test temperature was kept constant using a calorstat (Figure 3). In order to accurately reveal the size dependence of asphalt concrete, the constant strain rate method was adopted in this uniaxial compression experiment. The strain rates of these tests ranged from a quasi-static strain rate of 10^{-5} s^{-1} to the dynamic strain rates of 10^{-4} s^{-1}, 10^{-3} s^{-1}, and 10^{-2} s^{-1} (i.e., an earthquake) [26]. This test was set up in four groups according to the ratio of height to diameter: SE-1, SE-2, SE-3, and SE-4 (i.e., *L/D* = 0.5, 1, 1.5, and 2). The real-time force and displacement data of the test were acquired automatically through the acquisition module of the MTS system. The average value of the tested results of three specimens in each group was recorded if the difference between the average value and the maximum or minimum value of the tested result was less than 15%.

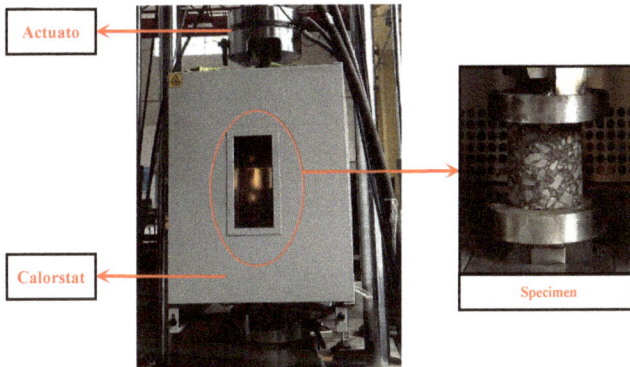

Figure 3. The dynamic loading system and calorstat.

3. Results and Analysis

3.1. Compressive Strength

Table 1 and Figure 4 show the average compressive strength of RCHAC specimens of different sizes under different loading rates. It can be seen that the compressive strength of the RCHAC decreases non-linearly with the increase in height, and the decrease trend is slower with the increase in specimen height. When the specimen height increases from 50 mm to 200 mm, the compressive strength decreases from 5.01 MPa to 2.30 MPa at the strain rate of 10^{-5} s^{-1}, and the compressive strength decreases from 35.98 MPa to 15.59 MPa at the strain rate of 10^{-2} s^{-1}. As shown in Figure 4, when the strain rate increases, the compressive strength decreases more significantly with the increase in size.

Table 1. Compressive strength of specimens with different sizes under different strain rates (MPa).

Height/mm	Strain Rate ($\dot{\varepsilon}$)/s^{-1}			
	10^{-5}	10^{-4}	10^{-3}	10^{-2}
50	5.01	10.22	26.44	35.98
100	3.40	4.33	12.88	18.22
150	2.25	3.43	8.83	14.76
200	2.30	4.43	7.91	15.59

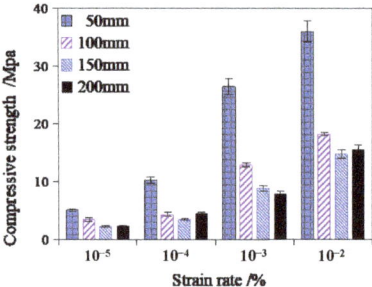

Figure 4. Test results of RCHAC compressive strength under different strain rates.

Meanwhile, the compressive strength of the RCHAC increases with the increase in strain rate. When the strain rate increases from 10^{-5} s^{-1} to 10^{-2} s^{-1}, the compressive strength of the specimen at the height of 50 mm increases from 5.01 MPa to 35.98 MPa, with an increase of 618.16%; the compressive strength of the 200-mm specimen increases by 577.83%, from 2.30 MPa to 15.59 MPa, showing an obvious strain rate enhancement effect. When the temperature of the experiments is above 0 °C, the viscous stress of asphalt increases rapidly with the increase in the strain rate, so the compressive strength of asphalt concrete is greatly enhanced with the increase in strain rate [27]. As shown in Figure 4, the compressive strength and strain rate of the RCHAC show a positive correlation trend, and this overall trend gradually weakens with the increase in the height of the specimen. Therefore, the variation law of the compressive strength of RCHAC of different sizes under dynamic load is the result of the coupling influence of the size effect and the strain rate effect. This is consistent with the research results on the size effect of the dynamic mechanical properties of rocks and concrete conducted by Mattia [28] and Milad [29].

3.2. Elastic Modulus

It can be seen from Table 2 and Figure 5 that when the height of the specimen increases from 50 mm to 200 mm and the strain rate is 10^{-5} s^{-1}, the elastic modulus increases from 0.51 GPa to 1.59 GPa. When the strain rate is 10^{-2} s^{-1}, the elastic modulus of the specimen increases from 5.08 GPa to 12.71 GPa, which means that the elastic modulus of the specimen has a size effect that gradually increases with the increase in its height. Additionally, Figure 5 shows that when the strain rate increases, the positive correlation between the elastic modulus and the specimen height becomes more obvious.

Table 2. Elastic modulus of specimens with different sizes under different strain rates (GPa).

Height/mm	Strain Rate ($\dot{\varepsilon}$)/s^{-1}			
	10^{-5}	10^{-4}	10^{-3}	10^{-2}
50	0.51	1.10	4.54	5.08
100	0.84	1.18	5.34	9.15
150	1.41	1.87	6.88	11.35
200	1.59	2.52	7.23	12.71

For the same specimen size, with the increase in strain rate, the elastic modulus also presents a non-linear increase, and the larger the strain rate, the more obvious the change trend. As can be seen from Table 2, as the strain rate increases from 10^{-5} s^{-1} to 10^{-2} s^{-1}, the elastic modulus of the specimen at the height of 50 mm increases from 0.51 GPa to 5.08 GPa, with an increase of 896.08%. The elastic modulus of the specimen at the height of 200 mm increased by 699.37% from 1.59 GPa to 12.71 GPa. The variation law of elastic modulus with strain rate shows a similar enhancement effect to the research results of Ning [18,30]. It is also shown that the strain rate effect of the elastic modulus of RCHAC decreases with an increase in dimension height. It can be known that the variation law of the elastic modulus of RCHAC is the result of the coupling effect between the size effect and the strain rate effect.

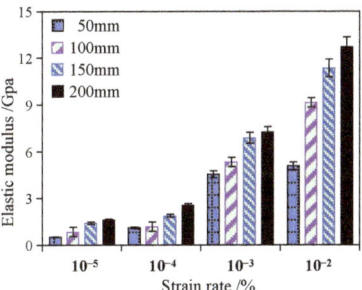

Figure 5. Test results of the elastic modulus of RCHAC under different strain rates.

3.3. Failure Modes

Figure 6 shows the failure mode of specimens with a height of 100 mm under different strain rates. It can be seen that the failure mode of specimens is greatly affected by the strain rate, which is also reflected in the study by Tekalur [27]. As can be seen from Figure 6, when the strain rate is 10^{-5} s^{-1}, there is no obvious damage to the aggregate on the surface of the specimen, only the small cracks caused by the debonding of the aggregate but no penetrating cracks. Due to the slow loading rate and relatively uniform distribution of stress inside the specimen, with the increase in stress, cracks first appear in the weak area, such as the interface between the aggregate and asphalt matrix, and then develop along the weak area around the aggregate. Meanwhile, the coarse aggregates extrude each other to gradually extrude the bonding material, the asphalt matrix. Therefore, the failure pattern of the specimen at a low strain rate is bond failure, which is mainly caused by the extrusion of the asphalt matrix and the debonding of the aggregate and matrix. With the increase in strain rate, some phenomena, such as asphalt matrix fracture and extrusion, coarse aggregate crushing, and aggregate vertical fracture, appear on the surface of the specimen. When the strain rate is 10^{-2} s^{-1}, multiple cracks on the surface of the specimen gradually connect through others, and multiple oblique cracks appear across the surface of the specimen. The above phenomenon occurs when the loading rate becomes faster and the internal stress distribution of RCHAC is more uniform. Therefore, although the weakest area has not reached the strength limit, many areas have reached the strength limit and cracks have occurred, and the generation and development of cracks is faster. The stress does not release along the shortest path of the weak area; it even releases through part of the aggregate to form cracks, and these cracks intersect each other to form multiple penetrating diagonal cracks.

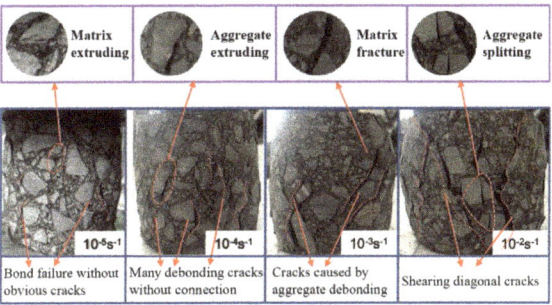

Figure 6. Failure modes of 100-mm-high specimens at different strain rates.

Figure 7 shows the failure mode of specimens with different heights at a strain rate of 10^{-3} s^{-1}. It can be seen that under the same loading rate, the failure mode of the specimen has a significant level of sensitivity to its size. When the height is 50 mm, part of the aggregate is fractured, and some of the aggregate is extruded out of the surface, and there are clear diagonal cracks on the surface of the specimen. At this time, the failure mode is mainly shear failure. As the height is increased, the proportion of influence of the specimen internal defects on the performance of RCHAC increases, and the influence of the structural effect (i.e., including lateral restraint and end friction effects) on the material decreases. The diagonal cracks on the surface of the specimen are distinctly reduced, and the phenomenon of aggregate crushing and extrusion is also gradually reduced. When the height is increased to 200 mm, the aggregate is complete without exhibiting obvious damage, and there is not any through cracks on the specimen; at this time, the failure mode is the bond failure caused by the debonding of the asphalt matrix and aggregate. With the increase in height, the damage degree of the specimen decreases obviously.

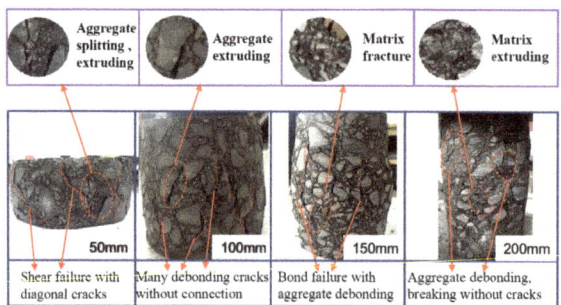

Figure 7. Failure modes of specimens of different heights at a 10^{-3} s^{-1} strain rate.

It can be seen from Figures 6 and 7 that the failure degree of the specimen decreases with the decrease in strain rate or the increase in specimen height. It shows that the failure process of the asphalt concrete specimen is not a single mode, and its failure mode is affected by the strain rate and specimen size.

3.4. Strain Energy Analysis of Failure Modes

According to the thermodynamics theory, the comprehensive effect of the energy conversion process is the driving factor of material failure. Assuming that the uniaxial compression test system is a closed system without heat exchange with the outside world, the total input strain energy, U, of the asphalt concrete specimen under uniaxial compression is as follows:

$$U = U^d + U^e \tag{1}$$

where U^d is the dissipative strain energy, which is used to induce internal damage and plastic deformation of the specimen, and it also contains viscous strain energy due to the influence of asphalt concrete viscosity. U^e is elastic strain energy, and the release of elastic strain energy stored in the specimen is the internal cause of cracks and failure.

Figure 8 shows the relationship between the energies in the stress–strain curve under uniaxial compression. The area enclosed by the curve and unloading modulus in the figure is the dissipated strain energy, and the triangular shaded area is the elastic strain energy. It should be noted that since the unloading test was not carried out in this experiment, the initial elastic modulus was used for energy calculation. Then, the energy of each part was calculated using Equations (1)–(3), as shown in Figure 9a–c.

$$U = \int_0^{\varepsilon_p} \sigma d\varepsilon \tag{2}$$

$$U^e = \frac{1}{2} \cdot \sigma_p \cdot \varepsilon^e \approx \frac{1}{2E_0} \cdot \sigma_p^2 \quad (3)$$

where ε_p is the peak strain; σ_p is the compressive strength; ε^e is the elastic strain; and E_0 is the modulus of elasticity.

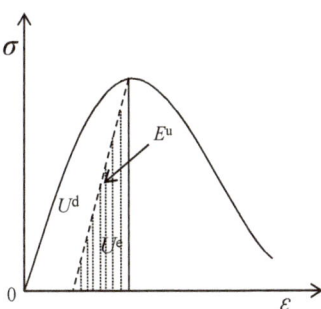

Figure 8. Dissipation strain energy and release strain energy under uniaxial compression.

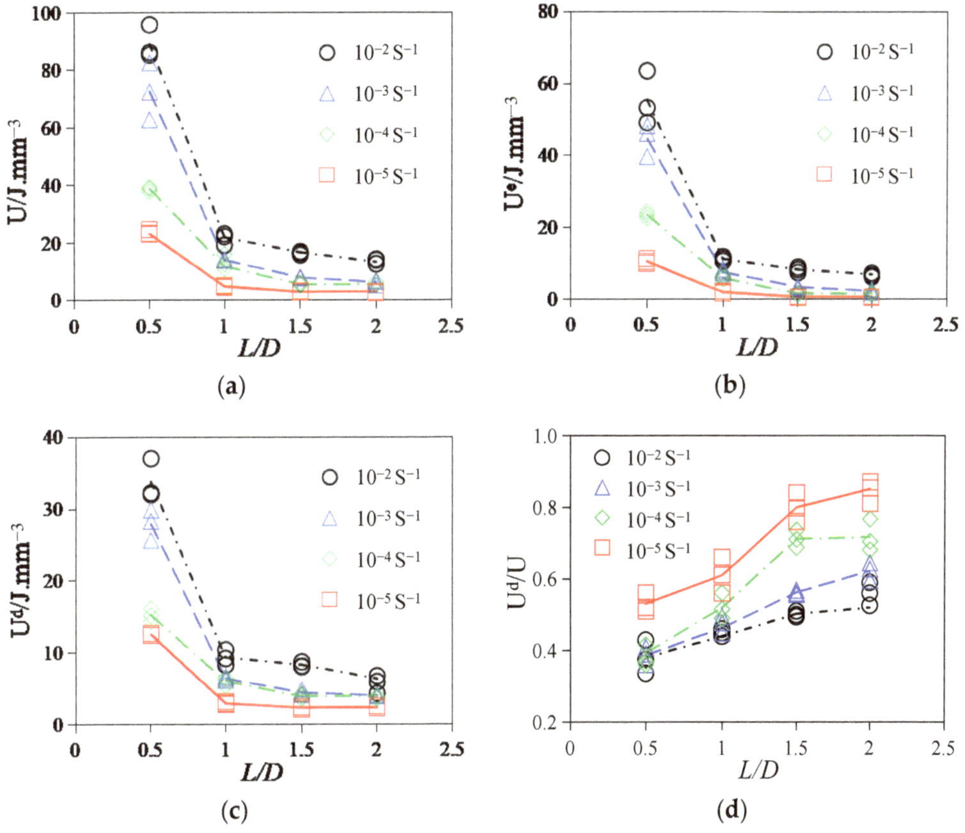

Figure 9. The size effect of various energies at different strain rates. (**a**) Total absorbed energy, U; (**b**) elastic strain energy, U^e; (**c**) dissipated strain energy, U^d; and (**d**) U^d/U.

It can be seen from Figure 9 that the variations in total absorption energy, U, dissipated strain energy, U^d, and elastic strain energy, U^e, in the process of compressive strength of all sizes of specimens are as follows: the U, U^d, and U^e of the specimens will increase with the increase in strain rates. Corresponding to the failure characteristics in Figures 6 and 7, it can also be seen that due to the increase in U per unit volume, a large number of cracks will be generated due to higher energy absorption, and the development of cracks needs to consume more energy.

Hence, the larger the strain rate, the more cracks will be generated during the final failure of the specimen, and the more distinct the brittle failure characteristics [31]. However, when the specimen size increases, U, U^d, and U^e all show a decreasing trend, and the amplitude of this decreasing trend becomes more obvious with the increase in strain rate. This is also consistent with the change rule of compressive strength in Figure 4, indicating that the strength of the specimen is related to the internal storage energy. The higher the value of U^e, the more energy is released when the specimen is damaged, and the bigger the strength [32]. Moreover, as the specimen increases and the strain rate decreases, U^e/U shows a gradually decreasing trend while U^d/U shows an increasing trend, indicating that the proportion of elastic strain energy used to release decreases at this time, and the degree of brittle failure of the specimen also decreases. An increase in the proportion of U^d indicates that the damage dissipation energy increases, the viscous strain energy increases, the rate of stress softening slows down, and the number of cracks decreases during failure. Thus, Figure 9 shows that the sensitivity of the asphalt concrete failure mode to size increases with the strain rate [33]. The tendency of the dissipated energy of RCHAC to decrease with increasing size is more pronounced for larger strain rates, and the slope of the change curve is larger [34]. When the specimen is large and the strain rate is low, the plastic damage of the specimen is greater, and the 'strength loss' increases due to the increase in the viscosity work of the asphalt concrete. It should be noted that 'loss of strength' does not mean 'holistic failure' [35]. It is because the proportion of viscous dissipation energy of the specimen increases that the release of residual elastic strain energy after the damage is not enough to break through the surface strain energy of the specimen and generate large cracks, so the damage to the specimen increases but the degree of failure decreases.; that is to say, when the specimen size is large and the strain rate is small, there is no distinct crack and the degree of failure is low.

4. The Dynamic Size Effect Theory

4.1. The Size Effect on Mechanical Properties

With great efforts to understand the size effect of quasi-brittle materials, many researchers have recognized that the size effect of quasi-brittle materials under compression is essentially a matter of material science [8], which is closely related to the material's aggregate particle size, shape, spatial distribution, pores, initial defects, and other micro- and meso-structural elements. Former researchers made a lot of explorations and published their corresponding important theories [35–37]. Currently, there are three main theories that describe the size effect: the size effect theory based on the weakest link and random strength theory, energy release, and multifractal theory. Among these, the size effect law based on the linear elastic fracture mechanics proposed by Bažant [26] has been proven to be able to describe the effect of the size effect on the properties of quasi-brittle materials. To further study the change law of the compressive strength and elastic modulus of RCHAC with height, the Bažant [8] size effect law is used to describe it:

$$\sigma_0 = \frac{B \cdot f_t}{\sqrt{1 + \frac{D}{D_0}}} \quad (4)$$

where D denotes the structural size; herein, it is the diameter of the cylinder specimen. σ_0 is the strength when the specimen size is D. f_t is the strength of the specimen at the height of 100 mm. B, D_0 is the empirical parameter, which is obtained by fitting and used to determine the value range of σ_0.

Based on the analysis of experimental data, Equation (4) can be modified to analyze the size effect of mechanical properties as follows:

$$F_c = \frac{\beta_c \cdot F_{c,100}}{\sqrt{1 + \frac{\gamma}{\gamma_c}}} \tag{5}$$

where γ is the height-to-diameter ratio of the cylindrical specimen, and the values in this experiment are 0.5, 1, 1.5, and 2. F_c is the mechanical parameter (e.g., compressive strength or elastic modulus) at any ratio of height to diameter. $F_{c,100}$ is the experimental mechanical parameter of the 100-mm-height specimen. β_c and γ_c are the fitting parameters corresponding to different F_c values.

Figure 10 shows the test values of compressive strength and the elastic modulus of RCHAC specimens under different strain rates, as well as the theoretical curves calculated using Equation (5). It is easy to understand from the slope changes of each curve in Figure 10 that with the increase in the height/diameter ratio of asphalt concrete specimens, the compressive strength tends to decrease while the elastic modulus tends to increase, and the trend of both tends to be flat. Therefore, the strain rate can enhance the dimensionality sensitivity of the asphalt concrete strength but suppress the dimensionality sensitivity of the elastic modulus [33,34]. Moreover, the strain rate can enhance the dimensionality sensitivity of the strength of RCHAC but suppress the dimensionality sensitivity of the elastic modulus.

In addition, as can be seen from the analysis in Figure 10, within the range of test loading rates, with the decrease in strain rate, the compressive strength and elastic modulus of specimens of the same size gradually decrease, while the peak strain gradually increases. As the strain rate further decreases, the variation trend of the size effect curve of the compressive strength and elastic modulus is more gradual.

As can be seen from Table 3, the mechanical parameters β_c and γ_c are closely related to the loading strain rate. Within the test conditions, with the increase in strain rate, each parameter presents a regular variation trend. It can be seen that Bažant's [8] size effect law under medium and low strain rate conditions cannot accurately describe the size effect of RCHAC. Therefore, it is necessary to modify the Bažant size effect law, considering the effect of strain rate, so as to make a more comprehensive analysis of the compressive performance of RCHAC.

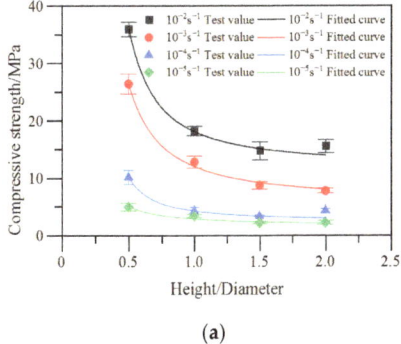
(a)

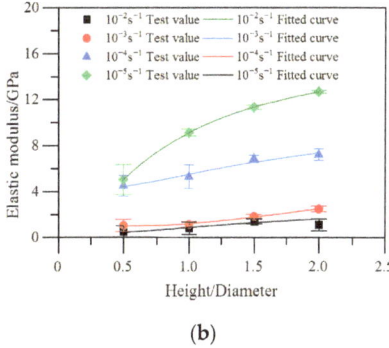
(b)

Figure 10. The experimental and theoretical values of the mechanical parameters of RCHAC of different sizes under different strain rates. (**a**) Compressive strength. (**b**) Elastic modulus.

Table 3. Fitting parameters of β and γ at different strain rates.

		Strain Rate/s^{-1}			
		10^{-5} s^{-1}	10^{-4} s^{-1}	10^{-3} s^{-1}	10^{-2} s^{-1}
Compressive strength	β_1	10.6836	5.3997	2.3286	1.7391
	γ_1	0.9548	0.8261	0.7876	0.6148
Elastic modulus	β_2	0.6667	0.7931	0.8142	0.8720
	γ_2	−2.9864	−2.8048	−2.4125	−2.4938

4.2. The Strain Rate Effect

The existing experimental results show that the dynamic mechanical properties of brittle or quasi-brittle materials, such as rocks and concrete, have a significant strain rate effect [26,38]. In this paper, by exploring the change law of compressive strength and the elastic modulus of RCHAC specimens under different strain rates (10^{-5} s^{-1}~10^{-4} s^{-1}), based on the research results of CEB-FIP [39] and Ning [40], the compressive dynamic increase factor (CDIF) about strain rate is introduced to show the strain rate effect of mechanical properties of RCHAC more intuitively and uniformly. CDIF (F_c) is defined as the ratio of the mechanical properties under the condition of the dynamic strain rate to the values under the condition of the 10^{-5} s^{-1} strain rate. Through calculation and analysis, it was found that the compressive strength CDIF (σ_0), the elastic modulus CDIF (E_0), and the logarithm of strain rate of each group of specimens show a non-linear relationship. After analyzing the test data, Equation (6) was used for fitting analysis as follows:

$$\text{CDIF}(F_c) = \frac{P_d}{P_s} = exp\left[\alpha \cdot lg\left(\frac{\dot{\varepsilon}_d}{\dot{\varepsilon}_s}\right)\right] \quad (6)$$

where F_c is the mechanical parameter; P_d and P_s are the mechanical parameters under dynamic loading and quasi-static loading, respectively; $\dot{\varepsilon}_d$ is the dynamic strain rate; $\dot{\varepsilon}_s$ represents the quasi-static strain rate 10^{-5} s^{-1}; and α is the material parameter obtained through fitting. The regression relationship between the compressive strength CDIF (σ_0), the elastic modulus CDIF (E_0), and the strain rate were obtained through fitting the above data.

The specific forms of Equation (1) for each mechanical property are as follows. The fitting parameters are shown in Table 4.

$$\text{CDIF}(\sigma) = \frac{\sigma_d}{\sigma_s} = exp\left[\alpha_1 \cdot lg\left(\frac{\dot{\varepsilon}_d}{\dot{\varepsilon}_s}\right)\right] \quad (7)$$

$$\text{CDIF}(E) = \frac{E_d}{E_s} = exp\left[\alpha_2 \cdot lg\left(\frac{\dot{\varepsilon}_d}{\dot{\varepsilon}_s}\right)\right] \quad (8)$$

Table 4. Fitting parameters of α at different heights.

		L/D			
		0.5	1	1.5	2
Compressive strength	α_1	0.6784	0.59145	0.63259	0.66595
	R^2	0.91164	0.94579	0.98534	0.99181
Elastic modulus	α_2	0.80165	0.80621	0.82631	0.74145
	R^2	0.85927	0.95991	0.95144	0.85222

Figure 11a,b show the variation curve of CDIF values calculated from the test data and the fitting result on the logarithm of strain rate. It can be seen from the distribution of CDIF values and the trend of the fitting curve that the compressive strength and elastic modulus of RCHAC specimens with different heights are significantly affected by the strain rate. As mentioned above, the CDIF values represent the increase or decrease in mechanical properties at dynamic loading rates and quasi-static loading rates. It can be seen from the calculation results and Figure 11 that under the condition of the dynamic loading rate, the compressive strength CDIF and elastic modulus CDIF are both greater than one, which indicates that the compressive strength and elastic modulus have significant strain rate enhancement effects. Meanwhile, the amplitude of compressive strength enhancement also increases significantly with the increase in strain rate.

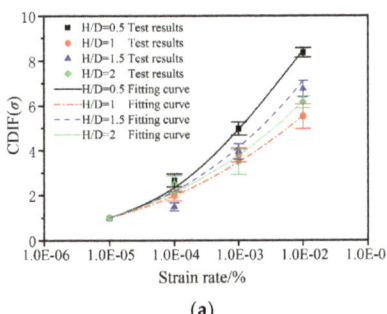 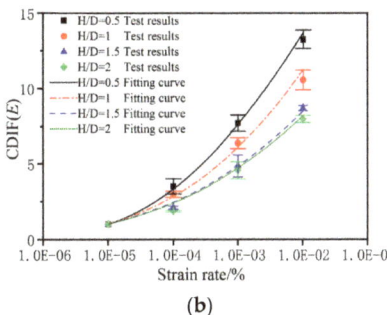

Figure 11. Trend regression curve of the CDIF with the strain rate of RCHAC under different heights. (**a**) The CDIF of compressive strength. (**b**) The CDIF of elastic modulus.

In addition, the dependence of CDIF values on the strain rate varies with different sizes; that is to say, the compressive strength and elastic modulus corresponding to the variation amplitude of the strain rate show size effects. As shown in Figure 11, the change rate of compressive strength and elastic modulus is the largest when the height is 50 mm. With the increase in the height, the change rate of each parameter decreases, and the strain rate effect decreases, which indicates that the strain rate effect of RCHAC has a significant size sensitivity in the strain rate range from $10^{-5}\ \mathrm{s}^{-1}$ to $10^{-2}\ \mathrm{s}^{-1}$.

4.3. The Dynamic Size Effect Model

Considering the influence of the strain rate on the mechanical properties and size effect of the specimen, the influence coefficient of the strain rate, $\varphi_{\dot{\varepsilon}}$, i.e., Equation (6), is introduced to modify Equation (5) to obtain Equation (9). Equation (9) is the theoretical model of the size effect of RCHAC considering the dependence of the strain rate under medium and low strain rates:

$$F_c = \frac{\beta_c \cdot F_{c,100}^{\dot{\varepsilon}_s}}{\sqrt{1 + \frac{\gamma}{\gamma_c}}} \cdot \varphi_{\dot{\varepsilon}} \qquad (9)$$

which is

$$F_c = \frac{\beta_c \cdot F_{c,100}^{\dot{\varepsilon}_s}}{\sqrt{1 + \frac{\gamma}{\gamma_c}}} \cdot exp\left(lg\frac{\dot{\varepsilon}_d}{\dot{\varepsilon}_s}\right) \qquad (10)$$

where $F_{c,100}^{\dot{\varepsilon}_s}$ denotes the mechanical index of asphalt concrete specimens with a height of 100 mm and under a quasi-static loading rate ($10^{-5}\ \mathrm{s}^{-1}$).

Equation (10) can be rearranged as follows:

$$Z = z_0 \cdot \frac{\beta_c}{\sqrt{1 + \frac{x}{\gamma_c}}} \cdot exp[\alpha \cdot (y + 5)] \qquad (11)$$

where $x = \gamma$, $y = \lg(\dot{\varepsilon}_d)$, $Z = F_c$, $z_0 = F_{c,100}^{\dot{\varepsilon}_s}$, and $\dot{\varepsilon}_s = 10^{-5}\ \mathrm{s}^{-1}$.

In order to further verify this theoretical model, the theoretical size effect model was used to calculate the compressive strength and elastic modulus under the experimental conditions and compared with the experimental results. In this theoretical model, z_0 is taken as the mechanical properties of the 100-mm-high specimens under the quasi-static loading rate of 10^{-5} s^{-1}, compressive strength $\sigma_0 = 3.4$ MPa, and elastic modulus $E_0 = 0.84$ GPa; the dynamic enhancement coefficient CDIF of each parameter can be obtained by fitting Equations (7) and (8). Then, the dynamic size effect model given in Equation (11) was used to calculate and predict the test data of the compressive strength and elastic modulus of different specimen heights and strain rates.

Figure 12 shows the predicted calculation results and test results of the modified size effect model. It is easy to know from Figure 12 that some theoretical results deviate slightly from the test values, but the variation trend of mechanical properties with height and strain rate is consistent with the analysis of the test results. The similarity between the theoretical values and experimental values preliminarily verify the rationality of the theoretical model, describing the size effect of asphalt concrete under different strain rates.

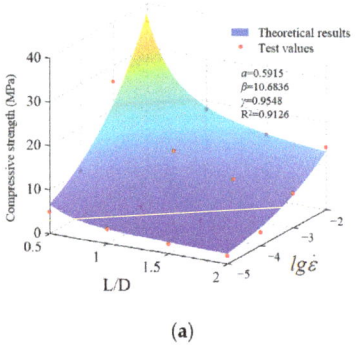
(a)

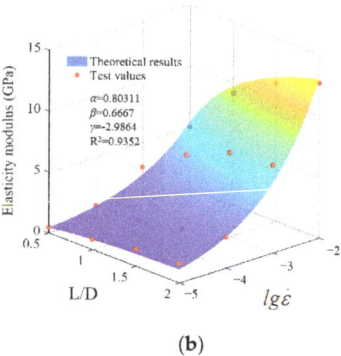
(b)

Figure 12. Theoretical and test values of the theoretical model of dynamic size effect. (**a**) Compressive strength. (**b**) Elastic modulus.

5. Conclusions

For studying the influence of the size effect and the strain rate effect on the mechanical properties and failure modes of RCHAC, a dynamic compression test was carried out under different strain rates. Moreover, a modified size effect theoretical model for RCHAC under different loading rates is proposed according to the influence mechanism of the strain rate and size effect. The following conclusions are obtained on the basis of the experimental studies:

(1) The strain rate and specimen size have coupling effects on the failure modes of RCHAC. When the strain rate is larger or the size is smaller, the damage degree becomes greater. When the strain rate is 10^{-3} s^{-1}, the specimen with a height-to-diameter ratio of 0.5 shows shear failure. When the strain rate decreases or the size increases, the failure modes of the specimen changes gradually from shear failure to aggregate fracture and bond failure. When the strain rate is 10^{-3} s^{-1}, the specimen with a height-to-diameter ratio of two has no obvious cracks, and only part of the asphalt matrix is extruded.

(2) When the ambient temperature is 5 °C, there is an obvious size effect on the mechanical properties of RCHAC when the specimen height ranges from 50 mm to 200 mm. When the specimen size increases, the compressive strength decreases non-linearly, and this decreasing trend gradually declines, while the elastic modulus increases non-linearly, and the increasing trend decreases gradually.

(3) The strain rate has an effect on the size effect of the mechanical properties of RCHAC. With the increase in strain rate, the variation trend of compressive strength and the elastic modulus with size is more significant.

(4) A dynamic size effect model considering strain rate enhancement is proposed, and the relationship between the dynamic loading rate, size, compressive strength, and elastic modulus is established, which can reasonably describe the size effect of dynamic compressive performance under strain rate effect.

(5) The proposed dynamic size effect model considering strain rate enhancement established the relationship between the dynamic loading rate, size, compressive strength, and elastic modulus, and it could reasonably describe the size effect of dynamic compressive performance under strain rates from 10^{-5} s^{-1} to 10^{-2} s^{-1}.

It is to be noted that the viscosity of asphalt concrete plays an important role in its mechanical properties with the increase in specimen size at room temperature. In this paper, the combined effect of the strain rate and size on the mechanical properties and failure modes of RCHAC have been explored, but the effect of the viscosity of RCHAC was not considered. It also can be known from the failure modes and energy characteristics that with the decrease in strain rate or increase in size, the bond failure of the specimen is more significant, which also shows that viscosity has an effect on the failure mode of the specimen. Therefore, the influence of viscosity should be considered in subsequent studies on the size effect of asphalt concrete.

Author Contributions: X.M. and Y.L. conceived and designed the experiments; X.M. and Z.N. performed the experiments; G.L. and J.D. analyzed the data; X.M. wrote the paper. All authors have read and agreed to the published version of the manuscript.

Funding: This research was funded by the National Natural Science Foundation of China: Yunhe Liu (No. 52039008) and Gang Liang (No. 52379133).

Institutional Review Board Statement: Not applicable.

Informed Consent Statement: Not applicable.

Data Availability Statement: Data are contained within the article.

Conflicts of Interest: The authors declare no conflicts of interest.

Nomenclature

Symbol	Paraphrase
U	Total input strain energy
U^d	Dissipative strain energy
U^e	Elastic strain energy
ε_p	Peak strain
σ_p	Compressive strength
ε^e	Elastic strain
D	Diameter of the specimen
f_t	Strength of the specimen with a height of 100 mm
γ	The ratio of height to diameter
F_c	The mechanical parameter
P_d	Dynamic mechanical parameters
P_s	Quasi-static mechanical parameters
$\dot{\varepsilon}_d$	Dynamic strain rate
$\dot{\varepsilon}_s$	Quasi-static strain rate
D_0, β_c, γ_c, and α	The fitting parameters

References

1. Wang, W.; Höeg, K. *The Asphalt Core Embankment Dam: A Very Competitive Alternative*; Modern Rockfill Dams: Beijing, China, 2009.
2. Hao, J.; Liu, Z.; Wang, Z. Development and prospect of hydropower project with asphalt concrete impervious elements in China. *J. Hydraul. Eng.* **2018**, *49*, 1137–1147.

3. Han, X.; Hu, Z.; Yu, L.; Pang, Y.; She, H.; Zhang, L.; Wang, X.; Qi, C. Dynamic Characteristics of Asphalt Concrete as an Impervious Core in Embankment Dams under Varying Temperatures and Stress States. *Materials* **2023**, *16*, 6529. [CrossRef] [PubMed]
4. Baziar, M.; Salemi, S.; Heidari, T. Analysis of earthquake response of an asphalt concrete core embankment dam. *Int. J. Civ. Eng.* **2006**, *4*, 192–210.
5. Wang, W.; Hu, K.; Feng, S.; Li, G.; Höeg, K. Shear behavior of hydraulic asphalt concrete at different temperatures and strain rates. *Constr. Build. Mater.* **2020**, *230*, 117022. [CrossRef]
6. Wang, W.; Höeg, K. Cyclic behavior of asphalt concrete used as impervious core in embankment dam. *J. Geotech. Environ. Eng. ASCE* **2011**, *137*, 536–544. [CrossRef]
7. Bažant, Z.P. Size Effect in Blunt Fracture: Concrete, Rock, Metal. *J. Eng. Mech.* **1984**, *110*, 518–535. [CrossRef]
8. Bažant, Z.; Planas, J. *Fracture and Size Effect in Concrete and Other Quasibrittle Materials*; CRC Press: Boca Raton, FL, USA, 1998; pp. 7–15.
9. Du, X.; Jin, L.; Li, D. A state-of-the-art review on the size effect of concretes and concrete structures (I): Concrete materials. *China Civ. Eng. J.* **2017**, *50*, 28–45.
10. Chowdari, S.G.; David, A.; Benjamin, F.B. Specimen size effect on dynamic modulus measurement of Cold recycled and full depth reclamation mixtures. *Constr. Build. Mater.* **2023**, *393*, 132095.
11. Hu, J.; Wu, X.; Hu, S. Dynamic mechanical behavior of EPS concrete. *J. Vib. Shock* **2011**, *30*, 205–209.
12. Wang, W.; Bai, Z.; Jiang, F. The Researches of engineering behaviors of asphalt concrete. *Pet. Asph.* **1997**, *4*, 21–25.
13. Kim, H.; Wagoner, M.P.; Buttlar, W.G. Numerical fracture analysis on the specimen size dependency of asphalt concrete using a cohesive softening model. *Constr. Build. Mater.* **2009**, *23*, 2112–2120. [CrossRef]
14. Liu, Y.; You, Z.; Zhao, Y. Three-dimensional discrete element modeling of asphalt concrete: Size effects of elements. *Constr. Build. Mater.* **2012**, *37*, 775–782. [CrossRef]
15. Haghighat, P.J.; Aliha, M.R.; Keymanesh, M.R. Evaluating mode I fracture resistance in asphalt mixtures using edge notched disc bend ENDB specimen with different geometrical and environmental conditions. *Eng. Fract. Mech.* **2018**, *190*, 245–258. [CrossRef]
16. Akhtarpour, A.; Khodaii, A. Experimental study of asphaltic concrete dynamic properties as an impervious core in embankment dams. *Constr. Build. Mater.* **2013**, *41*, 319–334. [CrossRef]
17. Nakamura, Y. Improvement of impervious asphalt mixture for high ductility against earthquake excitation. In Proceedings of the 4th International Conference Dam Engineering, Nanjing, China, 18–20 October; pp. 647–656.
18. Ning, Z.; Liu, Y.; Xue, X. Dynamic compressive behaviors of hydraulic asphalt concrete under different temperatures. *J. Hydroelectr. Eng.* **2019**, *38*, 24–34.
19. Chen, Y.; Jiang, T.; Huang, Z.; Fu, W. Effect of temperature on mechanical properties of asphalt concrete. *Rock Soil Mech.* **2010**, *31*, 92–96.
20. Albayati, A.H.; Ajool, Y.S.; Allawi, A.A. Comparative Analysis of Reinforced Asphalt Concrete Overlays: Effects of Thickness and Temperature. *Materials* **2023**, *16*, 5990. [CrossRef]
21. Krauthammer, T.; Elfahal, M.; Lim, J.; Ohno, T.; Beppu, M.; Markeset, G. Size effect for high-strength concrete cylinders subjected to axial impact. *Int. J. Impact Eng.* **2003**, *28*, 1001–1016. [CrossRef]
22. Elfahal, M.; Krauthammer, T.; Ohno, T.; Beppu, M.; Mindess, S. Size effect for normal strength concrete cylinders subjected to axial impact. *Int. J. Impact Eng.* **2005**, *31*, 461–481. [CrossRef]
23. Liang, C.; Li, X.; Zhang, H.; Li, S. Reserch on size effect of uniaxial compression properties of granite under medium and low strain rates. *Chin. J. Rock Mech. Eng.* **2013**, *32*, 528–536.
24. Jin, L.; Yu, W.; DU, X.; Zhang, S.; Li, D. Meso-scale modelling of the size effect on dynamic compressive failure of concrete under different strain rates. *Int. J. Impact Eng.* **2019**, *125*, 1–12. [CrossRef]
25. *DL/T 5362-2006*; Test Code for Hydraulic Bitumen Concrete. National Development and Reform Commission of the People's Republic of China: Beijing, China, 2018.
26. Du, X.; Wang, Y.; Lu, D. Nonlinear multiaxial dynamic strength criterion for concrete material. *J. Hydraul. Eng.* **2010**, *41*, 300–309.
27. Tekalur, S.A.; Shukla, A.; Sadd, M.; Lee, K.W. Mechanical characterization of a bituminous mix under quasi-static and high-strain rate loading. *Constr. Build. Mater.* **2009**, *23*, 1795–1802. [CrossRef]
28. Baldassari, N.; Monaco, A.; Sapora, A.; Cornetti, P. Size effect on flexural strength of notched and un-notched concrete and rock specimens by Finite Fracture Mechanics. *Theor. Appl. Fract. Mech.* **2023**, *125*, 103787. [CrossRef]
29. Milad, S.; Javad, A.; Ali, R.Z. Assessment of mode I fracture of rock-type sharp V-notched samples considering the size effect. *Theor. Appl. Fract. Mech.* **2021**, *116*, 103136.
30. Ning, Z.; Liu, Y.; Wang, Q.; Wang, W. Experimental study on the dynamic compressive behavior of asphalt concrete under different temperature. *J. Vib. Shock* **2021**, *40*, 243–250.
31. Wang, Q.; Li, Q.; Yin, X.; Xu, S. Structural size effect in the mode I and mixed mode I/II fracture of strain-hardening cementitious composites (SHCC). *Int. J. Solids Struct.* **2023**, *288*, 112628. [CrossRef]
32. Xie, H.; Ju, Y.; Li, L. Criteria for strength and structural failure of rocks based on energy dissipation and energy release principles. *Chin. J. Rock Mech. Eng.* **2005**, *17*, 3003–3010.
33. Jin, L.; Li, J.; Yu, W.; Du, X. Size effect modelling for dynamic biaxial compressive strength of concrete: Influence of lateral stress ratio and strain rate. *Int. J. Impact Eng.* **2021**, *156*, 103942. [CrossRef]

34. Yu, W.; Liu, J.; Du, X. Experimental investigation on splitting-tension failures of basalt fiber-reinforced lightweight aggregate concrete: Effects of strain rate and structure size. *J. Build. Eng.* **2023**, *68*, 105853. [CrossRef]
35. Xie, H.; Ju, Y.; Li, L.; Peng, R.D. Energy mechanism of formation and failure of rock masses. *Chin. J. Rock Mech. Eng.* **2008**, *9*, 6–17.
36. Weibull, W. The phenomenon of rupture in solids. *Proc. R. Swed. Inst. Eng. Res.* **1939**, *153*, 1–55.
37. Carpinteri, A.; Ferro, G. Size effects on tensile fractureproperties: A unified explanation based on disorder andfractality of concrete microstructure. *Mater. Struct.* **1994**, *27*, 563–571. [CrossRef]
38. Yan, D.; Li, H.; Liu, J.; Zheng, H. Dynamic properties of concrete specimens with different size. *J. Hydraul. Eng.* **2014**, *45* (Suppl. 1), 95–99.
39. Comite Euro-International Du Beton. *CEB-FIP Model Code 1990*; Redwood Books: Trowbridge, UK, 1993.
40. Ning, Z.; Liu, Y.; Wang, W. Compressive Behavior of Hydraulic Asphalt Concrete under Different Temperatures and Strain Rates. *J. Mater. Civ. Eng.* **2021**, *33*. [CrossRef]

Disclaimer/Publisher's Note: The statements, opinions and data contained in all publications are solely those of the individual author(s) and contributor(s) and not of MDPI and/or the editor(s). MDPI and/or the editor(s) disclaim responsibility for any injury to people or property resulting from any ideas, methods, instructions or products referred to in the content.

Article

Aging Behavior and Mechanism Evolution of Nano-Al₂O₃/Styrene-Butadiene-Styrene-Modified Asphalt under Thermal-Oxidative Aging

Zhiyuan Ji [1], Xing Wu [2], Yao Zhang [1,*] and Gabriele Milani [2]

[1] College of Architectural Science and Engineering, Yangzhou University, Yangzhou 225100, China; jzy@sinoroad.com
[2] Department of Architecture Built Environment and Construction Engineering, Politecnico di Milano, Piazza Leonardo da Vinci, 32, 20133 Milan, Italy; xing.wu@polimi.it (X.W.); gabriele.milani@polimi.it (G.M.)
* Correspondence: yaozhang@yzu.edu.cn

Abstract: The goal of this paper is to analyze the aging behavior and the mechanism evolution of nano-Al₂O₃ (NA)-reinforced styrene-butadiene-styrene (SBS) asphalt under different thermal-oxidative aging conditions. First, NA/SBS-modified asphalt and SBS-modified asphalt with different aging levels were prepared. Second, the viscosity and high temperature rheological performance of the specimens were tested and the property-related aging indexes were calculated and compared. Third, a Fourier transform infrared (FTIR) test of the specimen was conducted and the chemical group-related aging indexes were calculated and analyzed. Fourth, gel permeation chromatography (GPC) was used to analyze the molecular weight of the specimens under different aging levels. Then, an atomic force microscope (AFM) was adopted to analyze the microsurface morphology of different specimens. Finally, correlation analysis between property-related indexes and chemical group indexes was conducted. The results show that NA can enhance the thermal-oxidative aging resistance of SBS asphalt. NA can inhibit the increase in sulfoxide groups and the degradation of the SBS polymer with the increase in aging. NA can slow down the formation of large molecule during the aging process. The degree of change in both the bee structures and micromorphological roughness of NA/SBS asphalt is lower than that of SBS asphalt under different aging levels.

Keywords: aging behavior; mechanism evolution; nanoalumina; SBS asphalt; thermal-oxidative aging

1. Introduction

Asphalt concrete has excellent road performance and is widely used for highway pavement construction due to its convenient construction and maintenance [1,2]. Asphalt is a key material in asphalt concrete, which directly affects the performance of asphalt pavement [3]. However, during the utilization period of asphalt pavement, complex environmental conditions and traffic loads can lead to various road distresses. Researchers have indicated that thermal-oxidative aging causes irreversible damage to asphalt properties, and is one of the main reasons for road distresses [4]. Specifically, the thermal-oxidative effect could first decrease the performance of the asphalt, which decreases the adhesion ability and crack resistance of asphalt material [5]. Then, it induces other road distresses, such as cracking, etc.

SBS asphalt is widely used in the construction of high-grade road surfaces due to its excellent performance. However, SBS asphalt also undergoes noticeable aging under thermal-oxidative conditions [6]. Compared to the aging of base asphalt, the aging of SBS asphalt is more complex. It involves both the hardening phenomenon of base asphalt and the degradation of the SBS polymer [7]. The significant performance deterioration of SBS asphalt under thermal-oxidative aging conditions has been a concern among researchers.

Citation: Ji, Z.; Wu, X.; Zhang, Y.; Milani, G. Aging Behavior and Mechanism Evolution of Nano-Al₂O₃/Styrene-Butadiene-Styrene-Modified Asphalt under Thermal-Oxidative Aging. *Materials* **2023**, *16*, 5866. https://doi.org/10.3390/ma16175866

Academic Editor: Pengfei Liu

Received: 1 August 2023
Revised: 15 August 2023
Accepted: 25 August 2023
Published: 27 August 2023

Copyright: © 2023 by the authors. Licensee MDPI, Basel, Switzerland. This article is an open access article distributed under the terms and conditions of the Creative Commons Attribution (CC BY) license (https://creativecommons.org/licenses/by/4.0/).

There are studies indicating that the chemical groups in the polymer chains of SBS asphalt are highly sensitive to thermal oxidation conditions [4,8]. Hao et al. investigated the influence of aging conditions on the aging characteristics of SBS asphalt [9]. Cortizo et al. [10] believed that the formation of free radicals from chain scission could lead to an increase in polar compounds in SBS asphalt. Wei et al. [11] studied SBS asphalt from the perspective of microstructure and molecular weight, and found reactions such as oxidation, chain breakage, and large polymer cluster degradation in the aging process. Ruan et al. [12] believed that thermal-oxidative aging disrupted the polymer network structure, and as aging time increased, the polymer gradually degraded. Liu et al. [13] used the dynamic shear rheometer (DSR) test to test the rheological properties of aged SBS asphalt and characterized the changes in molecular weight before and after aging.

Considering the irreversible damage caused by asphalt aging on the performance of asphalt pavement, it is important to enhance its long-term resistance to thermal-oxidative aging. Numerous studies have utilized different modifiers to enhance the aging resistance of SBS asphalt, such as fibers, anti-rutting agents, etc. The application of nanomaterials in road engineering has been widely reported [14]. Sun and his colleagues believed that nanomaterials had the potential to be used as asphalt modifiers and conducted extensive experimental research on nanomaterial-modified asphalt, including the construction of test roads [15,16]. Ren et al. summarized and analyzed research into using nanomaterials in asphalt and attempted to modify bio-oil asphalt using five different nanomaterials. The experimental results showed improvements in the aging resistance of bio-oil asphalt [17]. Yadykova A. Y. et al. studied the influence of bio-oil and silica nanoparticles on the adhesion performance of asphalt. According to the adhesion performance test results, 5% bio-oil was recommended as the optimal mixing amount. It was found that hydrophobic silica could better improve asphalt adhesion performance than hydrophilic silica [18]. Yadykova A. Y. et al. also conducted a detailed study on bio-oil and nanomaterials, and their analysis revealed that the addition of 10% bio-oil could significantly improve the adhesion of asphalt to the maximum extent. Additionally, hydrophobic nanoclay enhanced the elasticity and adhesion of asphalt by inducing gelation [19]. Some researchers used metal oxide nanoparticles (nano-SiO_2, nano-TiO_2, nano-ZnO) to study the reinforcing effect of them on the aging resistance of SBS asphalt [20]. Zhang et al. [21] utilized rheological tests and aging tests and discovered that nano-SiO_2 had a positive effect on improving the rheological properties and aging resistance of asphalt. Nano-ZnO was reported multiple times to enhance the aging resistance of asphalt [22,23]. Li et al. [24] found that nano-ZnO could reduce the mass change and viscosity aging index during asphalt aging, which means it improved the aging resistance of asphalt. Wang et al. [25] reported excellent aging resistance properties of carbon nanotubes/SBS composite-modified asphalt.

Among all these nanomaterials, nano-Al_2O_3 (NA) has also been experimented with to modify asphalt due to its excellent high-temperature and chemical stability. Al-Mansob et al. [26] used NA as a modifier to enhance the properties of epoxy-modified natural rubber asphalt (ENRMA). Similarly, Ali and Shafabakhsh et al. [27,28] discovered that NA improved the high temperature properties and storage stability of asphalt. Bhata et al. [29] found that NA had the potential to enhance the aging resistance of SBS-modified asphalt. On the other hand, the potential of NA to modify polymer materials is attractive because NA plays a positive role in the modification and reinforcement of polymers, especially in the enhancement of polymer stability [30,31]. However, the current research on the antioxidant aging resistance of SBS asphalt modified by NA is not comprehensive, and the important mechanism of thermal-oxidative aging resistance of NA-reinforced SBS asphalt has been ignored.

Hence, it is of great significance to conduct a specialized study on the thermal-oxidative aging characteristics of NA/SBS modified asphalt, and to investigate the important aging strengthening mechanisms of NA in SBS asphalt. An increasing number of studies have adopted the approach of extending the aging test duration to examine the characteristics of asphalt thermal-oxidative aging [32]. Nagabhushanarao et al. [33] simulated long-term thermal-oxidative aging conditions by performing multiple cycles of the rolling thin film oven test (RTFOT). Ye et al. [34] conducted a comprehensive analysis of the effects of different durations and temperatures of the RTFOT on asphalt performance and concluded that aging had the most significant impact on asphalt fatigue performance. Based on a previous analysis, Yu et al. found that the prolonged aging time of the standard RTFOT could simulate long-term aging. Therefore, the standard aging time was extended to study the stability and aging characteristics of 15 asphalt samples under different aging times. Then, the destruction of the polymer network could be observed during the aging process [35]. Ibrahim, B. et al. extended the duration of the RTFOT test to 8 days to simulate the long-term aging of asphalt and studied the effect of antioxidants on asphalt performance. They recommended 10% crepe rubber and 2% trimethyl-quinoline as the optimal admixture combination based on the results in asphalt performance improvement [36]. Siddiqui et al. investigated the performance of four types of asphalt after undergoing multiple cycles of RTFOT aging. The results indicated that the asphalt aged through four cycles of the RTFOT exhibited similar rheological properties to those aged for an extended period of time [37]. At present, the approach of extending the aging time to study the aging characteristics of asphalt has become a common and effective research method.

Hence, in this study, we chose to perform multiple cycles of the RTFOT test (the first, second, third, fourth, and fifth cycles) to simulate the thermal-oxidative aging process from short-term aging to long-term aging.

Therefore, this paper first focuses on a detailed evaluation of the influence of NA on the aging property of SBS asphalt under different thermal-oxidative aging levels. Then, a comprehensive analysis of the thermal-oxidative aging mechanism of NA/SBS composite-modified asphalt is analyzed from the perspectives of its chemical functional groups, molecular weight, and micromorphology. Specifically, the NA/SBS-modified asphalt and SBS asphalt under different aging levels were prepared by changing the RTFOT cycles. Second, the rheological properties of the samples were tested and analyzed. Third, Fourier transform infrared spectroscopy (FTIR), gel permeation chromatography (GPC), and atomic force microscopy (AFM) were used to observe the chemical functional groups, molecular weight distribution, and micromorphology evolution of the tested asphalt under different thermal-oxidative aging levels. This paper could help to evaluate the reinforcing effect of NA on the long-term thermal-oxidative aging property of SBS asphalt and to reveal the anti-aging strengthening mechanism of NA in SBS asphalt.

The innovations of this study are as follows:

1. To evaluate the influence of NA on the thermal-oxidative aging properties of SBS asphalt, and to carry out research on the aging characteristics of NA/SBS-modified asphalt under long-term thermal-oxidative aging conditions.
2. To reveal the enhancement mechanism of NA/SBS under thermal-oxidative aging conditions from the perspective of chemical functional groups, molecular weight, and micromorphology.

2. Materials and Methods

2.1. Materials

The raw materials used to prepare the SBS asphalt and the NA/SBS-modified asphalt were base asphalt, linear SBS polymer, and nano-Al_2O_3 (NA). The base asphalt and SBS polymer were produced by China Petroleum & Chemical Corporation (Beijing, China), while the NA was manufactured by Zhejiang Hengna Co., Ltd. (Hangzhou, China). The physical properties of these materials are presented in Table 1.

2.2. Sample Preparation

The samples prepared in this paper can be divided into two categories, namely SBS asphalt (SBSAB) and NA/SBS composite-modified asphalt (NASBS). First, 5% SBS polymer (by weight of base asphalt) was added into the base asphalt to prepare the SBS asphalt. The linear SBS modifier was added into the hot base asphalt at 175 ± 5 °C. A high-speed shear mixer was used to mix them together at a rate of 6000 r/min for 1 h. Then, the SBS asphalt samples were aged using the rolling thin film oven test (RTFOT) under different standard aging cycles (cycles 0, 1, 2, 3, 4, and 5).

Table 1. Physical properties of the raw materials.

Material	Physical Property Index	Unit	Value or Characteristic
Base asphalt	Penetration (25 °C)	mm	6.7
	Softening point	°C	47
	Viscosity (135 °C)	Pa·s	0.52
	Ductility (10 °C)	mm	190
Nano-Al_2O_3	Grain size	nm	30
	Specific Surface Area	m^2/g	40–60
	Appearance	/	White powder solid
SBS	Appearance	/	Linear leaf
	Average molecule weight	g/mol	110,000
	Styrene content	wt%	30

In the research on using NA to modify other types of asphalt, the suggested NA content is 5% (by weight of SBS asphalt), because the NA particles might agglomerate with each other when their content is higher than 5% [38,39]. Therefore, this paper adopted 5% of the weight of SBS asphalt as the content of NA to prepare the NA/SBS-modified asphalt. NA was added into the SBS asphalt at 175 ± 5 °C, and they were mixed using a high-speed shear mixer at a rate of 6000 r/min for 1 h. After preparing the NASBS, the samples were also aged using the RTFOT under different standard aging cycles (cycles 0, 1, 2, 3, 4, and 5). The samples prepared in this paper are marked using codes to make it easier to understand. Samples' information is shown in Table 2.

Table 2. Asphalt sample information and code.

Asphalt Type	Aging Level	Code
SBS modified asphalt	Unaged	SBSAB
	Aging for 85 min	SR85
	Aging for 170 min	SR170
	Aging for 255 min	SR255
	Aging for 340 min	SR340
	Aging for 425 min	SR425
Nano-Al_2O_3/SBS composite modified asphalt	Unaged	NASBS
	Aging for 85 min	NSR85
	Aging for 170 min	NSR170
	Aging for 255 min	NSR255
	Aging for 340 min	NSR340
	Aging for 425 min	NSR425

2.3. Experimental Methods

2.3.1. Aging Procedure

The aging procedure adopted in this study is the rolling thin film oven test (RTFOT) aging method. It is a kind of thermal-oxidative aging procedure. During the aging procedure, aging cylinder bottles containing the test samples were placed in an oven, and the asphalt samples were ensured to maintain a thin film form. The temperature was set at 163 °C, and the flow rate of hot air was controlled within the range of 4000 mL/min ± 200 mL/min. The aging bottle was rotated at a speed of 15 r/min. The experimental procedures were conducted following ASTM D2872 [40].

The standard aging time of RTFOT aging is 85 min. Therefore, in order to study the aging behavior of the asphalt samples under different aging levels, the SBSAB and the

NASBS were aged under different RTFOT cycles. The number of RTFOT cycles was selected as 1, 2, 3, 4, and 5 loading cycles for both the SBSAB and the NASBS. The aging times for the different aging cycles were 85 min, 170 min, 255 min, 340 min, and 425 min, respectively.

In order to quantify the impact of thermal-oxidative aging on the properties of the asphalt samples, different aging indexes were used to characterize the aging behavior of the samples. The specific aging indexes are listed in the following sections.

2.3.2. Viscosity Test

The viscosity (η) of the asphalt samples under different aging levels was measured using a Brookfield rotational viscometer. This index evaluated the flowability of the asphalt samples. Experimental procedures were conducted following ASTM D4402 [41]. In each test, the asphalt samples were tested at 135 °C. The measurement device used for the viscosity testing were coaxial cylinders, with a specification of 27#. The shear rate was set to 50.00 s^{-1}. The aging index of η used in this study was the viscosity aging index (VAI), and it was calculated using Equation (1).

$$VAI = \frac{\eta_{Aged}}{\eta_{Unaged}} \quad (1)$$

2.3.3. DSR Test

A DSR test was used to evaluate the rheological properties of the asphalt samples. The asphalt samples were subjected to periodic shear loading during the test, which was able to provide the parameters of the asphalt's rheological properties to assess its deformability, etc. The rheological property indexes used in this paper were complex modulus (G^*) and rutting factor ($G^*/sin\delta$). The specific experimental procedures were listed in the specifications outlined in AASHTO T315 [42]. The angular frequency and strain parameters were set to 10 rad/s and 12%, respectively.

The aging indexes of G^* and $G^*/sin\delta$ are called the complex modulus aging index (CAI) and rutting factor aging index (RFAI), respectively. The aging indexes were calculated using the test data at 64 °C, and the indexes were calculated using Equations (2) and (3).

$$CAI = \frac{G^*_{Aged}}{G^*_{Unaged}} \quad (2)$$

$$RFAI = \frac{G^*/sin\delta_{Aged}}{G^*/sin\delta_{Unaged}} \quad (3)$$

2.3.4. FTIR Test

The aging of asphalt can also be reflected in a change in internal chemical bonds or functional groups [43,44]. The Fourier transform infrared (FTIR) test has been widely used in asphalt research to analyze the chemical components of asphalt materials and can help researchers understand the chemical functional group information and the molecular structure of asphalt. Therefore, a FTIR test was adopted in this study to obtain the spectrum of the unaged and aged asphalt samples. The FTIR spectrums were then analyzed using EZOMNIC version 7.3 software to extract the representative characteristic peak information and indexes. The spectrums were baseline-corrected using EZOMNIC version 7.3, and then the peak area tool in the software was used to capture the peaks at the specified positions. The software automatically captured the characteristic peaks and calculated their area. The detailed introduction of the aging indexes related to the FTIR test is listed in the discussion section.

2.3.5. GPC Test

Gel permeation chromatography (GPC) is an efficient liquid chromatography technique that could detect the molecular weight and accurately determine the molecular weight distribution of materials [45,46]. The principle of GPC analysis is to separate compounds using a gel column, where compounds with a higher molecular weight have faster penetration rates while compounds with a lower molecular weight have slower penetration rates. Compounds with a higher molecular weight are eluted first, and their signal appears first in the GPC test curve. Compounds with a lower molecular weight are eluted later, and their signal appears later. Asphalt and modified asphalt are a kind of complex blends composed of different molecules with different molecular weights. Each polymer with a different molecular weight inside the asphalt material has different contents. Therefore, GPC was used in this study to analyze the change in the molecular weight distribution of the asphalt materials before and after the aging procedure. Calibration is required for GPC testing, which involves testing standard samples with different molecular weights, constructing a standard curve based on the test results of the standard samples, and determining the equation of the standard curve.

In this study, the experiments were conducted at room temperature. As is shown in Figure 1, after obtaining the curve of the retention time and the signal strength, the curve is divided into three components when analyzing the GPC test result. The first part is the molecule of which the weight is higher than 19,000 Daltons, and it is identified as the SBS polymer. The second part is the molecule of which the weight is between 3000 Daltons and 19,000 Daltons, and it is identified as asphaltene [9,47].

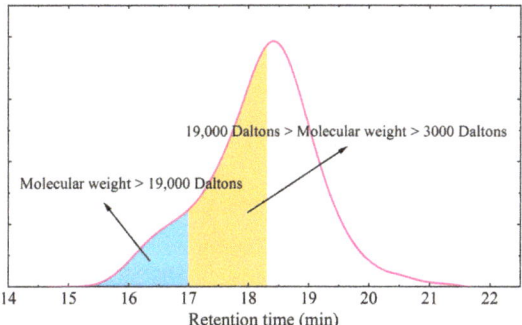

Figure 1. GPC test curve.

2.3.6. AFM Test

Atomic force microscope (AFM) is an analysis technique that allows for the observation of the surface microstructures of materials, and it is commonly used to study surface morphology and roughness. The principle of AFM analysis involves obtaining surface topography by measuring the interaction forces between a non-contact probe and the sample surface, thereby providing detailed information about the surface microstructure features [48]. The probe type is Tap300, with a needle tip curvature radius of less than 10 nm. The test is conducted in tapping mode.

AFM offers advantages such as high resolution, high sensitivity, and the ability to operate under ambient conditions without the need for a vacuum environment. AFM experiments are widely applied in fields such as materials science, biomedical research, and nanotechnology to investigate surface morphology, nanoscale structures, and the three-dimensional conformation of biomolecules.

In this study, AFM was conducted at room temperature, and the Nanoscope Analysis software version 1.5 was used for further analysis of the AFM images to obtain the surface micromorphology and the microsurface roughness data of the asphalt samples.

2.4. Experimental Design

The experimental design of this research is illustrated in Figure 2.

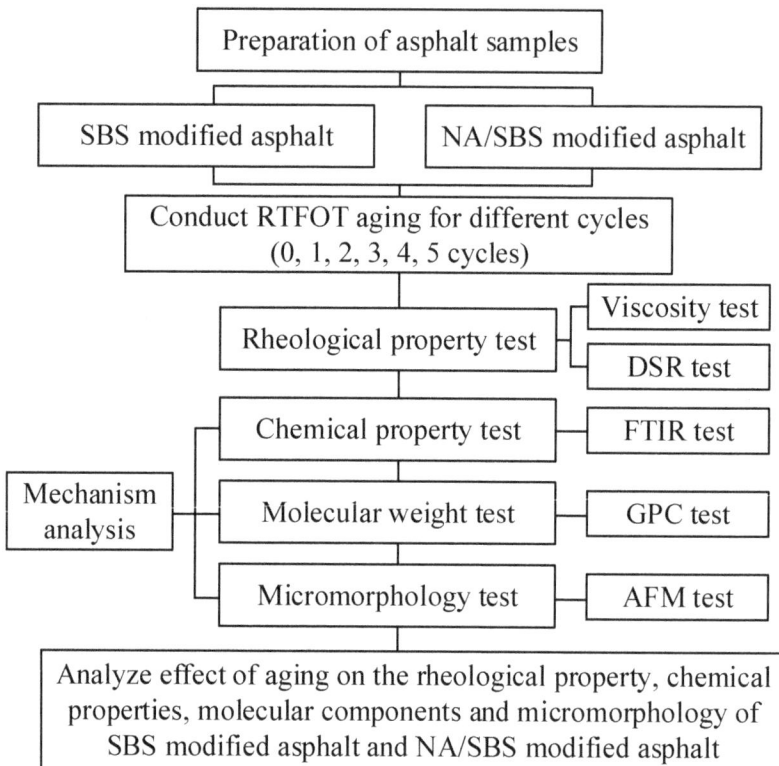

Figure 2. Flowchart of the experimental design.

3. Results and Discussion

3.1. Rheological Property Test Results

3.1.1. Viscosity Test Results

Figure 3 shows the viscosity and *VAI* values of the asphalt samples under different thermal-oxidative aging levels. As the aging time increases, both the SBSAB and NASBS asphalt samples exhibit an increasing trend in viscosity. The viscosity of SBSAB shows a maximum increase of 227%, while the viscosity of NASBS increases by 122%. Clearly, under the same duration of thermal-oxidative aging, NASBS exhibits a lower increase in viscosity compared to SBSAB, which is a positive outcome.

The *VAI* quantifies the extent of viscosity changes caused by thermal-oxidative aging. Prolonged aging leads to an increase in the *VAI* in both asphalt samples. However, the *VAI* growth rate of SBSAB is significantly higher than that of NASBS. Throughout the entire long-term aging process, the *VAI* values of NASBS are consistently lower than those of SBSAB in each aging cycle. A lower *VAI* value indicates a lower impact of aging on asphalt flowability and better anti-aging performance. This suggests that the addition of NA significantly enhances the thermal-oxidative aging resistance of SBSAB, and this positive effect becomes more pronounced when the thermal-oxidative aging level is higher.

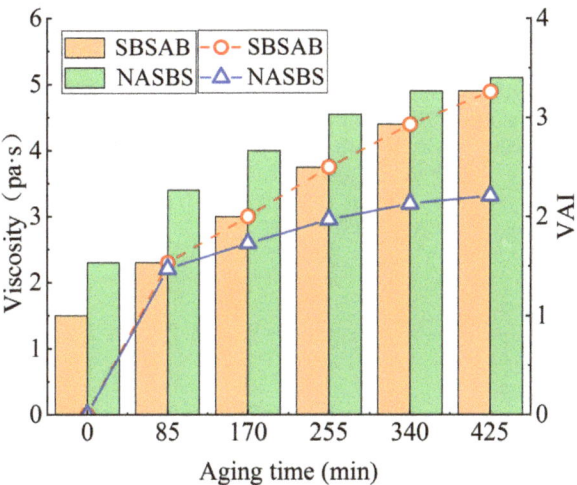

Figure 3. Viscosity (shear rate was set to 50.00 s^{-1}) and viscosity aging index.

3.1.2. DSR Test Results

Figure 4a,b show the complex modulus G^* of SBSAB and NASBS under different aging levels. The G^* values for both SBSAB and NASBS increase with the increase in aging time. The G^* value reflects the ability of asphalt materials to resist deformation, and the increase in G^* is mainly attributed to the conversion of small molecular substances into larger molecular substances under the influence of high temperature and oxygen, which leads to an increase in asphalt stiffness. Under the same thermal-oxidative aging time and test temperature (64 °C), the increasing extent of the G^* value of SBSAB is larger than that of NASBS. For example, the increases in the G^* value of SBSAB during the first and fifth aging processes are 115% and 399%, respectively. The corresponding increasing extents for NASBS are 53% and 161%. This indicates that the addition of NA could delay the hardening effect of SBS asphalt under aging conditions. After comparing with other references, it could also be noted that the anti-hardening effect of NA is more significant than that of mesoporous silica nanoparticles [49].

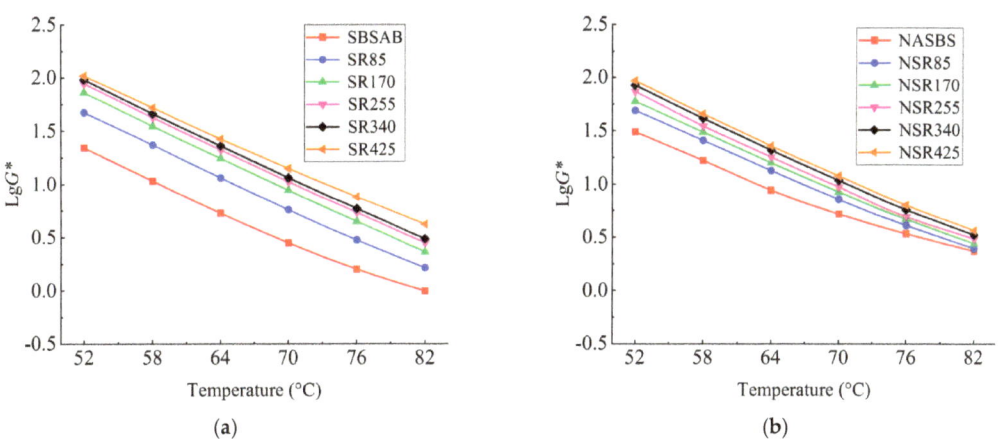

Figure 4. Complex modulus (angular frequency was set to 10 rad/s) of (**a**) SBS asphalt; (**b**) NA/SBS modified asphalt.

It can be observed from Figure 5 that the *CAI* values for SBSAB and NASBS both increase with the increase of aging time. The increasing extent of the *CAI* of SBSAB is more significant than that of NASBS. When the aging cycle changes from 1 to 5, the *CAI* value for SBSAB increases from 2.15 to 4.9. The increasing extent is 128%. In contrast, the *CAI* value for NASBS increases from 1.53 to 2.61, and the increasing extent is 71%. It can be seen that the increasing extent of *CAI* of NASBS is lower than that of SBSAB. Meanwhile, the *CAI* value for NASBS remains lower than that for SBSAB throughout the entire range of the aging process. This also indicates that NA has a positive effect on the thermal-oxidative aging resistance of SBS asphalt.

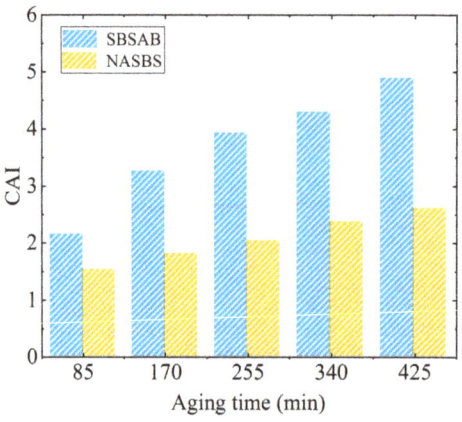

Figure 5. Complex modulus aging index.

Figure 6 illustrates the influence of thermal-oxidative aging time on the $G^*/sin\delta$ of SBSAB and NASBS. Under the same thermal-oxidative aging time, the $G^*/sin\delta$ values for both types of asphalt decrease as the test temperature increases. The $G^*/sin\delta$ shows an increasing trend as the thermal aging time increases under each temperature. The increasing extent of the $G^*/sin\delta$ of SBSAB is higher than that of NASBS. Thus, NASBS is more stable as the aging time increases.

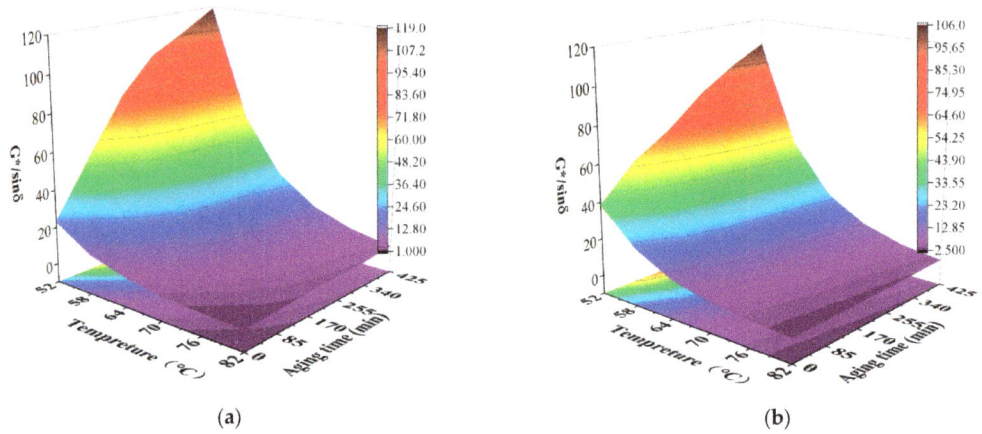

Figure 6. Rutting factor for (**a**) SBS asphalt; (**b**) NA/SBS modified asphalt.

As to the *RFAI* values shown in Figure 7, the *RFAI* values of both types of asphalt show a similar increasing trend as the aging time increases. It should be noted that the increasing extents of the *RFAI* of SBSAB and NASBS are 145% and 67%, respectively, when the aging time increases from 85 min to 425 min. Hence, the *RFAI* increasing extent of NASBS is much lower than that of SBSAB. Meanwhile, throughout the whole thermal-oxidative aging process, the *RFAI* value of NASBS remains lower than that of SBSAB, and the difference between them keeps increasing as the aging degree increases. In comparison with other research under the same level of aging, NA exhibits a more significant anti-aging effect compared to nano-zinc oxide and nanosilica particles [50,51].

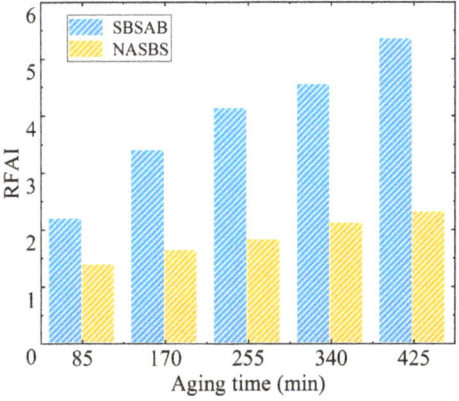

Figure 7. Rutting factor aging index.

Generally, the aging indexes of the rheological property tests showed that the anti-aging ability of NASBS was better than that of SBSAB, which indicated that NA could increase the anti-aging ability of SBS asphalt.

3.2. Mechanism Analyzing Test Results

Based on the analysis of the rheological property experimental data, it can be concluded that SBS asphalt modified with NA exhibits superior resistance to aging under thermal-oxidative conditions, and this positive effect becomes more pronounced with the increase in aging time. In the following sections, this paper focuses on investigating the underlying mechanisms of this phenomenon. It should be noted that three representative aging levels were adopted to analyze these mechanisms in SBSAB and NASBS, namely RTFOT aging cycles 0, 1, and 5.

3.2.1. FTIR Test Results

The Fourier transform infrared (FTIR) spectrums of SBSAB and NASBS under different aging cycles are shown in Figure 8a. The peak positions of the characteristic peaks in the spectrums of SBSAB and NASBS are basically the same, and there are no new absorption peaks formed. Specifically, the peaks at 1460 cm^{-1} and 1376 cm^{-1} are the in-plane bending vibration absorption peaks of aliphatic C-H groups. The characteristic absorption peak at 1030 cm^{-1} is the stretching vibration of the sulfoxide S=O group. The absorption peak at 966 cm^{-1} is the characteristic peak of the polybutadiene (PB) segment, and the absorption peak at 699 cm^{-1} is the characteristic peak of the polystyrene (PS) segment. Some chemical functional groups are affected by thermal-oxidative aging conditions. For instance, in the aging process, the peak absorbance at 1030 cm^{-1} increases dramatically, indicating that the amount of S=O functional groups in bitumen rises gradually. This change can be observed in the magnified portion of Figure 8 [52]. The SBS polymer contains unsaturated C=C bonds in its PB segment, which will degrade in a thermal-oxidative aging environment.

Aging affects both carbonyl and sulfone groups, but upon comparison, sulfone groups are more significantly influenced by aging and are easier to quantify for analysis [53].

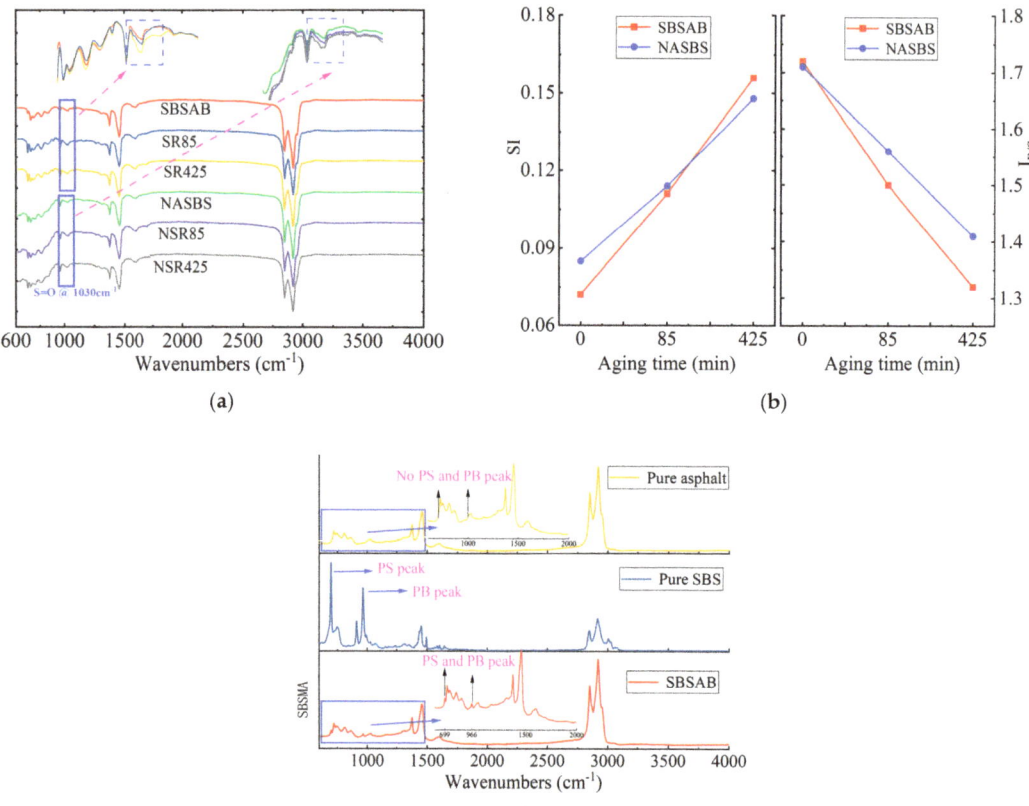

Figure 8. FTIR test results for (a) FTIR spectrums; (b) Sulfoxyl index and polymer index; and (c) SBS asphalt.

In analyzing SBS asphalt, some researchers adopt several aging indexes to quantitatively represent the aging degree of the asphalt. Sulfoxyl index (*SI*) and polymer index ($I_{B/S}$) are selected as indexes to quantify the content of sulfoxide groups and the cracking effect of the SBS modifier in the asphalt material [54,55]. The calculation equations are listed in Equations (4) and (5).

$$SI = \frac{A_{S=O}}{A_{1456 \text{cm}^{-1}}} \tag{4}$$

$$I_{B/S} = \frac{A_{PB}}{A_{PS}} = \frac{A_{966 \text{cm}^{-1}}}{A_{699 \text{cm}^{-1}}} \tag{5}$$

where $A_{S=O}$ represents the peak area centered at 1030 cm^{-1}; $A_{1456\text{cm}^{-1}}$ represents the area centered at 1456 cm^{-1}; A_{PB} represents the peak area at 966 cm^{-1}; and A_{PS} indicates the peak area at 699 cm^{-1}.

Figure 8b shows the *SI* values and $I_{B/S}$ values of SBSAB and NASBS at different thermal-oxidative aging degrees. The *SI* growth rate of SBSAB is significantly higher than that of NASBS under long-term thermal oxidation aging levels. After undergoing 85 min and 425 min of thermal-oxidative aging, the *SI* values of SBSAB increased by 54% and 117%, respectively. On the other hand, the *SI* values of NASBS experienced an increase of

34% and 74% under the corresponding thermal-oxidative aging time. This phenomenon indicates that the addition of NA inhibits the growth of sulfoxide groups in NASBS during thermal-oxidative aging.

The $I_{B/S}$ values of SBSAB and NASBS decreased significantly after thermal oxidation aging, which was caused by the degradation of the SBS. The rate of $I_{B/S}$ reduction in SBSAB was higher than that in NASBS. After undergoing 85 min and 425 min of thermal-oxidative aging, the $I_{B/S}$ values of SBSAB decreased by 13% and 23%, respectively. The $I_{B/S}$ values of NASBS reduced by 9% and 18% under the same thermal oxidation aging conditions, indicating that the SBS degradation in SBSAB was more severe during thermal oxidation aging. It can then be inferred that the NA in NASBS asphalt can alleviate the degradation of SBS under thermal oxidation aging conditions.

3.2.2. GPC Test Results

Figure 9 shows the gel permeation chromatography (GPC) test results of SBSAB and NASBS. As is mentioned before, the SBS polymer is the first to be detected in a GPC test, followed by asphaltene and other smaller components inside the asphalt material. There are mainly two peaks in Figure 9. Therefore, the first peak is the characteristic peak of the SBS polymer and the second peak represents the asphaltene in asphalt material.

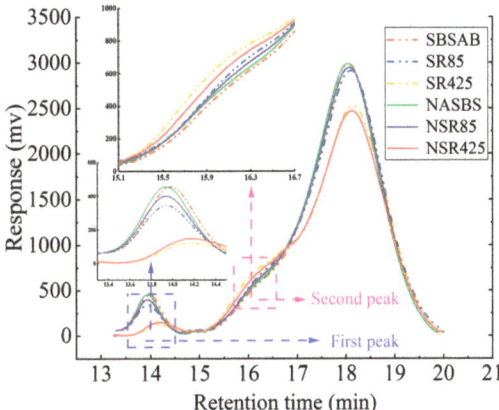

Figure 9. GPC test result.

It can be seen in Figure 9 that the first peak of SBSAB and NASBS decreases after the 85 min aging procedure. This means that the SBS polymer degrades during this period, decreasing its content. After being aged for 425 min, the decreasing extent of the first peak of SBSAB and NASBS is more obvious, which shows that the increase in aging time makes the degradation of the SBS polymers inside the asphalt more severe. However, it can be noticed that the decreasing degree of the first peak of NASBS is lower than that of SBSAB when the samples are aged for both 85 min and 425 min. These results show that NA can delay the degradation effect of the SBS polymer inside SBS asphalt under different thermal-oxidative aging conditions.

It can be seen in Figure 9 that the second peaks of unaged NASBS are higher than that of unaged SBSAB, which means NA increases the molecular weight of asphaltene in unaged SBSAB. After undergoing the thermal-oxidative aging process, the second peaks of SBSAB and NASBS increase, which means that the molecular weight of asphaltene increases. This is because the polar groups in asphalt are combined into large molecular glue substances as the aging time increases. It is also obvious that the increasing extent of NASBS is lower than that of SBSAB, which reveals that NA could decrease the sensitivity of SBS asphalt to thermal-oxidative aging.

In order to make a better comparison of the contents of different components inside the asphalt samples under different aging levels, this section also adopts the area ratio of the main components to reveal the change pattern in the different components. The calculation of area ratios is conducted by integrating the corresponding area under the GPC test curves. The area ratios represent the content of the corresponding components inside asphalt.

It can be seen in Figure 10 that the content of the SBS polymer in SBSAB decreases from 7.47% to 7.19% after 85 min of aging, and the decreasing extent is 3.75%. The corresponding decreasing degree of NASBS is 2.60%. The content of the SBS polymer in SBSAB decreases from 7.47% to 6.47% after 425 min of aging, and the decreasing extent is 13.39%. The corresponding decreasing degree of NASBS is 8.83% after undergoing 425 min of aging.

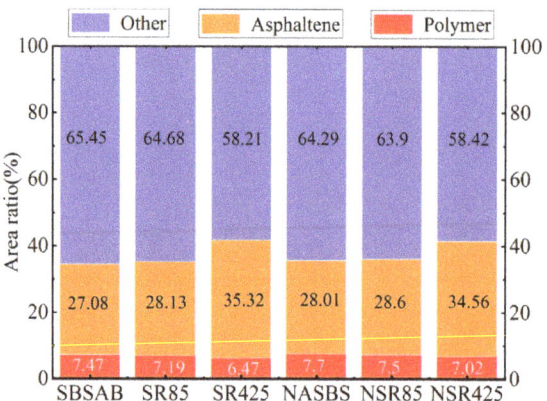

Figure 10. Area ratios of different components.

The content of asphaltene in SBSAB increases from 27.08% to 28.13% after 85 min of aging, and the increasing extent is 3.88%. The corresponding increasing degree of NASBS is 2.11%. The content of asphaltene in SBSAB increases from 27.08% to 35.32% after 425 min of aging, and the increasing extent is 30.43%. The corresponding increasing degree of NASBS is 23.38% after undergoing 425 min of aging.

This paper also uses an index called weight-average molecular weight (M_W) to represent the molecular weight of the SBS polymer and asphaltene inside different asphalt samples. M_W is a statistically averaged molecular weight based on the molecular weights of a sample. The calculation method can be referenced from Equation (6).

$$M_w = \sum_{i=1}^{n} \frac{w_i \times M_i}{w_i} \tag{6}$$

where M_W is the average molecular weight and w_i is the weight of molecular micelle M_i.

As is shown in Figure 11, during the whole aging process, the M_W of the SBS polymer inside different samples keeps decreasing. The decreasing extents of SBSAB after being aged for 85 min and 425 min are 4.67% and 36.38%, respectively. The related decreasing percentages of NASBS are 4.19% and 32.58%, respectively.

The M_W of asphaltene inside different samples keeps increasing as the aging time increases. The increasing extents of SBSAB after being aged for 85 min and 425 min are 11.83% and 46.47%, respectively. The related increasing percentages of NASBS are 7.72% and 39.60%, respectively. Therefore, it can be observed that the addition of NA delays the degradation of SBS polymers and the formation of macromolecular clusters under thermal-oxidative aging conditions [56]. Similarly, the conclusions drawn by Yan et al. support these findings [57].

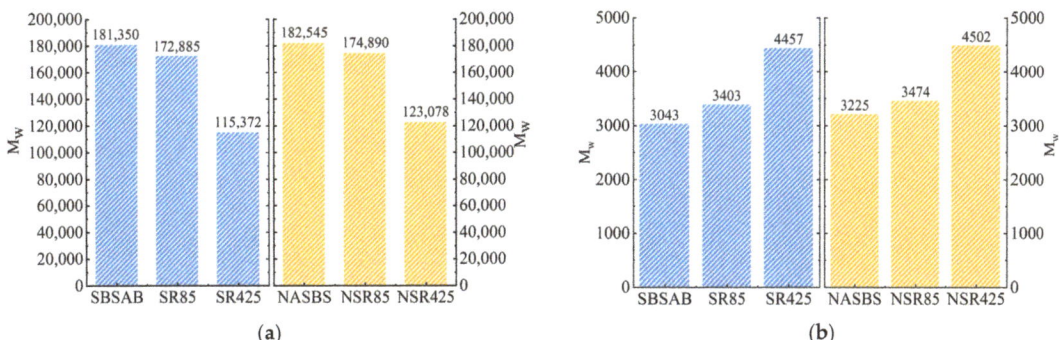

Figure 11. M_W calculation result for (**a**) SBS polymer; (**b**) asphaltene.

3.2.3. AFM Test Results

Figure 12 displays the 2D and 3D micromorphology of SBSAB under different aging times. The elliptical-shaped peaks marked in the figure are defined as the "bee structure". Currently, relevant research suggests that asphaltene is the main component of "bee structures" [58,59]. It can be observed that after aging for 85 min, the quantity of "bee structures" in SBSAB significantly increases, and the volume of these structures also increases. After aging for 425 min, the quantity of "bee structures" further increases, but the volume starts to decrease. The increase in the quantity of "bee structures" is attributed to the conversion of small molecular components of asphalt into larger molecular components dominated by asphaltene clusters under the thermal-oxidative aging condition. After long-term aging, the quantity of asphaltene continues to increase, but the volume starts to decrease, indicating that asphaltene clusters begin to undergo fragmentation under long-term thermal-oxidative aging conditions.

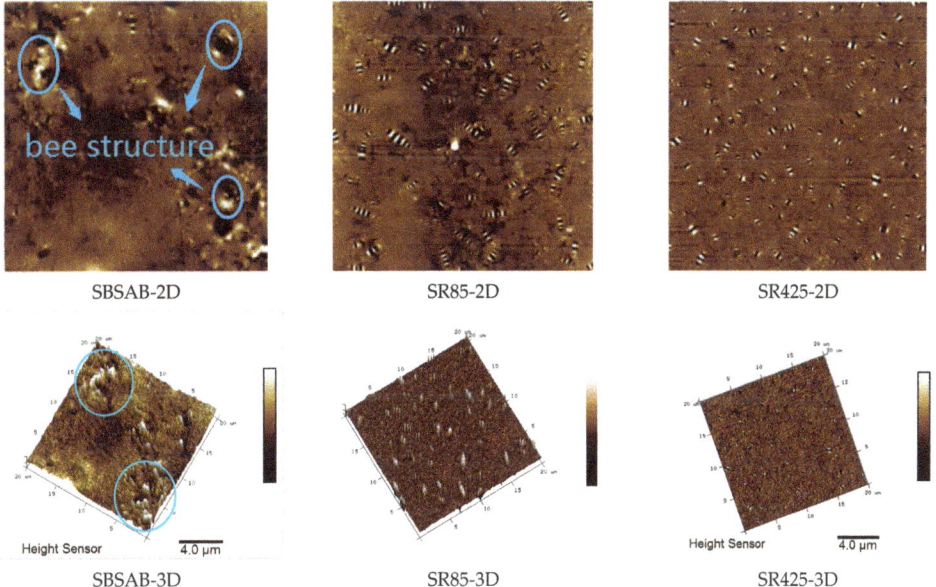

Figure 12. SBSAB micromorphology.

Figure 13 displays the 2D and 3D micromorphology of NASBS under different aging times. After aging for 85 min, the quantity of "bee structures" significantly increases, but the volume decreases slightly. After aging for 425 min, there is no significant increase in the quantity of "bee structures", but the volume decreases. By comparing the quantity and morphology changes in "bee structures" in the two asphalt binders, it can be concluded that NASBS with NA exhibits a more stable variation and stronger anti-aging ability under thermal-oxidative aging.

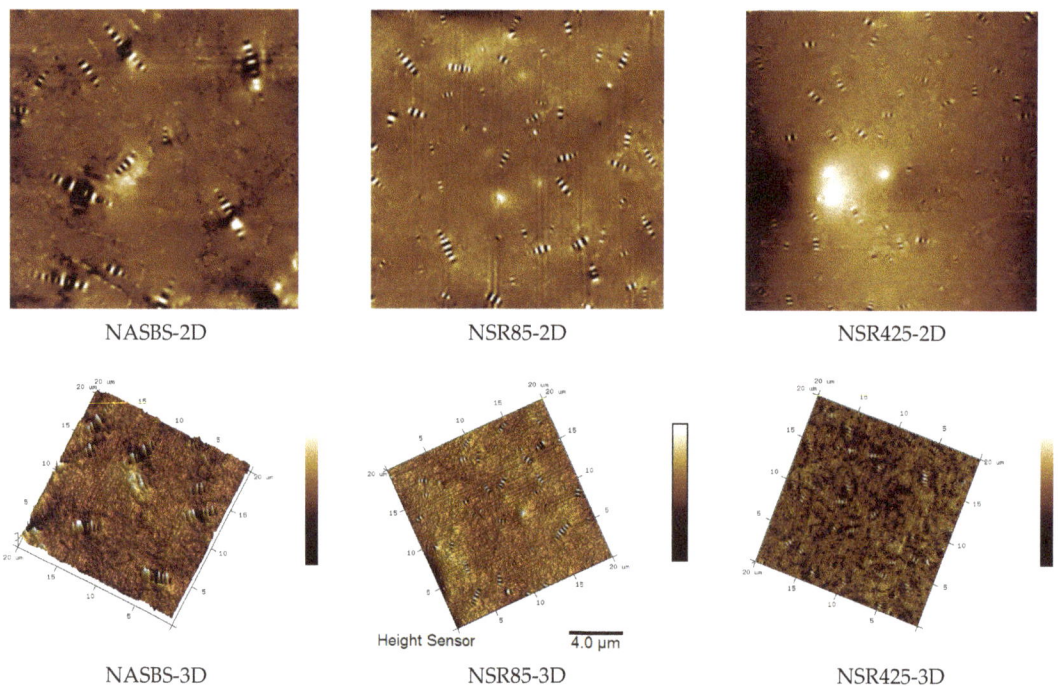

Figure 13. NASBS micromorphology.

In order to make a clearer comparison of the atomic force microscope (AFM) test results for SBSAB and NASBS, this section adopts roughness to compare the change in micromorphology under different aging times. The calculation method for maximum roughness (R_{max}) can be referenced from Equation (7).

$$R_{max} = R_P + R_V \tag{7}$$

where R_p is the height of the peak and R_v is the height of the valley.

It can be concluded from Figure 14 that after a short-term thermal-oxidative aging of 85 min, the roughness of SBSAB significantly increases. After undergoing 425 min of thermal-oxidative aging, the roughness decreases compared with the SBSAB aged for 85 min. This is because the content of asphaltene increases during the first 85 min of aging, but the asphaltene begins to fracture after 425 min of aging. The roughness of NASBS also increases under the influence of 85 min of thermal-oxidative aging, and it decreases when the aging time changes from 85 min to 425 min. The roughness maximum fluctuation extent of SBSAB is 73%, and that of NASBS is 21%. This indicates that under thermal-oxidative aging, the morphological changes in NASBS are more moderate compared to the morphological changes in SBSAB.

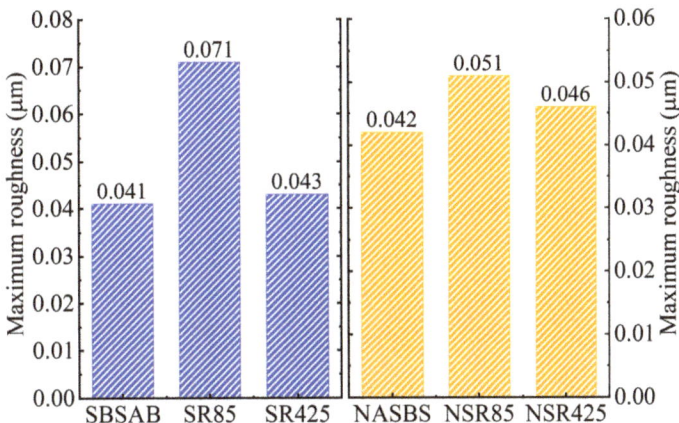

Figure 14. Roughness results.

3.2.4. Correlation Analysis between FTIR Indexes and Rheological Properties Indexes

FTIR test aging indexes have been used by other researchers to analyze the aging condition of asphalt materials and they are more accurate. Therefore, correlation analysis between the $I_{B/S}$ and rheological property aging indexes is analyzed in this section. The data are listed in Table 3.

Table 3. $I_{B/S}$ and rheological property aging indexes.

	Aging Time	$I_{B/S}$	VAI	CAI	RFAI
SBSAB	85 min	1.5	1.53	2.15	2.19
	170 min	1.47	2.00	3.27	3.4
	255 min	1.38	2.5	3.94	4.14
	340 min	1.313	2.93	4.3	4.56
	425 min	1.32	3.26	4.9	5.37
NASBS	85 min	1.56	1.47	1.53	1.38
	170 min	1.46	1.73	1.81	1.63
	255 min	1.45	1.97	2.04	1.82
	340 min	1.44	2.13	2.37	2.11
	425 min	1.41	2.21	2.61	2.31

The correlation analyzing the results between the $I_{B/S}$ and rheological property aging indexes of SBSAB and NASBS is illustrated in Figure 15a and 15b, respectively. The correlation between the $I_{B/S}$ and VAI is highest for SBSAB and NASBS, the decision coefficients (R^2) between the $I_{B/S}$ and VAI for SBSAB and NASBS being 0.9440 and 0.8579, respectively. The VAI represents the change in the flowability of the asphalt before and after thermal-oxidative aging. A lower VAI indicates better resistance to aging, and the $I_{B/S}$ to some extent reflects the polymer content of aged asphalt. Therefore, better resistance to aging corresponds to a higher remaining polymer content after aging.

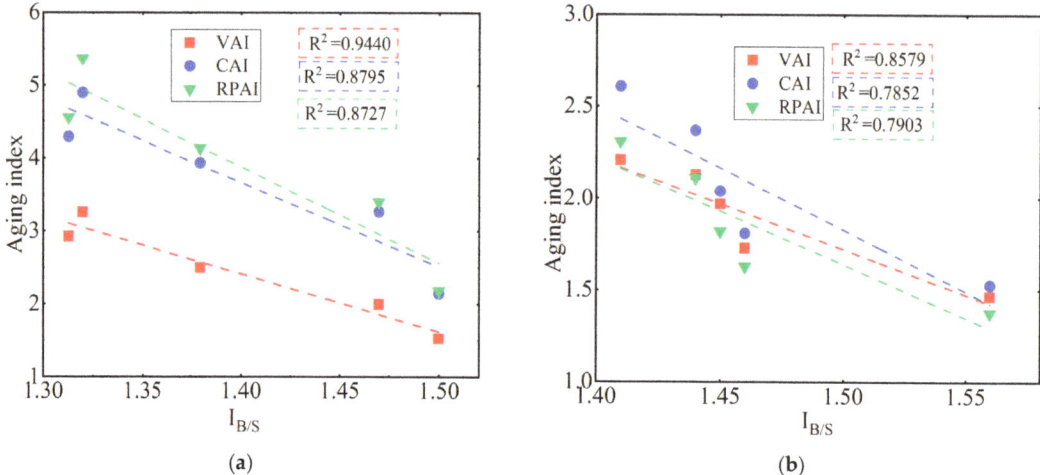

Figure 15. Correlation between the polymer index and rheological property aging indexes for (**a**) SBS asphalt; (**b**) NA/SBS modified asphalt.

4. Conclusions

In this study, a rolling thin film oven test (RTFOT) was used to investigate the thermal-oxidative aging performance of SBSAB and NASBS asphalt binders under different aging times. The changes in the asphalt binder properties before and after thermal-oxidative aging were evaluated and compared through viscosity and DSR tests. Furthermore, the anti-aging mechanism of NA in SBS asphalt was analyzed from the perspectives of chemical functional groups, molecular weight, and microstructure, etc., using FTIR, GPC, and AFM tests, respectively. The following conclusions were drawn:

(1) Based on the rheological property aging indexes *VAI, CAI, RFAI* under different aging times, it was found that NASBS exhibited significantly improved anti-thermal-oxidative aging performance compared to SBSAB.

(2) For both short-term aging for 85 min and long-term aging for 425 min, continuous exposure to thermal-oxidative aging accelerated the aging of SBS asphalt. NA lessened the aging of SBS asphalt.

(3) An analysis of the FTIR results revealed that continuous thermal-oxidative aging increased sulfoxide groups and caused a continuous degradation of SBS polymers. The addition of NA delayed the growth of sulfoxide groups and the degradation of SBS polymers, thus improving the anti-aging performance of SBS asphalt.

(4) GPC tests showed that both SBSAB and NASBS experienced a significant increase in the content of asphaltene and a decrease in the content of SBS polymers during thermal-oxidative aging. The addition of NA slowed down the conversion of small molecules to large molecules in the asphalt and hindered the degradation of SBS polymers.

(5) The volume of the "bee structure" in NASBS decreases and the quantity increased with the duration of thermal-oxidative aging. On the other hand, the volume of the "bee structure" in SBSAB showed an initial increase followed by a decrease, with an overall increase in quantity. The change in roughness of NASBS was less affected by thermal-oxidative aging compared to that of SBSAB.

Based on a summary of the current literature, this paper studied the influence of NA on the thermal-oxidative aging characteristics of SBS-modified asphalt. Mechanism analyses were conducted to understand the excellent anti-aging mechanism of NA/SBS asphalt, filling the research gap in the current literature. The improvement of the anti-aging performance of asphalt corresponds to better fatigue performance [60]. The excellent anti-aging performance of NA/SBS asphalt can effectively alleviate the diseases caused by the aging of SBS asphalt pavement and contribute to the construction of aging-resistant asphalt pavement. In the future, further experimental and engineering research is needed to explore the fatigue life and economic benefits of NA/SBS asphalt.

Author Contributions: Conceptualization, Z.J.; methodology, Z.J., X.W., G.M. and Y.Z.; software, Z.J.; validation, Z.J. and X.W.; formal analysis, Z.J.; investigation, Z.J. and X.W.; resources, Z.J. and Y.Z.; data curation, Z.J. and X.W.; writing—original draft preparation, Z.J. and X.W.; writing—review and editing, Z.J., X.W., G.M. and Y.Z.; visualization, Z.J. and X.W.; funding acquisition, Y.Z. All authors have read and agreed to the published version of the manuscript.

Funding: This research was funded by Innovation and Entrepreneurship Program of Jiangsu Province, grant number JSSCB20211065, and High-level Talent Introduction Project of Yangzhou University, grant number 137012062.

Institutional Review Board Statement: Not applicable.

Informed Consent Statement: Not applicable.

Data Availability Statement: The data presented in this study are available on request from the corresponding author.

Acknowledgments: The authors would like to thank Yangzhou University Test Center for providing experimental supports for this research.

Conflicts of Interest: The authors declare no conflict of interest.

References

1. Ren, J.; Xue, B.; Zhang, L.; Liu, W.; Li, D.; Xu, Y. Characterization and prediction of rutting resistance of rock asphalt mixture under the coupling effect of water and high-temperature. *Constr. Build. Mater.* **2020**, *254*, 119316. [CrossRef]
2. Zhang, Y.; Gu, Q.; Kang, A.; Ding, X.; Ma, T. Characterization of mesoscale fracture damage of asphalt mixtures with basalt fiber by environmental scanning electron microscopy. *Constr. Build. Mater.* **2022**, *344*, 128188. [CrossRef]
3. Ren, J.; Zang, G.; Xu, Y. Formula and pavement properties of a composite modified bioasphalt binder considering performance and economy. *J. Mater. Civil. Eng.* **2019**, *31*, 04019243. [CrossRef]
4. Sun, L.; Wang, Y.; Zhang, Y. Aging mechanism and effective recycling ratio of SBS modified asphalt. *Constr. Build. Mater.* **2014**, *70*, 26–35. [CrossRef]
5. Zhang, Z.; Han, S.; Han, X.; Cheng, X.; Yao, T. Comparison of SBS-modified asphalt rheological properties during simple-aging test. *J. Mater. Civil. Eng.* **2020**, *32*, 04020241. [CrossRef]
6. Zhang, D.; Zhang, H.; Shi, C. Investigation of aging performance of SBS modified asphalt with various aging methods. *Constr. Build. Mater.* **2017**, *145*, 445–451. [CrossRef]
7. Rivera, C.; Caro, S.; Arámbula-Mercado, E.; Sánchez, D.B.; Karki, P. Comparative evaluation of ageing effects on the properties of regular and highly polymer modified asphalt binders. *Constr. Build. Mater.* **2021**, *302*, 124163. [CrossRef]
8. Zhu, J.; Birgisson, B.; Kringos, N. Polymer modification of bitumen: Advances and challenges. *Eur. Polym. J.* **2014**, *54*, 18–38. [CrossRef]
9. Hao, G.; Huang, W.; Yuan, J.; Tang, N.; Xiao, F. Effect of aging on chemical and rheological properties of SBS modified asphalt with different compositions. *Constr. Build. Mater.* **2017**, *156*, 902–910. [CrossRef]
10. Cortizo, M.S.; Larsen, D.O.; Bianchetto, H.; Alessandrini, J.L. Effect of the thermal degradation of SBS copolymers during the ageing of modified asphalts. *Polym. Degrad. Stabil.* **2004**, *86*, 275–282. [CrossRef]
11. Wei, C.; Duan, H.; Zhang, H.; Chen, Z. Influence of SBS modifier on aging behaviors of SBS-modified asphalt. *J. Mater. Civil. Eng.* **2019**, *31*, 04019184. [CrossRef]
12. Ruan, Y.; Davison, R.R.; Glover, C.J. The effect of long-term oxidation on the rheological properties of polymer modified asphalts. *Fuel* **2003**, *82*, 1763–1773. [CrossRef]
13. Liu, G.; Nielsen, E.; Komacka, J.; Greet, L.; van de Ven, M. Rheological and chemical evaluation on the ageing properties of SBS polymer modified bitumen: From the laboratory to the field. *Constr. Build. Mater.* **2014**, *51*, 244–248. [CrossRef]
14. Hu, Z.; Xu, T.; Liu, P.; Oeser, M.; Wang, H. Improvements of developed graphite based composite anti-aging agent on thermal aging properties of asphalt. *Materials* **2020**, *13*, 4005. [CrossRef]
15. Sun, L.; Xin, X.; Ren, J. Inorganic nanoparticle-modified asphalt with enhanced performance at high temperature. *J. Mater. Civil. Eng.* **2017**, *29*, 04016227. [CrossRef]
16. Sun, L.; Xin, X.; Ren, J. Asphalt modification using nano-materials and polymers composite considering high and low temperature performance. *Constr. Build. Mater.* **2017**, *133*, 358–366. [CrossRef]
17. Ren, J.; Zang, G.; Wang, S.; Shi, J.; Wang, Y. Investigating the pavement performance and aging resistance of modified bio-asphalt with nano-particles. *PLoS ONE* **2020**, *15*, e0238817. [CrossRef]

18. Yadykova, A.Y.; Ilyin, S.O. Rheological and adhesive properties of nanocomposite bitumen binders based on hydrophilic or hydrophobic silica and modified with bio-oil. *Constr. Build. Mater.* **2022**, *342*, 127946. [CrossRef]
19. Yadykova, A.Y.; Ilyin, S.O. Bitumen improvement with bio-oil and natural or organomodified montmorillonite: Structure, rheology, and adhesion of composite asphalt binders. *Constr. Build. Mater.* **2023**, *364*, 129919. [CrossRef]
20. Chen, Z.; Zhang, H.; Zhu, C.; Zhao, B. Rheological examination of aging in bitumen with inorganic nanoparticles and organic expanded vermiculite. *Constr. Build. Mater.* **2015**, *101*, 884–891. [CrossRef]
21. Zhang, D.; Chen, Z.; Zhang, H.; Wei, C. Rheological and anti-aging performance of SBS modified asphalt binders with different multi-dimensional nanomaterials. *Constr. Build. Mater.* **2018**, *188*, 409–416. [CrossRef]
22. Zhang, H.; Zhu, C.; Chen, Z. Influence of multi-dimensional nanomaterials on the aging behavior of bitumen and SBS modified bitumen. *Petrol. Sci. Technol.* **2017**, *35*, 1931–1937. [CrossRef]
23. Xu, X.; Guo, H.; Wang, X.; Zhang, M.; Wang, Z.; Yang, B. Physical properties and anti-aging characteristics of asphalt modified with nano-zinc oxide powder. *Constr. Build. Mater.* **2019**, *224*, 732–742. [CrossRef]
24. Li, R.; Pei, J.; Sun, C. Effect of nano-ZnO with modified surface on properties of bitumen. *Constr. Build. Mater.* **2015**, *98*, 656–661. [CrossRef]
25. Wang, R.; Yue, M.; Xiong, Y.; Yue, J. Experimental study on mechanism, aging, rheology and fatigue performance of carbon nanomaterial/SBS-modified asphalt binders. *Constr. Build. Mater.* **2021**, *268*, 121189. [CrossRef]
26. Al-Mansob, R.A.; Ismail, A.; Rahmat, R.A.O.; Borhan, M.N.; Alsharef, J.M.; Albrka, S.I.; Karim, M.R. The performance of epoxidised natural rubber modified asphalt using nano-alumina as additive. *Constr. Build. Mater.* **2017**, *155*, 680–687. [CrossRef]
27. Ali, S.I.A.; Ismail, A.; Karim, M.R.; Yusoff, N.I.M.; Al-Mansob, R.A.; Aburkaba, E. Performance evaluation of Al_2O_3 nanoparticle-modified asphalt binder. *Road. Mater. Pavement* **2017**, *18*, 1251–1268. [CrossRef]
28. Shafabakhsh, G.; Aliakbari Bidokhti, M.; Divandari, H. Evaluation of the performance of SBS/Nano-Al_2O_3 composite-modified bitumen at high temperature. *Road. Mater. Pavement* **2021**, *22*, 2523–2537. [CrossRef]
29. Bhat, F.S.; Mir, M.S. A study investigating the influence of nano Al_2O_3 on the performance of SBS modified asphalt binder. *Constr. Build. Mater.* **2021**, *271*, 121499. [CrossRef]
30. Mallakpour, S.; Khadem, E. Recent development in the synthesis of polymer nanocomposites based on nano-alumina. *Prog. Polym. Sci.* **2015**, *51*, 74–93. [CrossRef]
31. Zhang, S.; Cao, X.Y.; Ma, Y.M.; Ke, Y.C.; Zhang, J.K.; Wang, F.S. The effects of particle size and content on the thermal conductivity and mechanical properties of Al_2O_3/high density polyethylene (HDPE) composites. *Express. Polym. Lett.* **2011**, *5*, 581–590. [CrossRef]
32. Cong, P.; Guo, X.; Mei, L. Investigation on rejuvenation methods of aged SBS modified asphalt binder. *Fuel* **2020**, *279*, 118556. [CrossRef]
33. Nagabhushanarao, S.S.; Vijayakumar, A.S. Chemical and rheological characteristics of accelerate aged asphalt binders using rolling thin film oven. *Constr. Build. Mater.* **2021**, *272*, 121995. [CrossRef]
34. Ye, W.; Jiang, W.; Li, P.; Yuan, D.; Shan, J.; Xiao, J. Analysis of mechanism and time-temperature equivalent effects of asphalt binder in short-term aging. *Constr. Build. Mater.* **2019**, *215*, 823–838. [CrossRef]
35. Yu, H.; Mo, L.; Zhang, Y.; Qi, C.; Wang, Y.; Li, X. Laboratory Investigation of Storage Stability and Aging Resistance of Slightly SBS-Modified Bitumen Binders. *Materials* **2023**, *16*, 2564. [CrossRef]
36. Ibrahim, B.; Wiranata, A.; Malik, A. The Effect of Addition of Antioxidant 1,2-dihydro-2,2,4-trimethyl-quinoline on Characteristics of Crepe Rubber Modified Asphalt in Short Term Aging and Long Term Aging Conditions. *Appl. Sci.* **2020**, *10*, 7236. [CrossRef]
37. Siddiqui, M.N.; Ali, M.F. Studies on the aging behavior of the Arabian asphalts. *Fuel* **1999**, *78*, 1005–1015. [CrossRef]
38. Adhikari, R.; Henning, S.; Lebek, W.; Godehardt, R.; Ilisch, S.; Michler, G.H. Structure and properties of nanocomposites based on SBS block copolymer and alumina. In *Macromolecular Symposia*; Wiley-VCH Verlag: Weinheim, Germany, 2005; Volume 231, pp. 116–124.
39. Charvani, S.; Reddy, S.S.K.; Narendar, G.; Reddy, C.G. Preparation characterisation of alumina nanocomposites. *Mater. Today Proc.* **2018**, *5*, 26817–26822. [CrossRef]
40. *ASTM D2872-19*; Standard Test Method for Effect of Heat and Air on a Moving Film of Asphalt (Rolling Thin-Film Oven Test). ASTM International: West Conshohocken, PA, USA, 2019.
41. *ASTM D4402-06*; Standard Test Method for Viscosity apparent Determination of Asphalt at Elevated Temperatures Using a Rotational Viscometer. ASTM International: West Conshohocken, PA, USA, 2006.
42. *AASHTO T 315*; Standard Method of Test for Determining the Rheological Properties of Asphalt Binder Using a Dynamic Shear Rheometer (DSR). American Association of State Highway and Transportation Officials (AASHTO): Washington, DC, USA, 2016.
43. Yao, H.; Dai, Q.; You, Z. Fourier Transform Infrared Spectroscopy characterization of aging-related properties of original and nano-modified asphalt binders. *Constr. Build. Mater.* **2015**, *101*, 1078–1087. [CrossRef]
44. Zhang, W.; Li, Q.; Wang, J.; Meng, Y.; Zhou, Z. Aging behavior of high-viscosity modified asphalt binder based on infrared spectrum test. *Materials* **2022**, *15*, 2778. [CrossRef]
45. Zhang, S.; Hong, H.; Zhang, H.; Chen, Z. Investigation of anti-aging mechanism of multi-dimensional nanomaterials modified asphalt by FTIR, NMR and GPC. *Constr. Build. Mater.* **2021**, *305*, 124809. [CrossRef]
46. Ying, G.; Fan, G.U.; Zhao, Y. Thermal oxidative aging characterization of sbs modified asphalt. *J. Wuhan Univ. Technol.-Mater. Sci.* **2013**, *28*, 88–91.

47. Daly, W.H.; Negulescu, I.; Balamurugan, S.S. *Implementation of GPC Characterization of Asphalt Binders at Louisiana Materials Laboratory*; Department of Transportation and Development: Louisiana, LA, USA, 2013.
48. Xu, M.; Yi, J.; Pei, Z.; Feng, D.; Huang, Y.; Yang, Y. Generation and evolution mechanisms of pavement asphalt aging based on variations in surface structure and micromechanical characteristics with AFM. Mater. *Today. Commun.* **2017**, *12*, 106–118.
49. Fini, E.H.; Hajikarimi, P.; Rahi, M.; Nejad, F.M. Physiochemical, Rheological, and Oxidative Aging Characteristics of Asphalt Binder in the Presence of Mesoporous Silica Nanoparticles. *J. Mater. Civil Eng.* **2016**, *28*, 04015133. [CrossRef]
50. Yunus, K.N.M.; Abdullah, M.E.; Ahmad, M.K.; Kamaruddin, N.H.M.; Tami, H.; Haryati, Y. Physical and rheological properties of nano zinc oxide modified asphalt binder. *MATEC Web Conferences* **2018**, *250*, 02004. [CrossRef]
51. Al-Sabaeei, A.M.; Napiah, M.B.; Sutanto, M.H.; Alaloul, W.S.; Usman, A. Influence of nanosilica particles on the high-temperature performance of waste denim fibre-modified bitumen. *Int. J. Pavement Eng.* **2020**, *9*, 1–14. [CrossRef]
52. Ren, S.; Liu, X.; Lin, P.; Jing, R.; Erkens, S. Toward the long-term aging influence and novel reaction kinetics models of bitumen. *Int. J. Pavement Eng.* **2022**. [CrossRef]
53. Hofko, B.; Porot, L.; Cannone, A.F.; Poulikakos, L.; Huber, L.; Lu, X. Ftir spectral analysis of bituminous binders: Reproducibility and impact of ageing temperature. *Mater Struct.* **2018**, *51*, 45. [CrossRef]
54. Kambham, B.S.; Ram, V.V.; Raju, S. Investigation of laboratory and field aging of bituminous concrete with and without anti-aging additives using FESEM and FTIR. *Constr. Build. Mater.* **2019**, *222*, 193–202. [CrossRef]
55. Yan, C.Y.; Huang, W.; Xiao, F.; Wang, L.; Li, Y. Proposing a new infrared index quantifying the aging extent of SBS-modified asphalt. *Road. Mater. Pavement* **2018**, *19*, 1406–1421. [CrossRef]
56. Zhao, X.; Wang, S.; Wang, Q.; Yao, H. Rheological and structural evolution of SBS modified asphalts under natural weathering. *Fuel* **2016**, *184*, 242–247. [CrossRef]
57. Yan, Z.; Junda, C.; Kai, Z.; Qinghai, G.; Hongmei, G.; Peng, X. Study on aging performance of modified asphalt binders based on characteristic peaks and molecular weights. *Constr. Build. Mater.* **2019**, *225*, 1077–1085.
58. Zhang, H.; Wang, Y.; Yu, T.; Liu, Z. Microstructural characteristics of differently aged asphalt samples based on atomic force microscopy (AFM). *Constr. Build. Mater.* **2020**, *255*, 119388. [CrossRef]
59. Zhang, H.L.; Wang, H.C.; Yu, J.Y. Effect of aging on morphology of organo-montmorillonite modified bitumen by atomic force microscopy. *J. Micros-Oxford.* **2011**, *242*, 37–45. [CrossRef] [PubMed]
60. Yan, Y.; Yang, Y.; Ran, M.; Zhou, X.; Guo, M. Application of infrared spectroscopy in prediction of asphalt aging time history and fatigue life. *Coating* **2020**, *10*, 959. [CrossRef]

Disclaimer/Publisher's Note: The statements, opinions and data contained in all publications are solely those of the individual author(s) and contributor(s) and not of MDPI and/or the editor(s). MDPI and/or the editor(s) disclaim responsibility for any injury to people or property resulting from any ideas, methods, instructions or products referred to in the content.

Article

Effects of the Mixing Process on the Rheological Properties of Waste PET-Modified Bitumen

Grzegorz Mazurek [1,*], Przemysław Buczyński [1], Marek Iwański [1], Marcin Podsiadło [1], Przemysław Pypeć [2] and Artur Kowalczyk [2]

[1] Department of Civil Engineering and Architecture, Kielce University of Technology, Al. Tysiąclecia Państwa Polskiego 7, 25-314 Kielce, Poland; p.buczynski@tu.kielce.pl (P.B.); miwanski@tu.kielce.pl (M.I.); mpodsiadlo@tu.kielce.pl (M.P.)

[2] TRAKT S.A., Szczukowskie Górki 1, 26-065 Piekoszów, Poland; przemyslaw.pypec@trakt.kielce.pl (P.P.); artur.kowalczyk@trakt.kielce.pl (A.K.)

* Correspondence: gmazurek@tu.kielce.pl

Abstract: This paper analyses the key findings of a study devoted to PET-modified bitumen. The research program was run according to the D-optimal experimental plan based on a factorial design. Five factors, i.e., the type of polymer (source), the type of bitumen (qualitative factors), PET amount, mixing rate, and mixing temperature (quantitative factors), controlled the bitumen–polymer mixing process. The experiment included a series of determinations of bitumen's rheological characteristics obtained by MSCR (Jnr, R) and $G^*/\sin(\delta)$ at 50 °C, 60 °C, and 70 °C. The low-temperature properties of the composite (critical temperature) were evaluated using a BBR test. The findings showed that bitumen modification with PET primarily reduced the creep susceptibility of the bituminous–polymer mixture. The low-temperature characteristics of the modified bitumen played a secondary but essential role. The amount of polymer and the mixing rate interacted with the temperature, significantly reducing the stiffness of the composite, while the type and amount of bitumen had a substantial effect on the results obtained in the BBR test. It is worth noting that when combining bitumen and plastomer, special attention should be paid to ensuring a high level of homogeneity of the mixture by controlling the parameters of the mixing process accordingly. The tests and analyses provided crucial models (GLM), which allowed for the prediction of the plastomer-modified bitumen's low- and high-temperature properties. The resulting relationships between factors and the identification of their impact on the bitumen properties enable a better understanding of the process of bitumen modification with PET. The conclusions presented here serve as a basis for future optimisation of the modified bitumen composition. The performed studies indicate that the use of >3% plastomer in bitumen 70/100 allows for a reduction in its susceptibility (MSCR) to below 0.5 kPa^{-1}, making it suitable for bituminous mixtures for high-traffic roads. No significant increase in critical temperature (BBR) was observed.

Keywords: waste polymers; bitumen modification; statistical modelling; rheological properties

Citation: Mazurek, G.; Buczyński, P.; Iwański, M.; Podsiadło, M.; Pypeć, P.; Kowalczyk, A. Effects of the Mixing Process on the Rheological Properties of Waste PET-Modified Bitumen. *Materials* **2023**, *16*, 7271. https://doi.org/10.3390/ma16237271

Academic Editors: Meng Ling, Yao Zhang, Haibo Ding and Yu Chen

Received: 23 October 2023
Revised: 13 November 2023
Accepted: 20 November 2023
Published: 22 November 2023

Copyright: © 2023 by the authors. Licensee MDPI, Basel, Switzerland. This article is an open access article distributed under the terms and conditions of the Creative Commons Attribution (CC BY) license (https://creativecommons.org/licenses/by/4.0/).

1. Introduction

Bitumen binder, a by-product of petroleum distillation, is a thermoplastic material that determines the durability of bituminous mixtures [1], whose rheological properties depend mainly on the characteristics of the bitumen used [2]. Most pavement distresses are caused by high and low temperatures, cyclic loading, and bitumen ageing. The most well-known way to improve the rheological and functional properties of bitumen for a high-quality binder is to add a polymer into its matrix [2,3]. Elastomers, e.g., styrene–butadiene–styrene (SBS), are the most commonly used [4] and effective polymers for bitumen modification, as confirmed by numerous studies. SBS-modified bitumen in bituminous mixtures improves the mix fatigue life and resistance to moisture and permanent deformation [5–7]. The molecular structure of SBS allows a cross-linked network to be formed via the process of polymer swelling and further absorption of the bitumen maltene fraction. Rigid polystyrene blocks in SBS are responsible for bitumen stiffness, while the polybutadiene blocks provide flexibility. In summary, the use of elastomers extends the viscoelastic range of the base bitumen, ensuring its high performance at low temperatures [8,9].

On the other hand, polymers from the plastomer group are a different class of bitumen modifiers. According to the literature, the primary effect of their inclusion in bitumen is an improved resistance to permanent deformation [10]. Thermoplastics are widely used in various industries (chemical, textile, food, etc.) due to their hardness and chemical inertness [11]. They are also incorporated as an additive into bituminous mixtures for bitumen modification. As highlighted by some researchers, the softening point of plastomers should be below the component mixing temperature, which is typically between 160 and 170 °C [12,13]. The use of a hard plastomer raises the bitumen softening point, reduces stability and rut resistance, and lowers the bituminous mix's ability to withstand low temperatures.

As of 2018, many countries producing significant amounts of plastomers (including PET) have imposed some restrictions on their import. As a result, special attention is now paid to plastic recycling. Using waste materials in bituminous mixtures allows for benefits such as reducing mix production costs and harnessing the potential of accumulated waste [14].

Different ways of recycling plastomers may include collection, sorting, and shredding. Chemical processes such as pyrolysis and gasification are also available. No less popular, however, is the mechanical method, which provides the material in the form of pellets of various particle sizes [15]. The final properties of polymer-modified bitumen depend on the type and properties of the added polymer and bitumen, the proportions used, and, importantly, the mixing process [16]. The polymer reactivity and chemical structure affect compatibility with bitumen, ultimately affecting the blend quality [17]. During bitumen modification with plastomers, segregation of the components may occur due to the high molecular weight and polarity of the polymer or insufficient maltene fraction in the bitumen [18]. The mixing process can also be affected by high mixing temperatures maintained for long periods, which causes additional bitumen ageing due to maltene fraction degradation or polymer oxidation [6]. As a result, the stability rapidly decreases when the blend is stored. One of the solutions to the low storage stability is to reduce the proportion of non-polar polymer chains by using, for example, butyl acrylate or reactive polymers [19,20]. As a result of the ageing-related higher mixing temperature as compared to ordinary elastomers, modification with plastomers provides additional polar groups and hence an improved bitumen–polymer compatibility [21]. Thus, the technological modification process is a crucial factor to account for during bitumen and plastomer homogenisation, as has been increasingly highlighted in publications. Naderi et al. [22] pointed out the complexity of the kinetics of polymer–bitumen mixing, which are significantly influenced by the polymer content, compatibilizer amount, mixing speed, temperature, and time, as well as the geometry of the mixer. The same conclusions were reported by researchers studying bitumen–SBS compatibility [23]. The ultimate properties of plastomer-modified bitumen will, therefore, depend on the type of bitumen (chemical composition) and available raw materials in the form of plastomers. Therefore, successful modification with plastomers is an iterative process by "trial and error".

Among a wide variety of plastomers on the market, there are several available in large quantities, e.g., polypropylene (PP), polyethylene terephthalate (PET), polyvinyl chloride (PVC), and high-density polyethylene (HDPE). In the present study, special attention was paid to PET-type plastomerx. PET is a polymer from which most water bottles and food packaging are made. Its beneficial effects of increasing the bitumen stiffness at high temperatures and thus its impact on the resistance to permanent deformation of bituminous mixtures have been confirmed in a few publications [24,25]. The wider use of PET is limited by its price. The price of waste PET is almost half that of SBS elastomer. Its base solubility is comparable to asphaltenes, so the mixing process will be significantly hindered compared to SBS. In addition, due to its lower solubility, PET will not produce the same cross-linked network as the elastomer under the same manufacturing conditions. However, paper [26] indicates that dosing as little as 2% PET significantly contributes to reducing mix deformation. On the other hand, a combination with a small amount of SBS (about 3%) results in the modified bitumen having properties comparable to that modified with SBS alone. The presence of a small amount of elastomer will compensate for the deficiency in elasticity characterised by bitumen ductility [26]. An excessive PET content reduces bitumen ductility. An increase in the PET content improves the stiffness and viscosity of bitumen and bituminous mixes [27,28]. Studies on PET implementation showed no evident mix stability losses or adverse impacts on the BBR critical temperature when the PET content was up to 6% [29].

PET has a high softening point, which may make its homogenisation with bitumen difficult compared to other plastomers [2]. This may also be due to its chemical inertness and high stiffness. However, on this topic, the opinions of researchers vary because the available comparisons refer to bitumen–polymer composites obtained in processes with different configurations of mixing parameters and using PET from various sources [30]. In the case of PET, the major factor that will significantly increase the homogeneity of the bitumen–plastomer mixture is the level of feed shredding. The literature reports adding PET into bitumen by incorporating > 1 mm flakes into bitumen mixtures [31,32] or incorporating it as a material with a high shredding level (<1.18 mm) added directly to the bitumen [33,34]. In the latter case, the plastomer dosage is usually up to 6% by weight of bitumen [35]. When using the second solution, much evidence indicates that the shredding effect is crucial for increasing the elasticity of the binder and the mix [36], as are factors such as the PET origin, contamination, and the chemical composition [30]. These parameters determine the quality of the mixture. Therefore, the mixing process is a critical issue, and a typical selection of mixing parameters in one case may need to be confirmed in experiments performed by researchers using different plastomers. Nevertheless, some aspects and conclusions can be generalised using an experimental design with many relevant factors.

It should be noted that the literature methods do not provide any information on the impact of the mixing process effects and usually consist of certain adopted values of mixing process factors without information on how their selection was determined. The results of the effectiveness of bitumen modification with a modifier are evaluated on the basis of the optimal configuration for SBS application in SBS-modified bitumen. Considering the different polarity and solubility of PET, bitumen modification with PET cannot be compared in the same way to bitumen modified with SBS. Therefore, without identifying the actual effects of the mixing process factors, it is impossible to take a position on the further strategy of modifying bitumen with PET. To the best of the authors' knowledge, very few comprehensive studies on PET-modified bitumen focusing on the mixing process with different plastomers and bitumens have been published so far. PET waste management is one of the biggest environmental challenges today, and promising test results should encourage further research [34,37]. Incorporating waste PET into bituminous mixtures has already attracted considerable interest [38]. In line with this trend, the present article provides information on the method of controlling the process of bitumen modification with PET, taking into account several variables/factors and rheological effects. The innovation of this operation was to perform mixing in a colloid mill rather than in a laboratory mixer. Thus, the analytical results are much closer to those obtained in real-life conditions. In addition, the identified relationships and their scale allow for effective optimisation of the PET and bitumen mixing process.

2. Materials and Methods
2.1. Bitumen

In the initial research phase, the bitumen to be modified with the chosen plastomers was selected and comprehensively evaluated [39]. Based on the conclusions from previous studies [34,40], two different types of bitumen, 50/70 (Orlen, Poland) and 70/100 (Orlen,

Poland), were chosen and subjected to basic rheological tests. The choice was defined by the need to reproduce different rheological states of the bitumen. The test results and 95% confidence intervals are compiled in Table 1.

Table 1. Paving bitumen results.

Features	Neat Bitumen		Standard
	50/70	70/100	
Penetration at 25 °C, 0.1 mm	60	91.5 ± 3.1	PN-EN 1426 [41]
Softening point $T_{R\&B}$, °C	48.6	44.7 ± 0.7	PN-EN 1427 [42]
Fraass breaking point, °C	−15	−13.4 ± 2.0	PN-EN 12593 [43]
$Jnr_{3.2kPa@60}$, kPa^{-1}	2.2 ± 0.3	5.7 ± 0.3	EN 16659 [44]
$R_{3.2kPa@60}$, %	1.7 ± 0.5	0	EN 16659 [44]
$T_{crit(Sm)}$ at Sm = 300 MPa	−16.6 ± 2.0	−16.9 ± 2.0	EN 14771 [45]
$T_{crit(m)}$ at m = 0.3	−15 ± 1.5	−16.2 ± 1.5	EN 14771 [45]

2.2. Waste Plastomers

The properties of the bitumen were modified by adding PET (polyethylene terephthalate) from two different suppliers. The plastomer used was a thermoplastic material with a crystalline structure. According to the suppliers' specifications, the PET had a maximum melting temperature of Tp = 256 °C and a maximum glass transition temperature of 75 °C.

The polymers were pulverized to maximise PET shredding. The samples shown in Figure 1 were then subjected to a granulation process so that the final polymer material had a grain size of <1 mm. The PET suppliers are uniquely identifiable by the plastomer colour resulting from the sorting process. Samples from supplier 1 were labelled SC1 (blue), and those from the second supplier were labelled SC2 (green).

(a) (b)

Figure 1. Test polymers; (a) supplier 1 (SC1); (b) supplier 2 (SC2).

2.3. Experimental Plan

The experimental plan required a customised approach. The sampling scheme was important for the prediction of dependent variables. It also influenced the formulation of conclusions. The starting topology of the plan included different values of bivariate and trivariate input quantities. These types of experimental plans combine fractional factorial design types $2^{(k-p)}$ and $3^{(k-p)}$ and were classified by Connor and Young for the US National Bureau of Standards [46]. The experimental domain, composed of the independent (controlled) variables listed in Table 2, served as the initial settings for the mixing process.

Table 2. Input variables (with abbreviations) and their levels in the process and bitumen and plastomer mixing.

Quantitative Variables	Levels	Qualitative Variables	Levels
Mixing speed (MS), rpm^{-1}	60, 1530, 3000	Plastomer type (PT)	SC1 (blue). SC2 (green)
Mixing temperature (MT), °C	160, 170, 180	Bitumen type (BT)	35/50 70/100
Plastomer content (PC), % (by bitumen mass)	3, 5, 7	-	-

Quantitative variables were entered at three levels while qualitative variables were at two levels. Additional coding was required for qualitative variables. This procedure was necessary to build a regression model determined based on the widely described generalised linear model (GLM) algorithm [47]. Accordingly, the type of polymer from supplier CS1 was given a code value of 0, while SC2 was given a value of 1. Similarly, in the case of the bitumen type, bitumen 50/70 was given a code value of 0, while bitumen 70/100 was given a value of 1. However, as the entire formula design would require 54 combinations, which would be inefficient under production conditions, the experimental design was altered to consider the adopted final mathematical model employed in subsequent calculations. The initial mathematical model was in the form of a polynomial function (1):

$$y = b_0 + b_1 x_1 + \cdots b_k x_k + b_{12} x_1 x_2 + \cdots + b_{k-1,k} x_{k-1} x_k + \cdots + b_{11} x_1^2 + \ldots + b_{kk} x_k^2 \quad (1)$$

where b_k—experimental coefficients, x_k—variables, y—dependent variable. Model (1) required some modifications, as two qualitative controlled variables could not be described by the quadratic terms of the equation. Therefore, for these variables, only the effect of the individual influence of a given variable on the dependent variable and their interactions with the other variables were included in the model. The modified model was further evaluated using D-optimal algorithms. Their task was to maximise the determinant of the $X^T X$ information matrix [48]. This was a key measure to minimise the volume of the total confidence set of all regression coefficients and increase the efficiency of inference about model parameters. In addition, the G-optimality criterion [49] was used (2):

$$\text{G-optimality} = 100 \times \text{square root}(p/N)^{1/2} / \Sigma_M \quad (2)$$

where p—the number of factor effects in the design, N—the number of required layouts, Σ_M—the maximum standard deviation of the predicted dependent variable value, with all proposed points considered. The G-optimality criterion minimises the largest value of the standard deviation of the determinable response surface (of the dependent variable determined from the approximating function). As a result, to meet the conditions for maximising the value of the G-optimality criterion, an experimental design containing 32 combinations was constructed, where 16 cases involved SC1 polymer while the remaining 16 used SC2. Each dependent feature was replicated at least twice, depending on the precision assigned by the standard requirements. It was very important to randomise the execution order of each combination in order to minimise the bias from the operator's precision in the execution of successive mixtures. Therefore, the order of execution of bitumen and plastomer mixtures was implemented in a coded form after randomisation. The experimental domain used in the study is shown in Table 3.

2.4. Sample Preparation

The test samples were mixed according to the plan adopted (Table 3) in a colloidal mixer under in situ conditions to obtain mixing conditions corresponding to those at the manufacturing plant. Based on the experiments presented in papers [39,40], a constant mixing time of 45 min was established for all formulations. A 15 min pre-mixing period was introduced before the main mixing period. After the mixing process, the material was left to cool. The tests started after a minimum of 24 h.

Table 3. Domain of experiment.

Case	Polymer Type (PT)	Bitumen Type (BT)	Mixing Speed rpm^{-1} (MS)	Polymer Content % (PC)	Mixing Temperature °C (MT)
1	SC1	50/70	60	3	180
2	SC1	50/70	60	5	170
3	SC1	50/70	60	7	160
4	SC1	50/70	1530	3	160
5	SC1	50/70	1530	7	180
6	SC1	50/70	3000	3	170
7	SC1	50/70	3000	3	180
8	SC1	50/70	3000	5	160
9	SC1	50/70	3000	7	160
10	SC1	70/100	60	3	160
11	SC1	70/100	60	5	160
12	SC1	70/100	60	7	180
13	SC1	70/100	1530	3	180
14	SC1	70/100	1530	7	160
15	SC1	70/100	3000	3	160
16	SC1	70/100	3000	7	180
17	SC2	50/70	60	3	180
18	SC2	50/70	60	5	170
19	SC2	50/70	60	7	160
20	SC2	50/70	1530	3	160
21	SC2	50/70	1530	7	180
22	SC2	50/70	3000	3	170
23	SC2	50/70	3000	3	180
24	SC2	50/70	3000	5	160
25	SC2	50/70	3000	7	160
26	SC2	70/100	60	3	160
27	SC2	70/100	60	5	160
28	SC2	70/100	60	7	180
29	SC2	70/100	1530	3	180
30	SC2	70/100	1530	7	160
31	SC2	70/100	3000	3	160
32	SC2	70/100	3000	7	180

2.5. Testing the Low-Temperature Creep of Bitumen (BBR)

The Superpave system uses a bending beam rheometer (BBR) to study the behaviour of bitumen at low temperatures. The test is performed according to EN 14771 [45]. To test the binder using the ATS (Rolling Meadows, IL, USA) BBR rheometer, beams made of a plastomer-modified mixture with dimensions of 125/12.5/6.25 mm were prepared from the original RTFO-aged and PAV-aged binders following the SHRP procedure. The binders were tested at -10 °C, -18 °C, and -26 °C. The beams were initially loaded with a force of 30 ± 5 mN. After that, a constant test load of 980 ± 50 mN was applied in the middle of the beam span. An extended unloading period was used. The total test time was 480 s, split into equal loading and unloading periods of 240 s. During the test, the beam deflection was continuously recorded to calculate the creep stiffness modulus (Sm) and the change in creep stiffness (m) after a loading time of 60 s. As a result, the minimum critical temperature $T_{crit(Sm)}$ was determined using the BBR test at Sm = 300 MPa and $T_{crit(m)}$ at m = 0.3.

2.6. Testing the High-Temperature Creep of Bitumen (MSCR)

A multi-stress creep recovery (MSCR) test was carried out according to EN 16659 [44] using a Discovery Hybrid Rheometer DHR-2, Hanover, MD, USA (DSR). It is a measuring system of parallel plates 25 mm in diameter with a gap of 1 mm. The test procedure involves repeatedly applying a load lasting 1 s to an RTFO-aged bitumen sample at constant stress. Each loading is followed by a recovery period of 9 s. A single creep recovery cycle lasting

10 s is repeated 10 times at a stress of 0.1 kPa, then further increased to 3.2 kPa. The measurement temperatures were 50 °C, 60 °C, and 70 °C.

2.7. Rutting Potential G*/sin(δ) in a DSR

Dynamic tests were performed according to the requirements of EN 14770 [50] using DSR. G*/sin(δ) measures the stiffness of the binder at high summer temperatures. Obtaining a high value of the G* modulus and a low value of phase angle δ is advisable to minimise permanent deformations. Such a relationship allows for a highly elastic response of the binder. The values depend largely on the temperature and frequency of the applied load. The frequency value of 1.96 Hz at which the G*/sin(δ) test was performed corresponded to an approximately 80 km/h vehicle speed, as per the provisions of the SUPERPAVE program (SHRP) [51]. The study used a measuring system of parallel plates with a diameter of 25 mm and a gap of 1 mm. The measurement temperatures were 50 °C, 60 °C, and 70 °C. In Poland, the recommended value of G*/sin(δ) is ≥1.0 kPa for unaged bitumen at the maximum pavement temperature.

3. Results

3.1. Evaluation of the Effect of Process Factors on Bitumen Properties

The first step in studying the influence of process factors on the properties of PET-modified bitumen was an analysis of variance (ANOVA), performed to determine the variation within the population that allows for searching for the parameters of model (1). In other words, an ANOVA determined whether there was at least one pair of results in the studied group with a significant difference at the assumed 95% probability level. Eleven dependent variables were used. The results are shown in Table 4.

Table 4. ANOVA for the complete model.

Dependent Variable	R^2	MSE [1]	F-Stat.	p-Value [2]
G*/sin(δ) 50 °C	0.66	39.66	12.57	0.00000
G*/sin(δ) 60 °C	0.50	13.61	6.43	0.00009
G*/sin(δ) 70 °C	0.36	22.74	3.69	0.00529
$T_{crit(Sm)}$, °C	0.44	1.08	5.08	0.00061
$T_{crit(m)}$, °C	0.45	14.02	5.41	0.00038
$Jnr_{3.2kPa@50}$, kPa^{-1}	0.57	4.51	8.71	0.00000
$Jnr_{3.2kPa@60}$, kPa^{-1}	0.54	0.29	7.67	0.00002
$Jnr_{3.2kPa@70}$, kPa^{-1}	0.53	0.01	7.26	0.00003
$R_{3.2kPa@50}$, %	0.37	2.84	3.79	0.00454
$R_{3.2kPa@60}$, %	0.43	0.33	4.83	0.00089
$R_{3.2kPa@70}$, %	0.34	17.88	3.38	0.00871

[1] root mean squared error. [2] variables with a p-value < 0.05 were insignificant.

The purpose of evaluating the complete model in Table 4 was to determine whether the choice of parameters was appropriate regarding their sensitivity to changing levels of controlled factors. The results in Table 4 indicate that at least one controlled factor in Table 2 significantly affected the dependent variable, providing a solid basis for finding a regression model. In other words, each factor of the mixing process causes a certain change in the bitumen characteristics shown in Table 4. However, the independent result of the influence of controlled factors on each selected dependent variable seems much more interesting. Before performing the main analysis of variance, the homogeneity of variance was assessed using the Levene test [52]. This test, developed in the 1960s, is less sensitive to the assumption of normality. Small sample sizes were used, which complied with standard requirements. As the samples were equinumerous in groups, it was reasonable to use the Levene test. In addition, prior to performing all analyses, outliers whose values were in the spread range defined by three standard deviations were eliminated. The ANOVA results for each characteristic are shown in Table 5.

Table 5. ANOVA results (parameterisation with sigma constraints).

Mixing Process Parameter	G*/sin(δ) 50 °C	G*/sin(δ) 60 °C	G*/sin(δ) 70 °C	$T_{crit(Sm)}$, °C	$T_{crit(m)}$, °C	$Jnr_{3.2kPa@70}$, kPa^{-1}	$Jnr_{3.2kPa@60}$, kPa^{-1}	$Jnr_{3.2kPa@50}$, kPa^{-1}	$Jnr_{3.2kPa@50}$, kPa^{-1}	$R_{3.2kPa@60}$, %	$R_{3.2kPa@70}$, %	$R_{3.2kPa@50}$, %
MS	N	N	N	N	S	S	S	S	S	S	S	S
PC	S	S	S	S	N	N	N	N	N	N	N	N
MT	N	N	N	S	N	N	N	N	N	N	N	N
PT	N	N	N	S	N	N	N	N	N	N	N	N
BT	S	S	N	S	S	S	S	S	S	N	S	N
PT*BT	N	N	N	N	N	N	N	N	N	N	N	N

A significant effect of a given factor of the mixing process is marked as "S", and the absence of any significant effect as "N". The ANOVA results in Table 5 confirm the findings presented in Table 4 and allow for some inference. The rotational speed and bitumen type significantly affect the variability of the results of the non-recoverable part of creep compliance, Jnr, and most cases of percent recovery, regardless of the test temperature. Also, the $T_{crit(m)}$ parameter determined in the BBR test indicates dependencies on the two mentioned factors. The type of bitumen caused significant changes in the set of G*/sin(δ) values at 50 °C and 60 °C. For this feature, the amount of polymer turned out to be important. The most significant number of essential factors in the mixing process shaped the variability of $T_{crit(Sm)}$. This feature is responsible for the behaviour of plastomer-modified bitumen at low temperatures. The bitumen type had a significant effect on bitumen properties. The lack of impact of the plastomer content on the Jnr level and the lack of a bitumen–plastomer-type interaction effect on the set of dependent variables were surprising. The analysis of variance revealed a strong effect of the mixing process on the improvement in the mix homogeneity. It can thus be expected that an increased mixing rate prevents the coagulation of plastomer particles, enhancing the interaction strength at the bitumen–plastomer phase boundary resulting from a larger PET–bitumen contact area. The validity of considering the mixing parameter effects was revealed when the BBR test results showed that the mix homogeneity was essential for obtaining a low critical temperature. The analysis of variance indicated a strong relationship between a low critical temperature and the mixing temperature. This is an important finding, as a high mixing temperature that is close to the PET melting temperature affects the increase in PET-modified mix homogeneity. However, the analysis of variance is a linear model with interactions between qualitative factors. Therefore, the suspected multicollinearity between variables and the lack of other interacting factors do not allow for a comprehensive decomposition of existing effects in the dataset. For these reasons, the next step was to perform a PCA.

3.2. Principal Component Analysis (PCA)

In practice, grouping variables is a desirable activity. Of the many clustering methods, a principal component analysis (PCA) is an effective technique for grouping and reducing variables. Its algorithm involves formulating mathematical models in linear equations with an idea similar to ANOVA analysis. It is essentially an orthogonal transformation of the observed results into a new set of uncorrelated variables (new principal components) [53]. In the present study, this approach was used to pre-classify the variables as a new space. The results of the projection of bitumen properties are shown in Figure 2.

The principal component analysis provided new variables (J1, J2) in a reduced number, which were a linear combination of the independent variables participating in the experiment. Standardised bitumen characteristics were used to build a new space and allowed for the detection of structure and general regularities in the current dataset. Using

Kaiser's criterion [54], three significant group variables explaining 86% of the variability of the entire set were established. However, only the first two variables, explaining 75% of the variability, were included in further analysis. The last new grouping variable only contributed 11% to the explanation.

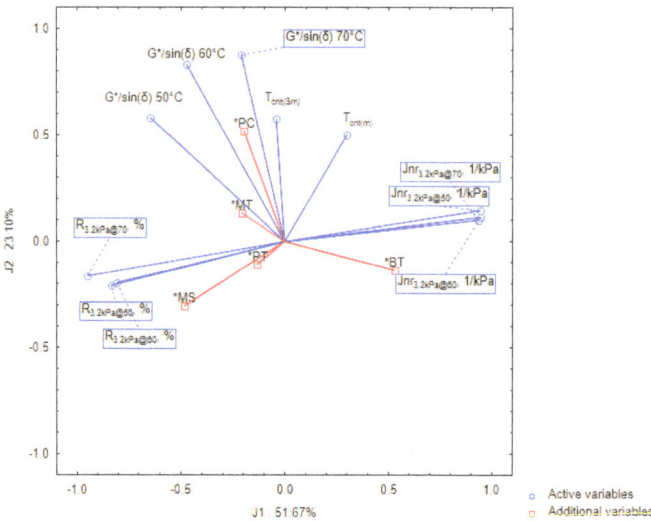

Figure 2. Group variable analysis of the properties of PET-modified bitumen.

The structure of variable distribution indicates the existence of two regularities. The first, labelled J1, explaining 52% of the variability of the entire set, includes the rheological properties of bitumen describing the behaviour of the modified bitumen at high temperatures. The formation of this variable was influenced by the results from the MSCR test (Jnr, R). The length of the vectors assigned to them indicates that an increase in Jnr correlates strongly with a decrease in R. This conclusion is consistent with the known state of knowledge, since a reduction in bitumen compliance also means a decline in its viscosity and a propensity of the bitumen structure towards a colloidal sol-type state. In establishing the relationships in the set, the effect of changing the measurement temperature was omitted, as the results in Figure 2 strongly correlate within the same characteristic regardless of the temperature type. Marked in red are the values assigned to the levels of controlling factors (independent variables). They did not participate in creating a set of new variables but were dropped into the existing principal component space. This allowed their correlation to be compiled into a new orthogonal result space. Comparing the vectors of dependent and independent variables, it should be concluded that the mixing rate and the type of bitumen are responsible for the change in the properties of modified bitumen at high temperatures. In the case of the Jnr characteristic, its increase was associated with using 70/100 bitumen. In contrast, the increase in mixing speed lowered it, probably due to the better homogeneity of the mixture. This key finding states that more than slow-speed mixing will be needed to ensure the high homogenisation of PET-modified bitumen. Thus, the values of elastic recovery, R, for the cases of mixtures prepared at low mixing speeds were reduced. The direction and level of the mixing temperature vector correlated with the first principal component, J1, suggesting that the applied mixing temperature range is responsible for the decrease in bitumen compliance and the increase in its elastic recovery to a lesser extent than the rotational speed. The modifier content and type had the lowest effect on Jnr and R.

The second component variable, J2, was identified based primarily on the variability of the $T_{crit(Sm)}$ and $T_{crit(m)}$ test results (BBR) and, to a lesser extent, of the $G^*/\sin\delta$ results. The vector assigned to the BBR features was orthogonal to Jnr and R vectors, so searching for the

J2 variable was justified. Its contribution to explaining the variability of the entire set was approximately 23%, i.e., half as high as the principal component J1. Therefore, the process of bitumen modification has a much smaller impact on the change in the low-temperature properties of the bitumen compared to its rheological properties at high temperatures. In the case of features $T_{crit(Sm)}$ and $T_{crit(m)}$, the amount of polymer showed the highest correlation with them, suggesting that an increase in the amount of plastomer leads to an increase in the critical temperature. In the case of the $G^*/\sin\delta$ feature, increasing the amount of plastomer stiffens the bitumen. The relationship between $T_{crit(Sm)}$ and $G^*/\sin\delta$ indicates that the increased stiffness of modified bitumen at high temperatures has an adverse effect on the increase in the critical temperature of the modified bitumen (tends to positive values). Therefore, the amount of plastomer requires special attention because, on the one hand, it reduces the bitumen deformation potential. Still, on the other hand, it makes it brittle at low temperatures. With respect to the mixing temperature, its increase contributes to a minor but unfavourable increase in the critical temperature, which may result from the progressive bitumen ageing process during mixing. An inverse relationship was observed between mixing speed and higher-penetration-grade bitumen 70/100. To a small extent, an increase in the mixing speed and the use of a high penetration grade reduced the critical temperature determined from the BBR test, which may be an indirect result of an increased level of mix homogeneity.

3.3. Modelling of Bitumen Properties Using a Generalised Linear Model (GLM)

The generalised model (GLM) was used to model the relationship between the characteristics of plastomer-modified bitumen and the mixing process variables. Its syntax allows for the quantitative and qualitative variables to be taken into account. Another notable advantage of GLM models is the inclusion of interaction effects and nonlinear terms in the function. Accordingly, model (1) was adopted as the research object function. Its form was introduced using an identity function binding the vector of independent variable values to the vector of expected values. This is because the distribution of the dependent variable, that is, the bitumen characteristics, had a normal distribution [47]. The generalised form of the adopted model was as follows (3):

$$g(Y) = X \cdot \beta + \varepsilon \quad (3)$$

where Y—vector of dependent variable values, X—matrix of independent variable values, ε—random component of residuals, β—vector of model coefficients, g—binding function. The last issue to consider was how to implement the qualitative variables. Accordingly, the qualitative variables were coded on a binary scale of 0–1 and were implemented in the regression model using the so-called "over-parametrised model". In this situation, a generalised inverse of the X'X matrix is required to determine the vector of model coefficients. Due to the modification of the initial design, the generalised inverse in the case of an incomplete-order matrix was calculated by simply intentionally zeroing the elements in the redundant rows and columns of the matrix [55].

From the fact that there is a high correlation between the results of some bitumen characteristics, depending on the test temperature (Figure 2), modelling results are presented only using selected characteristics such as:

- $G^*/\sin(\delta)$ at 60 °C;
- $T_{crit(Sm)}$ at Sm = 300 MPa;
- $Jnr_{3.2kPa@60}$, kPa^{-1};
- $R_{3.2kPa@60}$, %.

The optimal set of independent features describing the subsequent dependent variables was selected using the backward stepwise approach. This is a combination of the backwards elimination method and stepwise regression. Such a set may include significant variables with a *p*-value of < 0.05 and other variables selected to obtain the minimum value of the Akaike criterion (AIC) [56] and the minimum value of Pearson's scaled chi^2 statistic with respect to model (1). The first variable evaluated was the rutting coefficient $G^*/\sin(\delta)$. The optimal set of predictors is given in Table 6.

Table 6. $G^*/\sin(\delta)$ model parameters.

Independent Variable	Level	Coeff.	p-Value [2]	Scaled Chi^2 P.	R^2	RMSE [1]
Intercept		15.3593	0.313828			
BT*MS	70/100	0.0005	0.377883			
BT*MT	70/100	0.0934	0.274619			
PC		−7.4361	0.141744			
PC^2		0.9539	0.061419			
MS^2		−0.0001	0.426528			
MT^2		0.0001	0.668748	1.28	57.2	3.19
MS*PC		0.0004	0.174998			
BT*PC	70/100	0.9028	0.042943			
PT*MS	SC2 (green)	−0.0008	0.145242			
PT	SC1 (blue)	7.3626	0.001806			
BT	50/70	−18.5021	0.198334			
PT*PC	SC2 (green)	−1.7197	0.000036			

[1] root mean squared error. [2] variables with a p-value < 0.05 were insignificant.

The Level column represents the adopted reference value related to a given qualitative variable. When searching for a solution using another level of a qualitative variable, the value of this term in the equation will be zero. After the analysis, the final optimal model consisted of 12 parameters. The model explained approximately 52% of the variability in the dataset with an average estimation error of 3.19 kPa. This error is higher than the recommended higher minimum value of 1 kPa. Therefore, the minimum value will probably be obtained at a very low temperature, indicating that the contribution of a given bitumen to the mix deformation resistance will be considerable. The most significant impact was recorded for the plastomer type and the interaction with plastomer content in the mixture. In addition, the interaction between the polymer content and bitumen type also came into play. The number of interactions confirms the partial results given in Figure 2, thereby supporting the statement that the presence and quantity of PET are crucial from the perspective of controlling the $G^*/\sin(\delta)$ value. This is due to the fact that the plastomer stiffness is higher than that of bitumen. The effect of incorporating > 3% plastomer will be similar to that of a fine-grained filler, the only difference being that, to a small extent, the plastomer increases the elasticity, which will affect the elasticity of the mix. In the case of the $G^*/\sin(\delta)$ characteristic, its amount and type increased the high-temperature stiffness, which will have a measurably beneficial effect on the bituminous mixture in the summer season. Despite the chemical inertness of PET and its low solubility, the mentioned parameters of the mixing process play a special role in shaping the structure of PET-modified bitumen. Moreover, the interaction effect between the amount of polymer and the bitumen type suggests certain interactions in the bitumen–PET contact zone. The presented results complement the conclusions drawn from the ANOVA in Table 5. Extended findings, obtained in the context of relevant factors, result from the adopted model that takes into account interaction effects, additionally decomposing the total variability by adding new cognitive elements of the bitumen–plastomer mixing process. A visualisation of the fitting results against the cases defined in the experimental plan (Table 3) is presented in the variability diagram below (Figure 3).

From the model, it follows that the SC2 polymer (green) increases the average value of the $G^*/\sin(\delta)$ parameter. However, soft bitumen 70/100 significantly reduces the stiffness of the modified bitumen compared to bitumen 50/70. Another significant observation is that the $G^*/\sin(\delta)$ level increase is dependent on the PET content increase. Researchers studying other plastomers have reached the same conclusions [57].

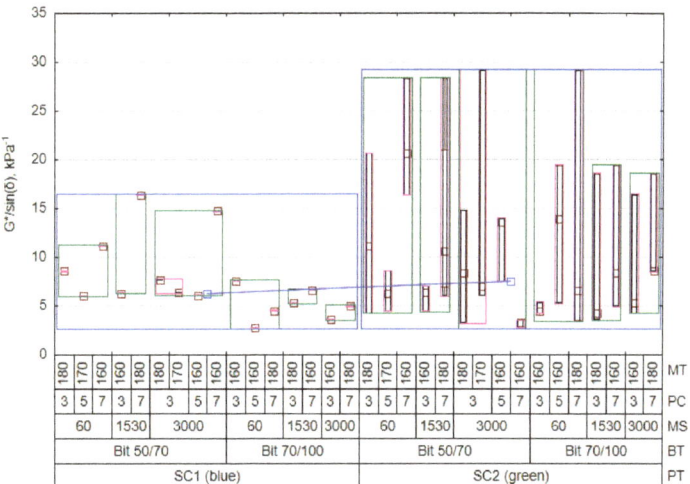

Figure 3. Variation in predicted G*/sin(δ) results.

The MSCR results complement the assessment of PET's effects on the properties of bitumen at high temperatures. As in the case of the G*/sin(δ) feature, an attempt was made to model the Jnr value using mixing process factors. A regression model was obtained with the same assumptions to predict the non-recoverable part of compliance $Jnr_{3.2kPa@60}$ at 60 °C, see Table 7.

Table 7. Model parameters for $Jnr_{3.2kPa@60}$.

	Level	Coeff.	p-Value [2]	Scaled Chi[2] P.	R^2	RMSE [1]
Intercept		0.488693	0.000000			
BT*PC	50/70	0.011806	0.017960			
MS*PC	50/70	0.000014	0.000374			
MS		−0.000110	0.000000	1.12	71	0.07
PC*MT	70/100	−0.000162	0.000516			
BT*MS	70/100	0.000032	0.000014			
BT*MT		−0.000957	0.000000			

[1] root mean squared error. [2] variables with a p-value < 0.05 were insignificant.

The creep of bitumen modified at a temperature of 60 °C strongly depended on the rotational speed. The remaining factors were strongly implicated in interactive effects. An interesting observation that emerged from Table 7 was the interaction between the type of bitumen and the amount of plastomer (BT*PC). Compared with G*/sin(δ), its statistical significance for the $Jnr_{3.2kPa@60}$ feature turned out to higher. This means that an increase in the amount of plastomer with a decrease in the penetration grade of the bitumen (bitumen 50/70 was specified as the reference) results in a rapid decrease in bitumen susceptibility, probably leading to an improved resistance of the mixture to permanent deformation. This proves that the resistance to deformation of bitumen with PET in cyclic creep testing is a complex phenomenon. Not accounting for the type of plastomer feature in the model is a symptomatic issue. With a coefficient of determination of $R^2 = 71\%$, a low estimation error of RMSE = 0.07 kPa^{-1} and a high value of fitting using Pearson's scaled chi[2] statistic, it was impossible to determine its influence on the model. Therefore, the level of PET homogenisation with bitumen was much more important than the contribution of plastomer stiffness to the change in the stiffness of the bitumen–plastomer blend. The prediction results corresponding to the cases adopted in the experiment are shown in Figure 4.

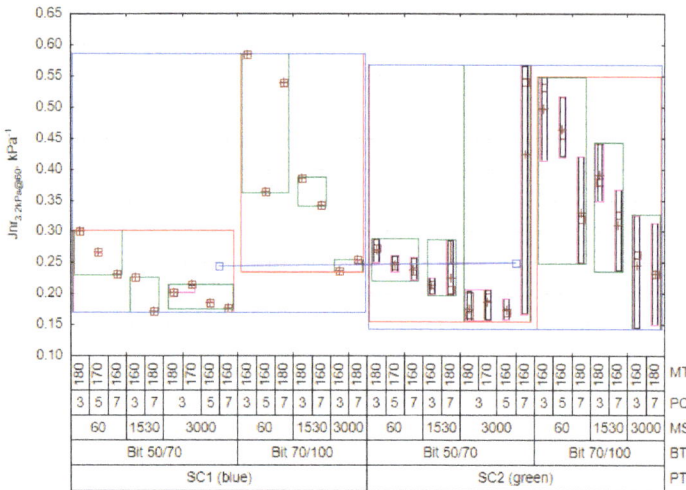

Figure 4. Variation in predicted Jnr$_{3.2kPa@60}$ results.

The results in Figure 4 show much more clearly that the PET effect causes a specific decrease in compliance, but it remains within the estimation error. The influence of a high rotational speed and the bitumen type on reducing Jnr$_{3.2kPa@60}$ was much more important. Also, despite interaction, a high mixing temperature increases the mixture's homogenisation level, resulting in reduced creep compliance of the bitumen. Since the PET softening point is high, its homogenisation with bitumen will depend on its value. As in the case of $G^*/\sin(\delta)$, an increase in the amount of plastomer resulted in an increased stiffness of the polymer–bitumen mixture. Another property determined during the MSCR test was the percent recovery R$_{3.2kPa@60}$, which characterises the modified bitumen's elasticity. The final results of assigning model weights are presented in Table 8.

Table 8. Model parameters for R$_{3.2kPa@60}$.

Independent Variable	Level	Coeff.	p-Value [2]	Scaled Chi^2 P.	R^2	RMSE [1]
Intercept		6.558793	0.000657			
BT*PC	70/100	−0.303740	0.210987			
BT*MT	70/100	0.023712	0.004149			
MS*MT		0.000029	0.000000			
BT*MS	70/100	−0.001201	0.000690	1.14	72	3.4
MS*PC		−0.000665	0.000317			
PC*MT		0.006114	0.004389			
PT*MT	SC2 (green)	0.002396	0.347372			
BT*PC	70/100	6.558793	0.000657			
PT*MT	70/100	−0.303740	0.210987			

[1] root mean squared error. [2] variables with p-value < 0.05 were insignificant.

Table 8 data indicate that the favourable increase in the R value is influenced by the rotational speed in interaction with an increase in the bitumen–plastomer mixing temperature. In this case, the interaction of the plastomer content with the type of bitumen did not cause a significant effect on the value of elastic recovery. The presence of this interaction in the model was solely due to the need to obtain a low estimation error. The amount of polymer present in the interacting relationships almost always results in a consistent decrease in elastic recovery. Thus, the amount of PET increases stiffness but at the same time decreases the elastic deformation capacity of the material, making it an elastic–brittle material with a possibly low LVE range. Unexpectedly, the interaction between the plastomer content and the increasing temperature level favourably increased the R level. The high R^2 and the low value of the RMSE error suggest that the adopted model properly describes the R variation using the adopted mixing process factors. The predicted value of recovery R against the assumed factor levels is shown in Figure 5.

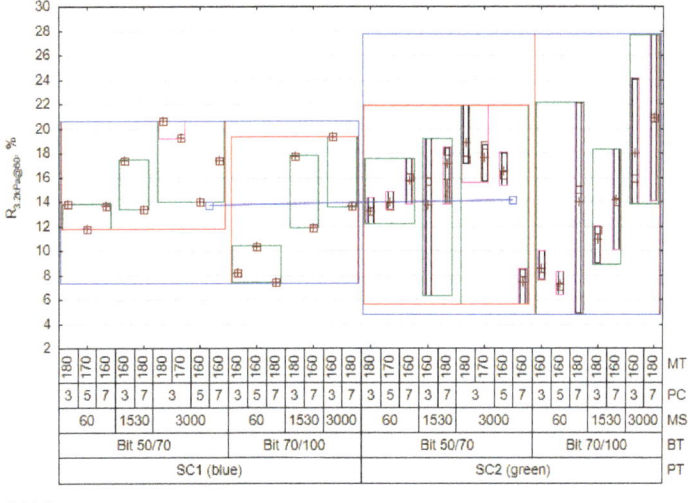

Figure 5. Variation in predicted $R_{3.2kPa@60}$ results.

As in the case of the $Jnr_{3.2kPa@60}$ feature, the plastomer type played a secondary role. A much greater variation was observed in relation to changes in rotational speed and plastomer content and the accompanying change in mixing temperature. Thus, the key to obtaining favourable viscoelastic properties is to ensure optimal mixing process parameters, with particular attention paid to mixing temperature, mixing speed, and plastomer content. In light of the conclusions obtained, forming a conclusion about the effectiveness of bitumen modification with PET without performing an experiment is impractical. An analysis of the mixing process should be the starting point for making decisions regarding further modifications with PET. Figure 6 illustrates the scatter of the results for $R_{3.2kPa@60}$ and $Jnr_{3.2kPa@60}$.

While observing the results in Figure 6, please note that almost all the results of PET-modified samples determined at 60 °C obtained compliance, allowing for extremely heavy traffic conditions (class E according to AASHTO M 332) [58]. The average value of $Jnr_{3.2kPa@60}$ for bitumen 50/70 was 2.2 kPa^{-1} and it was 5.7 kPa^{-1} for 70/100, but the elastic recovery value was low—below the solid line on the graph indicating an elastic response. Therefore, the modification of bitumen with PET cannot be considered the same as with SBS elastomer. Please note that the susceptibility of bitumen modified with plastomer SC2 (green) was the lowest (Figure 6) at the level of about 0.15 kPa^{-1}. Compared with commercial bitumen PmB 45/80–55 [59], the susceptibility value of the modified bitumen was lower by 0.05 kPa^{-1}. The elastic recovery was lower by about 30%, which shows that the modification did not obtain the same level of elasticity as with the use of SBS plastomer. Nevertheless, the elastic recovery R was higher than that obtained for neat bitumen and did not exceed 3% (Table 1). As mentioned, all the results are below the line separating bitumens modified with an acceptable elastomeric polymer. According to the provisions of AASHTO T350 [60], the presence of elastomer-modified bitumen is indicated by an R above the line. However, Morales et al. [61] point out a certain imperfection in the interpretation of the R parameter, which, in a non-obvious way, correlates with the resistance of bitumen mixtures to rutting. Therefore, modification results slightly below the line determined by the threshold function do not indicate that the bitumen in question should be disqualified as modified. However, the response during the MSCR test may be slightly different from other elastomer-modified bitumens. Evaluation of the MSCR results suggests that the effects of PET on bitumen are similar to those of fine filler. The difference between the two is due to the PET elasticity being greater than that of mineral aggregate. If the degree of PET grinding is increased, the bitumen and plastomer are mixed at a higher temperature, and if soft bitumen is used, a higher elastic recovery value can be obtained, probably due to an increase in the bitumen–PET compatibility. However, a high PET content reduces the properties related to the elasticity of the mixture (Figure 6).

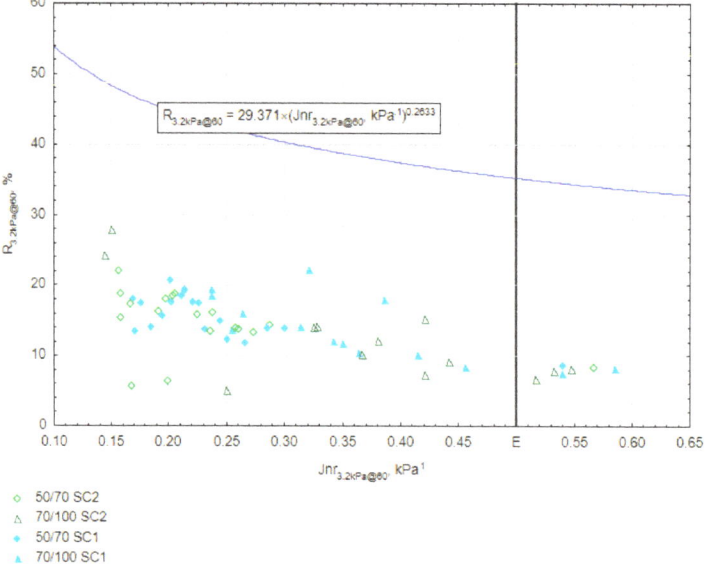

Figure 6. R vs. Jnr in MSCR.

The last parameter that characterises PET-modified bitumen is the critical temperature $T_{crit(Sm)}$, calculated for a stiffness of Sm = 300 MPa in the BBR test. This parameter is essential to complement the conclusions on the impact of PET on bitumen. It is vital to ensure that bitumen modification with a polymer does not worsen its properties at low temperatures, which would question the idea of bitumen modification. It should be emphasised that increasing the critical temperature has a considerable limiting effect on the effective viscoelastic range of the modified binder and hence its practical use. The results of modelling using the GLM algorithm are compiled in Table 9.

The best fit of the parameters in the adopted model (1) to the experimental results was obtained in the case of the $T_{crit(Sm)}$ feature. Many parameters from the selected optimal set of mixing process parameters considerably impacted the $T_{crit(Sm)}$ results. Most importantly, the nonlinear effect of temperature also played an important role in shaping the critical temperature prediction. For a given ($T_{crit(Sm)}$) feature, the interaction between low-penetration-grade bitumen (the base bitumen for this analysis is 70/100) and plastomer content will increase the $T_{crit(Sm)}$ value. Thus, this relationship is inversely proportional to the advantages of the plastomer content in the context of a decreased susceptibility. Moreover, both the polymer content and the bitumen amount directly influenced the $T_{crit(Sm)}$ property changes. The previous conclusions (Figure 2) are supported by the fact that the increase in the amount of polymer increases the critical temperature of the bitumen. The increase in mixing temperature was correlated with an increase in mix brittleness at low temperatures, probably due to bitumen ageing. The increased mixing speed in the interaction effects suggests an improvement in the characteristics of low-temperature bitumen–polymer mixtures. However, it should be emphasised that the presence of interaction effects indicates the complexity of the process factors' impact on the low-temperature properties of PET-modified binders. The results of predicting the critical temperature of the polymer–bitumen composites are presented in a variability graph (Figure 7).

Table 9. Model parameters for $T_{crit(Sm)}$.

	Level	Coeff.	p-Value [2]	Scaled. Chi[2] P.	R^2	RMSE [1]
Intercept		−313.552	0.000076			
PT*MT	SC2 (green)	−0.005	0.000147			
BT	50/70	−7.438	0.000000			
BT*PC	70/100	0.358	0.000000			
PC		0.150	0.000732	1.19	78.1	3.48
MT		3.379	0.000301			
MT^2		−0.010	0.000411			
BT*MT	70/100	0.034	0.000037			
BT*MS	70/100	−0.0001	0.000363			
PT*PC	SC2 (green)	0.153	0.000352			

[1] root mean squared error. [2] variables with p-value < 0.05 were insignificant.

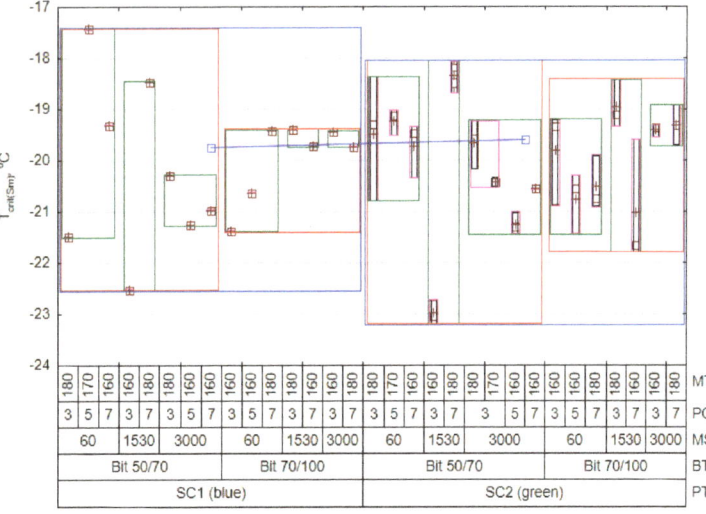

Figure 7. Variation in $T_{crit(Sm)}$ results.

An analysis of the results in Figure 7 indicates that the beneficial effect of the SC2 (green) polymer on reducing the bitumen deformability resulted in a slight undesirable increase in critical temperature (the line connecting the medians). An unexpectedly low $T_{crit(Sm)}$ was recorded for 1530 rpm^{-1}, 3% plastomer content, and 160 °C, in which bitumen 50/70 was used. This effect, repeated regardless of the source of PET, might be due to the low content of the modifier added to the bitumen at a minimal increase in its ageing level. In this case, the 1530 rpm^{-1} rotation level could evenly distribute the polymer in the bitumen. Regarding the results of $Jnr_{3.2kPa@60}$, it should be stated that the obtained resistance to deformation was moderate. However, regarding the material quality, an increase in the plastomer content resulted in a successive reduction in the bitumen resistance to low temperatures. Similar results were obtained by the authors of study [29], where it was demonstrated that a PET content increase to a level of 6% and enrichment with a compatibilizer does not imply any deterioration of the low-temperature properties of bitumen.

The results above show that it is important to keep in mind the limitations of the proposed model. The parameters were estimated based on the experimental plan given in Table 3. Despite meeting the postulate of normality of residuals for the constructed models, it should be taken into account that the approximation of the values of the dependent variables located at a large distance from the experimental domain may yield contradictory results. Therefore, the present modelling with GLM is valid for minor deviations from the range of the experimental domain.

4. Conclusions

This statistical-inference-based study provided a lot of important information about the process of mixing PET with bitumen. The key issue is that generalised conclusions about the effect of PET on bitumen modification cannot be drawn solely based on a modifier from one supplier. This study employed plastomers of two different colours from two sources. From the mixing process observations in the laboratory, it follows that the type of plastomer yielded different effects on the consistency of the mixture during mixing. So, the presence of fillers and pigment in the plastomer will make a huge difference. No comparative studies known to the authors take into account the variability due to the origin of the plastomer. It is widely accepted that contamination can lead to different results for bitumen characteristics, as confirmed by the tests included in this experiment. The level of compatibility of the same plastomer but from a different source was different, as observed in both susceptibility (MSCR) and critical temperature (BBR) tests. An important aspect that supports the representativeness of the results in this study was dealing with the results obtained from samples prepared in a colloid mill. This allowed us to include the effects related to imperfections accompanying the mixing process in the variability of the results.

Also, the consistency of bitumen with different proportions of asphaltene fraction versus maltene fraction repeatedly confirmed significant effects on the rheological properties of the modified bitumen. This finding is also supported by the results of the principal component analysis, according to which about 25% of the variation (Figure 2) cannot be explained by the mixing process factors adopted in the experiment. In this case, the chemical interaction, the mixer's geometry, and the shredding degree will effectively complement the influence of unexplained variability on the selected dependent characteristics. We should also keep in mind the mixing time, which, for technical reasons, still needs to be evaluated. Its value was assumed based on previously performed exploratory studies based on the Plackett–Burman elimination plan [40]. Nevertheless, given the current state of knowledge, the mixing time factor will be included in subsequent experiments on optimising the mixing process. The confirmed significant impact of PET on bitumen properties allows for the formation of the detailed conclusions below. These detailed conclusions can be the argument for undertaking further implementation measures:

- A principal component analysis of the set variability structure indicated the existence of two dominant factors, explained in approximately 75% by the selected tests on

bitumen. The first factor is the bitumen rutting resistance, explained in 52%. The low-temperature properties of PET-modified bitumen can be explained in approximately 23% by the research methods used;
- The vectors assigned to the variables in the PCA and the ANOVA results indicate that a high mixing speed and the bitumen type significantly reduce the compliance of plastomer-modified bitumen. However, the mixing rate (MS) and the use of high-penetration bitumen (PT) minimise this effect;
- A detailed analysis of modelling using GLM indicates that the mixing speed (MS) is a crucial factor for obtaining a low bitumen deformability (MSCR) and a low critical temperature (BBR). This result was probably related to the increase in the homogeneity of the mixture, to which, in the case of PET-modified bitumen, special attention should be paid;
- The amount of plastomer (PC) > 3% was by far the most important factor for ensuring a high $G^*/\sin\delta$ value and a low Jnr, but it had a negative impact on the critical temperature;
- The temperature of mixing (MT) > 1500 rpm^{-1} aids bitumen–PET homogenisation. However, in the case of low-temperature properties, it slightly increased the critical temperature, which might be related to an accelerated ageing process. The low mixing rate does not ensure a satisfactory quality and standard characteristics of the plastomer-modified bitumen;
- The plastomer type (PT) was the least significant factor. However, its impact was complex and involved in numerous interactions;
- The presence of pigment and impurities strongly affects the final quality of the plastomer-modified bitumen;
- Significant bitumen–plastomer interactions suggest different compatibility levels between these components resulting from plastomer solubility differences. Therefore, research should be continued;
- The bitumen type (BT) was the most important factor for low-temperature characteristics and ensuring a high homogenisation of the mixture. The use of lower-viscosity bitumen significantly increased the level of its deformation. To ensure a high resistance of the modified bitumen to deformation, the use of an increased amount of plastomer is required;
- The results of the R–Jnr relationship proved that adding a plastomer allows for the compliance to be reduced to the level of >0.5 kPa^{-1}. Therefore, the obtained modified bitumen can be used in extremely heavy traffic conditions.

It should be strongly emphasised that the modification of bitumen with PET does not expand the bitumen viscoelastic range to the level obtained by the modification with SBS. These are two different types of polymers. The lower compatibility between bitumen and PET requires further research, with a focus on the contact phenomena between bitumen and PET. The occurrence of significant interactions in the developed models suggests that the bitumen–PET interaction begins during mixing. Therefore, in the future, studies involving an evaluation of the chemical composition of the plastomer will be implemented. Further optimisation of the mixing process supported by the selection of a suitable compatibilizer will definitely increase the efficiency of PET use. The presented detailed feasibility study of bitumen modification with PET, taking into account the consistency of the input bitumen and the type of plastomer, confirms the possibility of its implementation and is an innovative measure that should be further pursued.

Author Contributions: Conceptualization, G.M. and M.I.; Methodology, P.B. and A.K.; Validation, G.M., P.B. and M.P.; Formal analysis, G.M.; Investigation, P.B., M.P. and A.K.; Resources, A.K.; Data curation, P.B. and M.P.; Writing—original draft, G.M.; Writing—review & editing, M.I.; Visualization, G.M.; Supervision, M.I. and P.P.; Project administration, M.I. and P.P.; Funding acquisition, P.P. All authors have read and agreed to the published version of the manuscript.

Funding: The research was carried out as part of Project No. RPSW.01.02.00-26-0011/22, "Development of an innovative bitumen binder modified with waste PET polymer (Poly(ethylene terephthalate))", co-financed under the Regional Operational Program of the Świętokrzyskie Region for 2014–2020 under the European Regional Development Fund, Priority Axis 1 Innovation and Science, Measure 1.2 Research and development in the Świętokrzyskie entrepreneurship sector.

Institutional Review Board Statement: Not applicable.

Informed Consent Statement: Written informed consent has been obtained from the patient(s) to publish this paper.

Data Availability Statement: Data available on request from the corresponding author.

Conflicts of Interest: The authors declare no conflict of interest.

References

1. Christensen, D.W.; Bonaquist, R. Use of Strength Tests for Evalu-Ating the Rut Resistance of Asphalt Concrete. *J. Assoc. Asph. Paving Technol.* **2002**, *71*, 697–711.
2. Nizamuddin, S.; Boom, Y.J.; Giustozzi, F. Sustainable Polymers from Recycled Waste Plastics and Their Virgin Counterparts as Bitumen Modifiers: A Comprehensive Review. *Polymers* **2021**, *13*, 3242. [CrossRef] [PubMed]
3. Zhu, J.; Birgisson, B.; Kringos, N. Polymer Modification of Bitumen: Advances and Challenges. *Eur. Polym. J.* **2014**, *54*, 18–38. [CrossRef]
4. Read, J.; Whiteoak, D.; Hunter, R.N. *The Shell Bitumen Handbook*, 5th ed.; Thomas Telford: London, UK, 2003; ISBN 978-0-7277-3220-0.
5. Islam, S.S.; Singh, S.K.; Ransinchung, G.D.; Ravindranath, S.S. Effect of Property Deterioration in SBS Modified Binders during Storage on the Performance of Asphalt Mix. *Constr. Build. Mater.* **2021**, *272*, 121644. [CrossRef]
6. Airey, G.D. Styrene Butadiene Styrene Polymer Modification of Road Bitumens. *J. Mater. Sci.* **2004**, *39*, 951–959. [CrossRef]
7. Han, S.; Niu, D.Y.; Liu, Y.M.; Chen, D.; Liu, D.W. Analysis on the Impact of the Type and Content of SBS on the Performance of the Modified Asphalt Mixture. *Adv. Mater. Res.* **2014**, *919*, 1079–1084. [CrossRef]
8. Dong, F.; Yu, X.; Wang, T.; Yin, L.; Li, N.; Si, J.; Li, J. Influence of Base Asphalt Aging Levels on the Foaming Characteristics and Rheological Properties of Foamed Asphalt. *Constr. Build. Mater.* **2018**, *177*, 43–50. [CrossRef]
9. Airey, G.D. Rheological Evaluation of Ethylene Vinyl Acetate Polymer Modified Bitumens. *Constr. Build. Mater.* **2002**, *16*, 473–487. [CrossRef]
10. Li, B.; Li, X.; Kundwa, M.J.; Li, Z.; Wei, D. Evaluation of the Adhesion Characteristics of Material Composition for Polyphosphoric Acid and SBS Modified Bitumen Based on Surface Free Energy Theory. *Constr. Build. Mater.* **2021**, *266*, 121022. [CrossRef]
11. Behnood, A.; Modiri Gharehveran, M. Morphology, Rheology, and Physical Properties of Polymer-Modified Asphalt Binders. *Eur. Polym. J.* **2019**, *112*, 766–791. [CrossRef]
12. Padhan, R.K.; Sreeram, A. Enhancement of Storage Stability and Rheological Properties of Polyethylene (PE) Modified Asphalt Using Cross Linking and Reactive Polymer Based Additives. *Constr. Build. Mater.* **2018**, *188*, 772–780. [CrossRef]
13. Airey, G. Rheological Properties of Styrene Butadiene Styrene Polymer Modified Road Bitumens★. *Fuel* **2003**, *82*, 1709–1719. [CrossRef]
14. Lecomte, M.; Hacker, S.; Teymourpour, P.; Bahia, H. Use of Plastomeric Additives to Improve Mechanical Performance of Warm Mix Asphalt. In Proceedings of the 6th Eurasphalt & Eurobitume Congress, Prague, Czech Republic, 1–3 June 2016.
15. Ragaert, K.; Delva, L.; Van Geem, K. Mechanical and Chemical Recycling of Solid Plastic Waste. *Waste Manag.* **2017**, *69*, 24–58. [CrossRef] [PubMed]
16. McNally, T. Introduction to Polymer Modified Bitumen (PmB). In *Polymer Modified Bitumen*; Elsevier: Amsterdam, The Netherlands, 2011; pp. 1–21, ISBN 978-0-85709-048-5.
17. Singh, B.; Kumar, L.; Gupta, M.; Chauhan, G.S. Polymer-Modified Bitumen of Recycled LDPE and Maleated Bitumen. *J. Appl. Polym. Sci.* **2013**, *127*, 67–78. [CrossRef]
18. Khakimullin, Y.N. Properties of Bitumens Modified by Thermoplastic Elastomers. *Mech. Compos. Mater.* **2000**, *36*, 417–422. [CrossRef]
19. Giavarini, C.; De Filippis, P.; Santarelli, M.L.; Scarsella, M. Production of Stable Polypropylene-Modified Bitumens. *Fuel* **1996**, *75*, 681–686. [CrossRef]
20. Pyshyev, S.; Gunka, V.; Grytsenko, Y.; Bratychak, M. Polymer Modified Bitumen: Review. *ChChT* **2016**, *10*, 631–636. [CrossRef]
21. Liu, P.; Lu, K.; Li, J.; Wu, X.; Qian, L.; Wang, J.; Gao, S. Effect of Aging on Adsorption Behavior of Polystyrene Microplastics for Pharmaceuticals: Adsorption Mechanism and Role of Aging Intermediates. *J. Hazard. Mater.* **2020**, *384*, 121193. [CrossRef]
22. Naderi, K.; Jonas, C.; Carbonneau, X. Investigating the Link between the Chemical Composition of Bitumen and the Kinetics of the Styrene-Butadiene-Styrene Swelling Process. *Road Mater. Pavement Des.* **2023**, *24*, 263–278. [CrossRef]
23. Polacco, G.; Filippi, S.; Merusi, F.; Stastna, G. A Review of the Fundamentals of Polymer-Modified Asphalts: Asphalt/Polymer Interactions and Principles of Compatibility. *Adv. Colloid Interface Sci.* **2015**, *224*, 72–112. [CrossRef]
24. El-Naga, I.A.; Ragab, M. Benefits of Utilization the Recycle Polyethylene Terephthalate Waste Plastic Materials as a Modifier to Asphalt Mixtures. *Constr. Build. Mater.* **2019**, *219*, 81–90. [CrossRef]

25. Choudhary, R.; Kumar, A.; Murkute, K. Properties of Waste Polyethylene Terephthalate (PET) Modified Asphalt Mixes: Dependence on PET Size, PET Content, and Mixing Process. *Period. Polytech. Civ. Eng.* **2018**, *62*, 685–693. [CrossRef]
26. Imanbayev, Y.; Bussurmanova, A.; Ongarbayev, Y.; Serikbayeva, A.; Sydykov, S.; Tabylganov, M.; Akkenzheyeva, A.; Izteleu, N.; Mussabekova, Z.; Amangeldin, D.; et al. Modification of Bitumen with Recycled PET Plastics from Waste Materials. *Polymers* **2022**, *14*, 4719. [CrossRef]
27. Agha, N.; Hussain, A.; Ali, A.S.; Qiu, Y. Performance Evaluation of Hot Mix Asphalt (HMA) Containing Polyethylene Terephthalate (PET) Using Wet and Dry Mixing Techniques. *Polymers* **2023**, *15*, 1211. [CrossRef] [PubMed]
28. Majka, T.M.; Ostrowski, K.A.; Piechaczek, M. Research on the Development of a Way to Modify Asphalt Mixtures with PET Recyclates. *Materials* **2023**, *16*, 6258. [CrossRef] [PubMed]
29. Beyza İnce, C.; Geckil, T. Effects of Recycled PET and TEOA on Performance Characteristics of Bitumen. *JCE* **2022**, *74*, 105–114. [CrossRef]
30. Mashaan, N.; Chegenizadeh, A.; Nikraz, H. Laboratory Properties of Waste PET Plastic-Modified Asphalt Mixes. *Recycling* **2021**, *6*, 49. [CrossRef]
31. Soltani, M.; Moghaddam, T.B.; Karim, M.R.; Baaj, H. Analysis of Fatigue Properties of Unmodified and Polyethylene Terephthalate Modified Asphalt Mixtures Using Response Surface Methodology. *Eng. Fail. Anal.* **2015**, *58*, 238–248. [CrossRef]
32. Brasileiro, L.; Moreno-Navarro, F.; Tauste-Martínez, R.; Matos, J.; Rubio-Gámez, M. Reclaimed Polymers as Asphalt Binder Modifiers for More Sustainable Roads: A Review. *Sustainability* **2019**, *11*, 646. [CrossRef]
33. Silva, J.D.A.A.E.; Rodrigues, J.K.G.; De Carvalho, M.W.; Lucena, L.C.D.F.L.; Cavalcante, E.H. Mechanical Performance of Asphalt Mixtures Using Polymer-Micronized PET-Modified Binder. *Road Mater. Pavement Des.* **2018**, *19*, 1001–1009. [CrossRef]
34. Domínguez, F.J.N.; García-Morales, M. The Use of Waste Polymers to Modify Bitumen. In *Polymer Modified Bitumen*; Elsevier: Amsterdam, The Netherlands, 2011; pp. 98–135, ISBN 978-0-85709-048-5.
35. Sojobi, A.O.; Nwobodo, S.E.; Aladegboye, O.J. Recycling of Polyethylene Terephthalate (PET) Plastic Bottle Wastes in Bituminous Asphaltic Concrete. *Cogent Eng.* **2016**, *3*, 1133480. [CrossRef]
36. Almeida E Silva, J.D.A.; Lopes Lucena, L.C.D.F.; Guedes Rodrigues, J.K.; Carvalho, M.W.; Beserra Costa, D. Use of Micronized Polyethylene Terephthalate (Pet) Waste in Asphalt Binder. *Pet. Sci. Technol.* **2015**, *33*, 1508–1515. [CrossRef]
37. Poulikakos, L.D.; Kakar, M.R.; Piao, Z. Urban Mining for Low-Noise Urban Roads towards More Sustainability in the Urban Environment. *Road Mater. Pavement Des.* **2023**, *24*, 309–320. [CrossRef]
38. Wang, J.; Yuan, J.; Xiao, F.; Li, Z.; Wang, J.; Xu, Z. Performance Investigation and Sustainability Evaluation of Multiple-Polymer Asphalt Mixtures in Airfield Pavement. *J. Clean. Prod.* **2018**, *189*, 67–77. [CrossRef]
39. Mazurek, G.; Podsiadło, M. Optimisation of Polymer Addition Using the Plackett-Burman Experiment Plan. *IOP Conf. Ser. Mater. Sci. Eng.* **2021**, *1203*, 022003. [CrossRef]
40. Mazurek, G.; Šrámek, J.; Buczyński, P. Composition Optimisation of Selected Waste Polymer-Modified Bitumen. *Materials* **2022**, *15*, 8714. [CrossRef]
41. EN 1426; Bitumen and Bituminous Binders—Determination of Needle Penetration. European Commission: Brussels, Belgium, 2015.
42. EN 1427; Bitumen and Bituminous Binders—Determination of the Softening Point—Ring and Ball Method. European Commission: Brussels, Belgium, 2015.
43. EN 12593; Bitumen and Bituminous Binders—Determination of the Fraass Breaking Point. European Commission: Brussels, Belgium, 2015.
44. EN 16659; Bitumen and Bituminous Binders—Multiple Stress Creep and Recovery Test (MSCRT). European Commission: Brussels, Belgium, 2015.
45. EN 14771; Bitumen and Bituminous Binders—Determination of the Flexural Creep Stiffness—Bending Beam Rheometer (BBR). European Commission: Brussels, Belgium, 2015.
46. Anderson, V.L.; McLean, R.A. *Design of Experiments: A Realistic Approach*, 1st ed.; CRC Press: Boca Raton, FL, USA, 2018; ISBN 978-1-315-14103-9.
47. Nelder, J.A.; Wedderburn, R.W.M. Generalized Linear Models. *J. R. Stat. Society. Ser. A* **1972**, *135*, 370. [CrossRef]
48. Box, G.E.P.; Cox, D.R. An Analysis of Transformations. *J. R. Stat. Soc.* **1964**, *26*, 211–252. [CrossRef]
49. Dykstra, O. The Augmentation of Experimental Data to Maximize |X'X|. *Technometrics* **1971**, *13*, 682. [CrossRef]
50. EN 14770; Bitumen and Bituminous Binders—Determination of Complex Shear Modulus and Phase Angle—Dynamic Shear Rheometer (DSR). European Commission: Brussels, Belgium, 2012.
51. Stuart, K.D.; Izzo, R.P. Correlation of superpave g*/sin delta with rutting susceptibility from laboratory mixture tests (with discussion and closure). *Transp. Res. Rec.* **1995**, *1492*, 176–183.
52. Montgomery, D.C. *Design and Analysis of Experiments*, 8th ed.; John Wiley & Sons, Inc.: Hoboken, NJ, USA, 2013; ISBN 978-1-118-14692-7.
53. Miller, J.N.; Miller, J.C. *Statistics and Chemometrics for Analytical Chemistry*, 6th ed.; Prentice Hall: Harlow, UK, 2010; ISBN 978-0-273-73042-2.
54. Kaiser, H.F. The Varimax Criterion for Analytic Rotation in Factor Analysis. *Psychometrika* **1958**, *23*, 187–200. [CrossRef]
55. Dempster, A.P. *Elements of Continuous Multivariate Analysis*; Addison-Wesley Pub. Co.: Reading, MA, USA, 1969; ISBN 978-0-201-01485-3.
56. Akaike, H.; Parzen, E.; Tanabe, K.; Kitagawa, G. *Selected Papers of Hirotugu Akaike*; Springer International Publishing: Cham, Switzerland, 1998; ISBN 978-1-4612-1694-0.

57. García-Morales, M.; Partal, P.; Navarro, F.J.; Martínez-Boza, F.; Gallegos, C.; González, N.; González, O.; Muñoz, M.E. Viscous Properties and Microstructure of Recycled Eva Modified Bitumen. *Fuel* **2004**, *83*, 31–38. [CrossRef]
58. *AASHTO M 332*; Standard Specification for Performance-Graded Asphalt Binder Using Multiple Stress Creep Recovery (MSCR) Test. American Association of State Highway and Transportation Officials: Washington, DC, USA, 2022.
59. Błażejowski, K.; Wójcik-Wiśniewska, M.; Baranowska, W.; Ostrowski, P. *Bitumen Handbook*; ORLEN Asfalt sp. z o.o.: Płock, Poland, 2022.
60. *AASHTO T 350*; Standard Method of Test for Multiple Stress Creep Recovery (MSCR) Test of Asphalt Binder Using a Dynamic Shear Rheometer (DSR). American Association of State Highway and Transportation Officials: Washington, DC, USA, 2019.
61. Morales, R.; Bahia, H.; Swiertz, D. WHRP (0092-14-20) TPF-5 (302) Pooled Fund Project. In *Modified Binder (PG+) Specifications and Quality Control Criteria*; Wisconsin Department of Transportation: Milwaukee, WI, USA, 2016.

Disclaimer/Publisher's Note: The statements, opinions and data contained in all publications are solely those of the individual author(s) and contributor(s) and not of MDPI and/or the editor(s). MDPI and/or the editor(s) disclaim responsibility for any injury to people or property resulting from any ideas, methods, instructions or products referred to in the content.

Article

A Fractional Creep Constitutive Model Considering the Viscoelastic–Viscoplastic Coexistence Mechanism

Jia Zhao [1], Weigang Zhao [2], Kaize Xie [2,*] and Yong Yang [2]

[1] School of Civil Engineering, Shijiazhuang Tiedao University, Shijiazhuang 050043, China; 220190114@student.stdu.edu.cn

[2] School of Safety Engineering and Emergency Management, Shijiazhuang Tiedao University, Shijiazhuang 050043, China; zhaoweig2002@163.com (W.Z.); yangy@stdu.edu.cn (Y.Y.)

* Correspondence: kzxie1988@stdu.edu.cn

Abstract: In order to improve the accuracy and universality of the nonlinear viscoelastic-plastic mechanical behavior characterization method of asphalt mixture, a new criterion for the division of the creep process of materials was established based on the strain yield characteristics, and the coexistence mechanism of Viscoelastic–Viscoplastic strain was proposed in the subsequent yield phase; then, a viscoelastic element was constructed in the form of a parallel connection of two fractional viscoelastic elements based on fractional calculus theory, and its mathematical equations were derived; with novel viscoelastic elements, a constitutive model characterizing the whole creep process of asphalt mixtures was developed and its analytical expression was derived. The laboratory short-term creep test of Cement and Asphalt Mortar (CA mortar) and the simulation test data of asphalt mixtures from the references were used to verify the constitutive model. The results show that the creep constitutive model of asphalt mixture established in this paper has excellent fitting accuracy for different phases of the creep process of asphalt mixture under different stress levels, where the minimum fitting correlation values R^2 for CA mortar, asphalt mixture (applied to pavement engineering), and asphalt sand are 0.9976, 0.981, and 0.979, respectively. Therefore, this model can be used to provide a theoretical reference for the study of the characterization of the mechanical behavior of asphalt materials.

Keywords: constitutive models; Viscoelastic–Viscoplastic coexistence mechanism; fractional calculus; asphalt mixtures; creep test

1. Introduction

Due to the presence of asphalt matrix components in asphalt mixtures, which can provide excellent bonding and vibration reduction performance for engineering structures, they are widely used in engineering fields such as road pavement engineering and high-speed railway ballastless track engineering. However, the asphalt mixture exhibits complex time-varying mechanical behavior characteristics under long-term load, which poses a great challenge for structural design [1–3]. The establishment of a reasonable, accurate, and universal constitutive model of asphalt mixture has become a research hotspot in the field of engineering, which is of great significance for studying the evolution of the mechanical properties of materials and optimizing structural design methods.

As a typical viscoelastic-plastic material, the viscous mechanical behavior of asphalt mixture has a significant temporal correlation. For verifying and predicting the time-varying characteristics of the deformation of asphalt mixture under prolonged-term stress, the creep testing is an essential research measure [4–7]. According to the change characteristics of the whole creep process of asphalt mixture and the classification method of the creep process of viscoelastic materials in most studies, the creep process of asphalt mixture is divided into three phases: the decay creep phase, the stable creep phase, and the accelerated creep phase. For describing the strain changes of the creep process of the asphalt mixture, based on the total strain decomposition method, researchers usually use strain decomposition ideas to describe the different phases of the creep process of asphalt mixture using appropriate mechanical properties elements such as elastic, plastic, and viscoelastic or viscoplastic elements. With different combinations, constitutive models suitable for describing the stress–strain curve properties of the studied materials at different creep phases can be constructed. For instance, Zhang [8] developed a modified Burgers model to describe the nonlinear viscoelastic properties of rubber asphalt materials by concatenating one nonlinear modified viscous element, one elastic element, and one Kelvin model; Xiao [9] concatenated a nonlinear rheological element in series to the classical Burgers model to form a new constitutive model to describe the creep behavior of asphalt mixtures with large porosity; Yin [10] used the modified Burgers model, which combines the classical Burgers model with one Maxwell model and one fractional Kevin model in series, to analyze the dynamic creep performance of asphalt mixtures of the pavement. Xu [11] realized the study of transplanting the series Viscoelastic–Viscoplastic constitutive model to finite element calculation, and analyzed the mechanical behavior simulation and calculation of the asphalt pavement structure. Zhang [12] developed a creep model of asphalt mixture by using a three-element linear viscoelastic model in series with a viscoplastic element, and verified the model by repeated laboratory tests. During service, the creep behavior of the asphalt mixture is susceptible to load or environmental factors. Therefore, researchers have further carried out the creep model research of asphalt mixture under the influence of different conditions (temperature, stress level, water, etc.). An [13] carried out creep tests of asphalt mixtures under the influence of different temperatures and stress levels, and used a viscoplastic creep model in series with a nonlinear viscoplastic element and the Burgers model to achieve high-precision data fitting, and used this model to predict the creep behavior of asphalt mixtures under complex conditions. Zhou [14] used the Burgers model to characterize the viscoelastic behavior of asphalt mixture, studied the influence of two kinds of water effects, the freeze–thaw cycle and elevated-temperature water immersion, on the parameters of the creep model of asphalt mixture, and made a preliminary qualitative analysis of the law. Li [15,16] studied the influence of temperature and load stress on the creep deformation rate of asphalt mixture with different grades through static creep experiments, and used the Burgers model to characterize the viscoelastic energy of asphalt mixture, and analyzed the shift law of parameters in the model under different temperature and stress levels. The results show that the increase in temperature and stress level has a clear effect on the viscoelasticity of the asphalt mixture, with the temperature factor having more influence on the overall consistency of the model parameters. Zhao [17] analyzed the influence of viscoelastic short-term and long-term creep compliance of asphalt mixture under different confining pressure and temperature conditions. It can be seen that the creep test is an indispensable test method for constructing the constitutive model of asphalt mixture.

The construction methods for the creep constitutive model of asphalt mixture can be broadly divided into three categories. The first is the direct data fitting method, that is, according to the stress–strain data obtained from a static or quasi-static creep test of the asphalt mixture, after dimensionless processing, the geometric characteristics of the data curve are analyzed, and the appropriate mathematical function is selected for data fitting, and the creep constitutive model of the asphalt mixture is constructed. In this process, the quantitative relationship between the external factors and the coefficients of the chosen function can be analyzed and constitutive equations can be established to predict the mechanical behavior of asphalt mixtures under complex conditions [18–20]. However, while these methods can achieve higher fitting accuracy for test data, the constitutive equations established lack physical meaning and are difficult to apply to other types of asphalt mixtures. The second is to develop an appropriate constitutive model based on thermodynamics and the theory of internal variables, taking into account the dissipative mechanisms induced by irreversible displacements. This kind of method satisfies the basic laws of thermodynamics. By introducing a set of internal variables characterizing the internal structural changes of the material during viscoelastic deformation, the internal state of the material and the external response are linked through mathematical derivation and the irreversible energy dissipation process inside the material is realized. This approach has found numerous applications in the study of constitutive models of concrete materials that take into account performance degradation, and in the study of rheological constitutive models of rock materials. Currently, it is also being applied to the study of creep constitutive models for asphalt mixtures. This approach clearly expresses the shift of the internal state of matter away from logic and has a clear physical meaning. However, due to the unfixed potential function form of the thermodynamic state variables involved in the modeling process and the large number of internal variables, the mathematical derivation workload is heavy and most of the internal variables cannot be verified in the test, which makes it difficult to be widely promoted in practical applications [21–24]. The third (mechanical component combination) is widely used at present, which applies some mechanical components that can characterize the elasticity, plasticity, viscoelasticity, or viscoplasticity of the asphalt mixture, through a reasonable combination way, to establish the constitutive model that can adapt to the mechanical behavior of asphalt mixture at different phases of creep process. At first, most of these approaches were aimed at constructing constitutive models capable of characterizing the mechanical behavior of linear viscoelasticity [25,26]. As nonlinear problems became more and more prominent and intractable in the engineering field [27], some researchers began to study the nonlinear mechanical behavior of asphalt mixture under complex conditions through the measures of the nonlinear correction of the components in the viscoelastic constitutive model or the establishment of new elements adapted to nonlinear mechanical behavior combined with relevant nonlinear theories. In this regard, Zhang [28] introduced the strain equicalent principle of damage mechanics, establishing a type of nonlinear viscoelastic-plastic damage constitutive model. Zhang [29] considered the nonlinearity induced by the hardening and damage mechanisms of asphalt mixtures throughout the creep process and constructed a fractional creep damage model based on the fractional calculus theory to describe the complex viscoelastic mechanical behavior of asphalt mixtures in terms of fractional Abel elements. An [30] constructed a nonlinear viscoelastic-plastic creep model of the asphalt mixture with six mechanical parameters in series with the Burgers model and nonlinear viscoplastic elements, which can better describe the accelerated creep phase of asphalt mixture. Chen [31] used the Schapery model, the Findley model, and the multiple creep integral model to characterize the nonlinear viscoelastic properties of asphalt mixture. Liu [32] used the Weibull function to describe the distribution of internal defects in asphalt mixture, and introduced it into the nonlinear adhesive pot element of the modified Burgers model to establish the viscoelastic damage model of asphalt mixture. The results show that this model can fully reflect the creep characteristics of asphalt mixture in three phases. For another instance, the external factors affecting the material parameters related to asphalt mixture were introduced, and the nonlinear function (power function or exponential function) relationship between them and the material parameters of the constitutive model was constructed, so as to modify the material parameters in the established linear constitutive model, and realize the expression of the nonlinear viscoelastic mechanical behavior of the asphalt mixture constitutive model [33,34].

Recently, based on fractional order theory, a nonlinear viscoelastic constitutive model has been developed by scholars to characterize asphalt mixtures. Firstly, fractional mechanical components (such as the Abel adhesive pot) between linear elasticity and linear viscoelasticity has been developed by using fractional derivative theory, which realized the characterization of the nonlinear viscoelastic mechanical behavior of asphalt mixture. Then, according to the differences in mechanical behavior of materials with different types of components, and combining some linear mechanical elements, constitutive models suitable for the corresponding asphalt mixture were established [35–37]. However, most of the current models only achieve a characterization of the viscoelastic or viscoplastic mechanical behavior of a particular asphalt mixture, but do not uniformly characterize the mechanical and deformation behavior of a composite containing an asphalt matrix. On the one hand, this is related to the ability of constitutive models with different structural forms to characterize certain mechanical behaviors. On the other hand, this may also be related to the lack of uniform and accurate qualitative understanding of the deformation of the asphalt mixtures at different phases in the stress deformation process.

Motivated by the above discussion, in this paper, we establish a different criterion for dividing the creep process of asphalt mixtures according to the mechanical yield of the material, and propose a mechanism for the coexistence of viscoelastic and viscoplastic strains in the subsequent yield phase. In order to improve the generality of the viscoelastic characteristic, a special viscoelastic element consisting of two fractional viscoelastic elements in parallel has been proposed. Based on this, a constitutive model that characterizes the whole creep process of asphalt mixtures was developed. The correctness and applicability of the developed model is verified by short-term creep tests on CA mortar (one type of asphalt mixture) in the laboratory and test data on asphalt mixtures in the fitting reference example. The results can provide a new research perspective for analyzing the phase division of the creep development of asphalt mixture, which is different from traditional theories, and provide a new theoretical model reference for the subsequent study of the Viscoelastic–Viscoplastic mechanical behavior of asphalt mixture under the influence of damage and multiple factors.

2. Phases Division and Strain Property of Creep Process

The three phases of creep are mainly divided according to the different strain rate characteristics of the different phases, as shown in Figures 1 and 2. In the decay creep phase, the strain rate gradually decreases from an initially high value. It enters the stable creep phase when the strain rate does not change. In the stable creep phase, the creep rate of the asphalt mixture remains essentially constant, and it enters the accelerated creep phase when the strain rate starts to increase. It is well known that the key to determining the dividing points of the different phases is the strain rate of the stable creep phase. When the strain rate and time curve are calculated with strain-time test data, the strain rate entering the stable creep phase is an approximation rather than a constant value. In this way, a judgment error will be introduced in determining the time points at which the different phases enter, resulting in an inaccurate constitutive model.

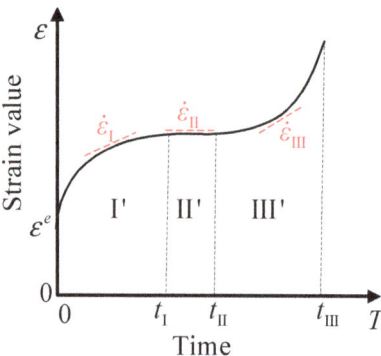

Figure 1. Classical three phases of creep.

Moreover, it is known that, according to the creep test results of viscoelastic-plastic materials in literature, without considering the influence of temperature and loading rate, when the loading stress is large, the time of each creep phase, especially the time of the stable phase, will be sharply reduced, so the stable creep phase is not obvious, and even may not appear. However, most scholars conduct modeling studies by separately matching different creep phases with appropriate mechanical models. This well-established model does not effectively explain the disappearance of the stable creep phase.

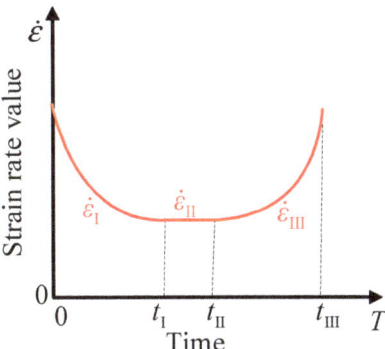

Figure 2. Variation of strain rate in three phases of creep.

Also, in some papers [24,30], the acceleration phase is described using a viscoplastic model. However, such a description directly restricts the strain properties of the accelerated creep phase to the viscoplastic strain, which lacks a criterion of discrimination.

To this end, starting from the study of the strain properties in the whole creep process of viscoelastic-plastic materials in this paper, we divide the phases into the instantaneous elastic phase, the pre-strain yield phase, and the subsequent strain yield phase by using the strain yield values. The strain properties of the corresponding phases are elastic, viscoelastic, and viscoelastic-plastic strains, respectively. The judgment on the strain properties of the subsequent yield phase can be understood as follows: the stress–strain relation of the material during the loading process has a one-to-one correspondence, provided that the temperature and loading rate are constrained. However, the deformation of viscoelastic-plastic materials is time-dependent due to the presence of viscous effects. The mechanism of viscoelastic strain does not disappear immediately when the material undergoes yielding and plastic deformation. Instead, it persists under the influence of plastic deformation through a different mechanism than the viscoelastic deformation prior to yielding. This can be defined as the "inertia effect of viscoelastic deformation". As a result, both viscoelastic and viscoplastic strains coexist during the subsequent yielding phase of viscoelastic-plastic materials. Therefore, the strain properties at different phases of the creep process will be shown in Figure 3.

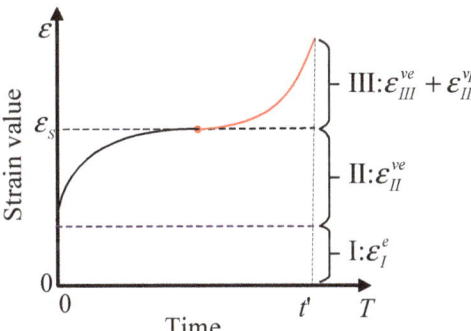

Figure 3. New three phases of creep and corresponding stain properties.

3. A Fractional Viscoelastic Element

Maxwell, Kelvin, and other linear models composed of Hooke elements and Newton elements in different combinations are constantly used to describe the viscoelastic mechanical behavior of materials. However, the viscoelastic deformation of asphalt mixtures at larger stresses exhibits pronounced nonlinear features. Therefore, based on the superiority of fractional calculus theory in describing nonlinear physical features, a different viscoelastic element with two fractional viscoelastic elements in parallel is proposed, as shown in Figure 4. η_i, γ_i represent the viscoelastic coefficient and derivative order of the viscoelastic element; the subscript i is the number of elements of the two fractional viscoelasticity.

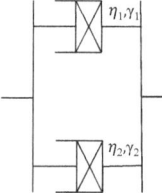

Figure 4. Fractional viscoelastic element.

According to the definition of the fractional derivative operator of the Riemann–Liouville type, the constitutive equation for the fractional viscoelastic element can be written as follows:

$$\sigma = \eta \frac{d^\gamma \varepsilon}{dt^\gamma} \tag{1}$$

where σ and ε are stress and strain respectively; t represents time.

When $\gamma = 0$, Equation (1) can be simplified as $\sigma = \eta\varepsilon$, which shows the time-independent features of the Hooke element; when $\gamma = 1$, Equation (1) becomes $\sigma = \eta\dot{\varepsilon}$, showing that the stress is proportional to the strain rate which is the feature of the Newton element; when $0 < \gamma < 1$, fractional viscoelastic elements are a class of nonlinear elements characterized as intermediate between elastic and Newtonian elements; when $\gamma > 1$, the more complex nonlinear mechanical behavior of the accelerated rheology can be expressed.

By adjusting the value of η_i and γ_i of the two fractional viscoelastic elements in Figure 4, the fractional viscoelastic element can be flexibly changed between elasticity, viscoelasticity, and viscosity. Viscous elements have a richer combinatorial form and greater generality compared with single viscoelastic elements or viscoelastic elements combined with linear elements. Its specific constitutive equations are derived as follows:

$$\sigma = \sigma_1 + \sigma_2 \tag{2}$$

$$\varepsilon = \varepsilon_1(t) = \varepsilon_2(t) \tag{3}$$

where σ and ε are the total stress and strain of the viscoelastic element, respectively; σ_1, ε_1 and σ_2, ε_2 represent the stress and strain of viscoelastic element 1 and element 2, respectively.

According to Equation (1):

$$\sigma_1 = \eta_1 \frac{d^{\gamma_1}\varepsilon_1(t)}{dt^{\gamma_1}} \tag{4}$$

$$\sigma_2 = \eta_2 \frac{d^{\gamma_2}\varepsilon_2(t)}{dt^{\gamma_2}} \tag{5}$$

It can be derived from Equation (1) to Equation (5):

$$\sigma = \eta_1 \frac{d^{\gamma_1}\varepsilon_1(t)}{dt^{\gamma_1}} + \eta_2 \frac{d^{\gamma_2}\varepsilon_2(t)}{dt^{\gamma_2}} \tag{6}$$

Equation (7) can be obtained by performing the Laplace transformation on both sides of Equation (6):

$$\tilde{\sigma}(s) = \eta_1\left[s^{\gamma_1}\tilde{\varepsilon}_1(s) - s^{\gamma_1-1}\varepsilon_1(0)\right] + \eta_2\left[s^{\gamma_2}\tilde{\varepsilon}_1(s) - s^{\gamma_2-1}\varepsilon_1(0)\right] \quad (7)$$

Assuming that the value of the stress σ applied is constant σ_0, and considering the initial state $\varepsilon_1(0) = 0$, Equation (7) can be transformed into Equations (8) and (9).

$$\frac{\sigma_0}{s} = (\eta_1 s^{\gamma_1} + \eta_2 s^{\gamma_2})\tilde{\varepsilon}_1(s) \quad (8)$$

$$\tilde{\varepsilon}_1(s) = \frac{\sigma_0}{s(\eta_1 s^{\gamma_1} + \eta_2 s^{\gamma_2})} \quad (9)$$

To obtain the inverse Laplace transform of Equation (9), the definition of the Mittag–Leffler function is as follows:

$$E_{a,b}(z) = \sum_{j=0}^{\infty} \frac{z^j}{\Gamma(ak+b)}, \quad a > 0, \, b > 0 \quad (10)$$

where $\Gamma()$ is the Gamma function, defined as:

$$\Gamma(z) = \int_0^{\infty} e^{-t} t^z \mathrm{d}t, \quad \mathrm{Re}(z) > 0 \quad (11)$$

According to the Laplace transformation formula of the Mittag–leffler function:

$$\int_0^{\infty} e^{-pt} t^{ak+b-1} E_{a,b}^{(k)}(nt^a)\mathrm{d}t = \frac{k! p^{a-b}}{(p^a - n)^{k+1}}, \quad \mathrm{Re}(s) > |n|^{\frac{1}{a}} \quad (12)$$

where $E_{a,b}^{(k)}(nt^a)$ is the kth derivative of the Mittag–leffler function. So, the Laplace inverse transformation of $\tilde{\varepsilon}_1(s)$ can be derived, and contacting with the Equation (3), the result is obtained as Equation (13).

$$\varepsilon(t) = \frac{\sigma_0 t^{\gamma_2}}{\eta_2} E_{\gamma_2-\gamma_1,\gamma_2+1}\left(-\frac{\eta_1}{\eta_2} t^{\gamma_2-\gamma_1}\right) \quad (13)$$

This formula is the constitutive model of the fractional viscoelastic element proposed in this paper. For convenience of application, the first two terms of the Mittag–Leffler function are expressed approximately, so that Equation (13) can be simplified as follows:

$$\varepsilon(t) = \frac{\sigma_0 t^{\gamma_2}}{\eta_2}\left[\frac{1}{\Gamma(\gamma_2+1)} + \frac{-\frac{\eta_1}{\eta_2} t^{\gamma_2-\gamma_1}}{\Gamma(2\gamma_2 - \gamma_1 + 1)}\right] \quad (14)$$

4. Nonlinear Creep Constitutive Model of Asphalt Mixture

Due to the presence of asphalt binder, asphalt mixtures can essentially be classified as a type of organic–inorganic composite material with nonlinear viscoelastic-plastic mechanical characteristics. It is often assumed that viscoelastic-plastic materials undergo viscoelastic creep deformation only at low stress. However, during creep beyond the yield strength of the material, as described in Section 1, viscoelastic strain continues to occur in the asphalt mixture. Moreover, a mechanism different from viscoelastic deformation prior to yielding occurs due to the effect of plastic deformation. This means that during the subsequent yielding phase of the asphalt mixture, two types of deformation coexist: viscoelastic and viscoplastic deformation.

Based on the above views, in this paper, a creep constitutive model of asphalt mixture referring to the Burgers model and the Nishihara model is established, as shown in Figure 5.

In the model, the instantaneous elastic strain is represented by a Hooke element, the viscoelastic strain is represented by a viscoelastic element, and the viscoelastic-plastic strain is represented by a viscoelastic element attached to a fractional viscoplastic element. The specific constitutive equations are derived as follows.

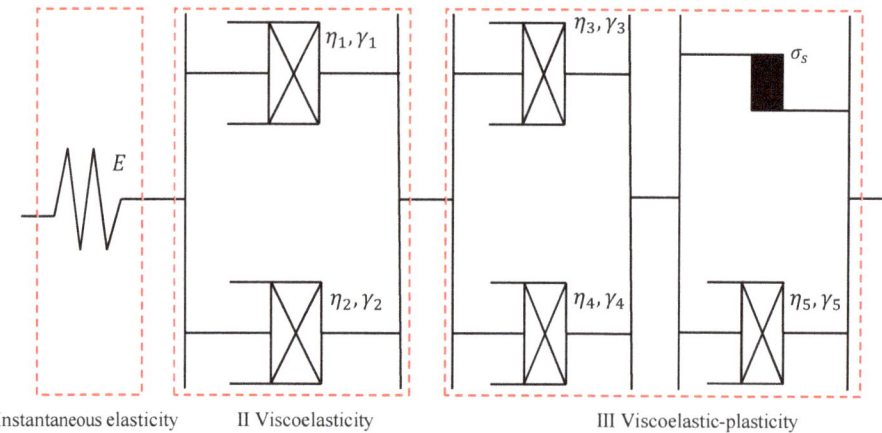

Figure 5. Nonlinear creep constitutive model.

For the creep process of asphalt mixtures, when applying the constant stress σ_0, the different components of the strain responses of the nonlinear creep constitutive model can be described as follows:

(1) Instantaneous elasticity

The constitutive equation of the instantaneous elastic model can be obtained from the generalized Hooke law:

$$\varepsilon_I = \frac{\sigma_0}{E} \tag{15}$$

where E is the elastic modulus that is corrected considering the amount of plastic deformation caused by the micro-defects existing in the initial structure.

(2) Viscoelasticity

Applying the derivation results of the fractional viscoelastic element constitutive equation in Section 3, the constitutive equation of the viscoelastic model (part II of nonlinear creep constitutive model) is obtained, denoted as Equation (16).

$$\varepsilon_{II}(t) = \frac{\sigma_0 t^{\gamma_2}}{\eta_2}\left[\frac{1}{\Gamma(\gamma_2+1)} + \frac{-\frac{\eta_1}{\eta_2}t^{\gamma_2-\gamma_1}}{\Gamma(2\gamma_2-\gamma_1+1)}\right] \tag{16}$$

where ε_{II} is the strain in part II of nonlinear creep constitutive model.

(3) Viscoelastic-plasticity

The strain in part III of the nonlinear creep constitutive model consists of two parts, related by a series relation: the viscoelastic strain due to the effects of plastic deformation and the viscoplastic strain in the subsequent strain-yielding phase.

The constitutive equation of the viscoelastic element in the viscoplastic element is expressed as Equation (17):

$$\sigma_5 = \eta_5 \frac{d^{\gamma_5}\varepsilon_5(t)}{dt^{\gamma_5}} \tag{17}$$

where σ_5 and $\varepsilon_5(t)$ are the stress and strain on the viscoelastic element, respectively; η_5 is the parameter related to material properties; γ_5 is the corresponding derivative order.

Assuming that the yield stress of the material is σ_s, the viscoplastic constitutive equation of the nonlinear creep constitutive model can be written as follows:

$$\sigma_5 = \langle \sigma_0 - \sigma_s \rangle = \eta_5 \frac{d^{\gamma_5} \varepsilon_5(t)}{dt^{\gamma_5}} \tag{18}$$

where $\langle \rangle$ is the Heaviside function. When $\sigma_0 - \sigma_s \leq 0$, the value is 0; when $\sigma_0 - \sigma_s \geq 0$, the value is $\sigma_0 - \sigma_s$.

Applying the Laplace transformation to Equation (18), then taking the inverse Laplace transform, the constitutive equation of the viscoplastic element can be obtained:

$$\varepsilon_5(t) = \frac{\langle \sigma_0 - \sigma_s \rangle t^{\gamma_5}}{\eta_5 \Gamma(\gamma_5 + 1)} \tag{19}$$

In order to characterize the Viscoelastic–Viscoplastic coexistence mechanism during creep, a fractional viscoelastic element with different parameters from the viscoelastic constitutive equation in part II is taken in the model. According to Equation (16), the strain can be expressed as follows:

$$\varepsilon'(t) = \frac{\langle \sigma_0 - \sigma_s \rangle t^{\gamma_4}}{\eta_4} \left[\frac{1}{\Gamma(\gamma_4 + 1)} + \frac{-\frac{\eta_3}{\eta_4} t^{\gamma_4 - \gamma_3}}{\Gamma(2\gamma_4 - \gamma_3 + 1)} \right] \tag{20}$$

where $\varepsilon'(t)$ is the strain representing the coexistence mechanism of viscoelasticity and viscoplasticity during the third phase; η_3 and η_4 are the material parameters of two viscoelastic elements; γ_3 and γ_4 are the corresponding derivative order.

Combining with Equations (19) and (20), the strain of part III of the nonlinear creep constitutive model is as follows:

$$\varepsilon_{III}(t) = \varepsilon'(t) + \varepsilon_5(t) = \frac{\langle \sigma_0 - \sigma_s \rangle t^{\gamma_4}}{\eta_4} \left[\frac{1}{\Gamma(\gamma_4 + 1)} + \frac{-\frac{\eta_3}{\eta_4} t^{\gamma_4 - \gamma_3}}{\Gamma(2\gamma_4 - \gamma_3 + 1)} \right] + \frac{\langle \sigma_0 - \sigma_s \rangle t^{\gamma_5}}{\eta_5 \Gamma(\gamma_5 + 1)} \tag{21}$$

Combining with Equations (15), (16) and (21), the constitutive equation of the creep process constructed is obtained, which is divided into two situations specifically, namely:

$$\begin{cases} \varepsilon = \varepsilon_I + \varepsilon_{II} = \frac{\sigma_0}{E} + \frac{\sigma_0 t^{\gamma_2}}{\eta_2} \left[\frac{1}{\Gamma(\gamma_2+1)} - \frac{\frac{\eta_1}{\eta_2} t^{\gamma_2 - \gamma_1}}{\Gamma(2\gamma_2 - \gamma_1 + 1)} \right], & \sigma_0 < \sigma_s \\ \varepsilon = \varepsilon_I + \varepsilon_{II} + \varepsilon_{III} = \frac{\sigma_0}{E} + \frac{\sigma_0 t^{\gamma_2}}{\eta_2} \left[\frac{1}{\Gamma(\gamma_2+1)} - \frac{\frac{\eta_1}{\eta_2} t^{\gamma_2 - \gamma_1}}{\Gamma(2\gamma_2 - \gamma_1 + 1)} \right] \\ + \frac{\langle \sigma_0 - \sigma_s \rangle t^{\gamma_4}}{\eta_4} \left[\frac{1}{\Gamma(\gamma_4+1)} - \frac{\frac{\eta_3}{\eta_4} t^{\gamma_4 - \gamma_3}}{\Gamma(2\gamma_4 - \gamma_3 + 1)} \right] + \frac{\langle \sigma_0 - \sigma_s \rangle t^{\gamma_5}}{\eta_5 \Gamma(\gamma_5 + 1)}, & \sigma_0 \geq \sigma_s \end{cases} \tag{22}$$

5. Model Parameters Determination and Verification

5.1. Determination Method for Model Parameters

In order to obtain the model parameters, the constitutive equation can be simplified as follows:

$$\begin{cases} \varepsilon = \sigma_0 \left(\frac{1}{E} + \frac{t^{\gamma_2}}{a} - b t^{2\gamma_2 - \gamma_1} \right) & \sigma_0 < \sigma_s \quad (23a) \\ \varepsilon = \sigma_0 \left(\frac{1}{E} + \frac{t^{\gamma_2}}{a} - b t^{2\gamma_2 - \gamma_1} \right) + \langle \sigma_0 - \sigma_s \rangle \left(c t^{\gamma_4} - d t^{2\gamma_4 - \gamma_3} + f t^{\gamma_5} \right) & \sigma_0 \geq \sigma_s \quad (23b) \end{cases}$$

where $a = \eta_2 \Gamma(\gamma_2 + 1)$, $b = \frac{\eta_1}{\eta_2^2 \Gamma(2\gamma_2 - \gamma_1 + 1)}$, $c = \frac{1}{\eta_4 \Gamma(\gamma_4 + 1)}$, $d = \frac{\eta_3}{\eta_4 \Gamma(2\gamma_4 - \gamma_3 + 1)}$, $f = \frac{1}{\eta_5 \Gamma(\gamma_5 + 1)}$.

When σ_0 is less than σ_s, the Equation (23a) is used to fit the test data with the L-M algorithm and the corresponding parameter E, a, b, γ_1, γ_2 can be obtained directly. Then η_1, η_2 can be obtained by the conversion formulas.

When σ_0 is more than σ_s, segmented fitting is required. According to the material creep test data, the first term of the corresponding constitutive equation in Equation (23b) is fitted by using the test data from 0 to t_1 corresponding to the strain dividing point. The corresponding parameter $E, a, b, \gamma_1, \gamma_2$ is obtained using the method applied in the condition $\sigma_0 < \sigma_s$. In this case, the instantaneous elastic strain and viscoelastic strain of the creep model can be obtained by fitting the results data. However, considering that the viscoelastic strain will not disappear and can continue to develop in the subsequent yield phase, the viscoelastic-plastic strain should be as Equation (24).

$$\varepsilon_{III} = \varepsilon - (\varepsilon_I + \varepsilon_{II}), t_1 \leq t \leq t' \quad (24)$$

At this time, the obtained parameter values are fixed, and the viscoelastic-plastic strain data in the creep process are calculated using Equation (24), then the second term of the corresponding constitutive equation in Equation (23b) is applied to fit. The parameters $c, d, f, \gamma_3, \gamma_4, \gamma_5$ are obtained, and η_3, η_4, η_5 are obtained by the conversion formulas. The process of determining model parameters is shown in Figure 6.

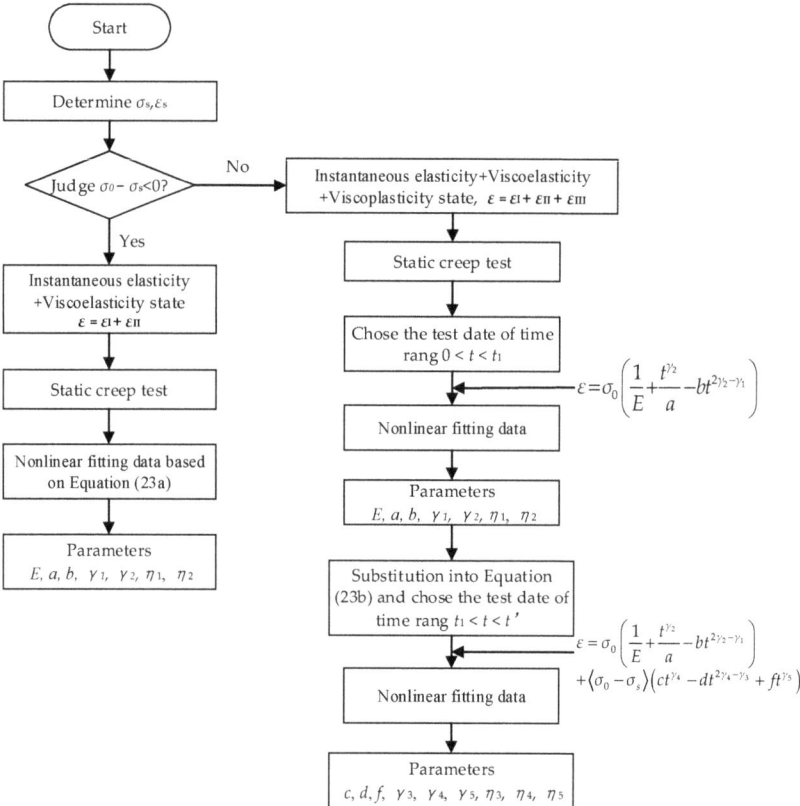

Figure 6. Model parameter determination flowchart.

If the yield stress σ_s is known, the uniaxial compression test is not carried out; namely, the yield strain cannot be determined. For this case, an optimization algorithm was proposed that can quickly determine the yield strain throughout creep without the need for uniaxial compression tests and improve the efficiency of the study.

By setting the number of measuring points on the whole curve as N, and the time corresponding to the Nth measuring point as t_n, then the minimum error between the fitting results and the test results is taken as the target to determine the strain yield point. The optimized model established is shown in Equation (25):

$$\min f = \sum_{i=1}^{n} \left\{ \frac{\sigma_0}{E} + \frac{\sigma_0 t_i^{\gamma_2}}{\eta_2} \left[\frac{1}{\Gamma(\gamma_2+1)} + \frac{-\frac{\eta_1}{\eta_2} t_i^{\gamma_2-\gamma_1}}{\Gamma(2\gamma_2-\gamma_1+1)} \right] - \varepsilon_i \right\}^2 \\ + \sum_{j=n+1}^{N} \left\{ \begin{array}{l} \frac{\sigma_0}{E} + \frac{\sigma_0 t_j^{\gamma_2}}{\eta_2} \left[\frac{1}{\Gamma(\gamma_2+1)} + \frac{-\frac{\eta_1}{\eta_2} t_j^{\gamma_2-\gamma_1}}{\Gamma(2\gamma_2-\gamma_1+1)} \right] + \\ \frac{\sigma_0-\sigma_s t_j^{\gamma_4}}{\eta_4} \left[\frac{1}{\Gamma(\gamma_4+1)} + \frac{-\frac{\eta_3}{\eta_4} t_j^{\gamma_4-\gamma_3}}{\Gamma(2\gamma_4-\gamma_3+1)} \right] + \frac{\sigma_0-\sigma_s t_j^{\gamma_5}}{\eta_5 \Gamma(\gamma_5+1)} - \varepsilon_j \end{array} \right\}^2 \quad (25)$$

In the test curve of Figure 7, the yield strain appears in the surrounding region of strain transforming, so simplifying the calculation method of Equation (25) is as follows.

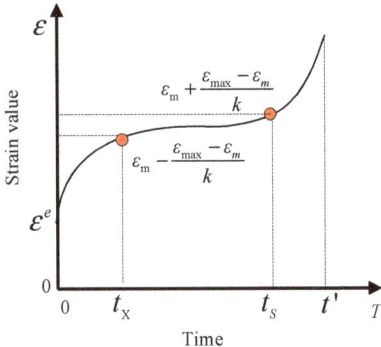

Figure 7. Yield strain determination method.

In the range 0 to t_n, first, the forward difference of the strain (the time interval is only related to the test frequency and is a constant more than 0) can be calculated and the time tm corresponding to the minimum difference value and strain ε_m can be determined; second, the possible range of yield strain as Equation (26) can be determined.

$$\varepsilon_m - \frac{\varepsilon_{max} - \varepsilon_m}{k} \leq \varepsilon_s \leq \varepsilon_m + \frac{\varepsilon_{max} - \varepsilon_m}{k} \quad (26)$$

where ε_{max} is the maximum strain measured in Figure 6 and k is the empirical coefficient. The smaller the value, the larger the range determined, and the longer the optimization time. The recommended value ranges from 3 to 5 as the actual fitting experience.

Solving Equation (25) could use some intelligent optimization algorithms such as the Genetic Algorithm or the Ant Colony Algorithm. The optimized results not only include yield strain point and corresponding time, but also include specific fitting parameters.

5.2. Model Verification

Two types of test data are adopted for verification. One is the short-term creep test of CA mortar in the laboratory, and the other is the test data from the references [29,38].

(1) Laboratory creep test

CA mortar is the composition material of the ballastless tracks for high-speed railways. According to Xie's test research [39], it is known that under the condition of small stress, it takes 3 years or more to obtain the complete creep process of CA mortar. So, the short-time creep test of CA mortar in laboratory is carried out based on Xu's research [40].

The test of raw materials contain the early strength of Portland cement (P·II 52.5R), fine sand (the maximum particle size not exceeding 1.18 mm), water, SBS-modified emulsified asphalt (the physical properties showed in Table 1), water reducer, defoamer (silicone defoamer), aluminum powder (mass fraction ≥ 85%), and the other additives (early strength agent). The mix proportion of CA mortar is shown in Table 2, where the actual weights of the materials can be weighed according to the proportional relationship shown in the table.

Table 1. Properties of asphalt emulsion.

De-Emulsification Speed	Evaporation Residue			
	Residue Content/%	Penetration Index/ (0.1 mm)	Softening Point/°C	Ductility (25 °C)/cm
Slow-breaking	58.5	62.3	45.5	130

Table 2. Quality mix proportion of CA mortar.

Material	Cement	Fine Sand	Emulsified Asphalt	Water	Water Reducer	Defoamer	Aluminum Powder
Weight parts	366.7	733.3	515	50	12.5	1.5	0.05

The preparation of CA mortar material: Firstly, pour water, SBS emulsified asphalt, and water reducing agent into a mixer and mix at a speed of 60 r/min for 30 s. Then, add cement, fine sand, defoamer, and aluminum powder, mix at a speed of 60 r/min for 60 s, then mix at a speed of 120 r/min for 180 s, and then mix at a speed of 60 r/min for 60 s. The preparation of CA mortar is complete.

The preparation process of the test was:

a. Pouring the freshly mixed CA mortar into cylindrical specimens with the size of Φ50 × 100 mm;
b. Placing in the environment box with relative humidity of (65 ± 5)% and curing for 24 h;
c. Removing from the testing mold and curing in standard curing room for 28 days, then the test specimens were completed as shown in Figure 8;
d. After completing the above steps, taking out specimens and carrying out the creep loading test of CA mortar.

Figure 8. CA mortar specimens.

The creep test used the electronic universal testing machine made in China shown in Figure 9. The measurement range of the machine is 50 kN and the accuracy is ±0.5% of the value displayed.

Figure 9. Electronic universal testing machine.

The test was carried out at an environment temperature of 25 °C.

The loading stress levels were set at 0.1 MPa and 0.5 MPa, and the data acquisition frequency was 0.04 Hz. Before starting the normal test, 0.005 MPa should be preloaded to eliminate the gap between the test instrument and the specimen. The stress of 0.005 MPa should be maintained for 1 min and then the set stress level should be loaded quickly and maintained for 1.0 h. In order to ensure the stability of the test data, the test data with large discreteness were removed, and the number of specimens was increased to ensure that each stress level had three groups of qualified data. Then, the data curve changes smoothly for the fitting of the constitutive model.

According to the regulation [41] and the compressive strength value of CA mortar measured in the previous experiment, the 28-day compressive strength of CA mortar is 1.8 MPa, and it is pointed out in reference [42] that the recommended yield strength of general viscoelastic materials is 40% of the compressive strength value. So, if the minimum yield value of CA mortar is calculated in accordance with the specification, the yield stress value of CA mortar is 0.72 MPa. Because the stress values applied in the specimens are all smaller than 0.72 MPa, the corresponding equation of the case $\sigma_0 < \sigma_s$ in the established constitutive Equation (23a,b) was used to fit the test data. The test data and model fitting results are shown in Figure 10. The fitting parameters are shown in Table 3. In order to increase the reliability of the model validation process in this article and the impact of different models on the characterization of mechanical properties of the same material, this article compared the short-term creep test data of CA mortar in reference [40]. The specimen size used in this experiment is Φ50 × 50 mm (Φ50 × 100 mm in this article), the test temperature is controlled at 25 °C (the same as the test environment temperature in this article), and the loading stress levels are 0.1 MPa and 0.5 MPa, respectively (the same as the test loading stress size in this article). The loading process also involves preloading with a stress load of 0.005 MPa for 10 min, then quickly applying it to the required loading stress value and maintaining it for 1 h. In reference [40], the Burgers model and the four-element five-parameter model were used to fit experimental data, and parameter E, which also characterizes the instantaneous elastic deformation of the CA mortar creep process, was compared, as shown in Table 4.

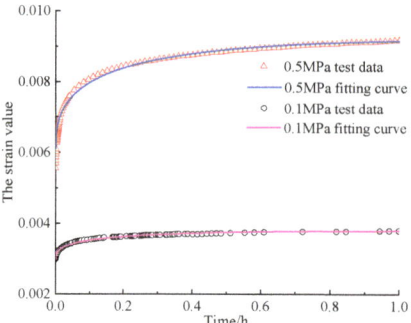

Figure 10. Short-term creep test data and fitting results of CA mortar.

Table 3. Fitting values of model parameters under different stress levels.

Stress/MPa	E/MPa	γ_1	γ_2	a	b	R-Squared
0.1	33.46	0.5047	0.5095	1.1646	0.8505	0.9985
0.5	81.7285	0.5335	0.5364	0.998	0.9959	0.9976

Table 4. Parameters of CA mortar creep test model in reference [40].

Stress/MPa	Elastic Modulus/MPa (Burgers Model)	Elastic Modulus/MPa (Four-Element Five-Parameter Model)
0.1	33.326	35.49
0.5	84.549	90.39

Figure 10 shows that the viscoelastic-plastic constitutive model established in this paper can be used to fit the test data under different stresses during the short-term creep process of CA mortar, and the goodness of fit is above 0.99. Meanwhile, compared with the CA mortar elastic model established in the literature in Table 4, it can be seen that the fitting results are consistent under the condition of 0.1 MPa, but the elastic modulus obtained under 0.5 MPa stress is slightly smaller, with 3.33% deviation from Burgers model and 9.58% deviation from the four-element, five-parameter model. On the one hand, it may be caused by different test loading conditions; on the other hand, it is due to the differences in the physical expressions of material deformation by different constitutive models.

(2) Model adaptability verification

In order to verify the adaptability of the model to asphalt mixture, the creep test data of different components of asphalt mixture in references [29,38] were fitted with the established model, and the fitting results were shown in Figures 11 and 12.

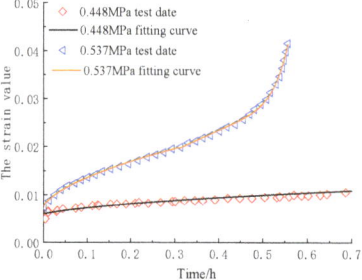

Figure 11. Fitting of creep test results of asphalt mixture.

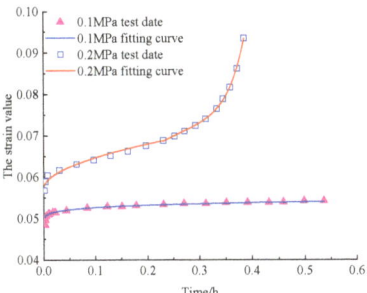

Figure 12. Fitting of creep test results of asphalt sands.

According to the reference corresponding to the test data, the yield strength of asphalt mixture and asphalt sand are 0.358 MPa and 0.05 MPa, respectively. The asphalt mixture in Figure 11 at a stress value of 0.537 MPa and the asphalt sand in Figure 12 at a creep value of 0.2 MPa both demonstrated obvious viscoelastic-plastic deformation phases. Therefore, the test data should be fitted by using the constitutive equation parameter determination method corresponding to the case of $\sigma_0 \geq \sigma_s$ in Section 5.1. However, the creep deformation rate of asphalt mixture under a stress value of 0.448 MPa and asphalt sand under a stress value of 0.1 MPa was slow, and the strain value did not reach the yield strain value during the test time. So, the test data should be fitted with the first two terms of the constitutive equation corresponding to the case of $\sigma_0 \geq \sigma_s$.

As the results shown in Figures 11 and 12 and Table 5 demonstrate, the fractional-order creep model established considering the Viscoelastic–Viscoplastic coupling mechanism in this paper has a good fitting effect at different phases of the creep process under different stress conditions. And it also shows that the dividing method based on the strain yield proposed in this paper is also feasible.

Table 5. Fitting results of creep test parameters of different asphalt mixtures by this model.

Data Sources	Stress	E	γ_1	γ_2	a	b	γ_3
Reference [29]	0.448	69.053	0.385	0.382	0.952	1.037	
	0.537	72.264	0.483	0.478	0.811	1.186	35.492
Reference [38]	0.1	2.799	0.0472	0.0471	0.822	1.027	
	0.2	3.488	0.527	0.526	0.9465	0.9318	26.948

Date Source	Stress	γ_4	γ_5	c	d	f	R^2
Reference [29]	0.448	-	-	-	-	-	0.981
	0.537	18.02	0.5464	2625.16	0.737	1.154	0.999/0.998
Reference [38]	0.1	-	-	-	-	-	0.979
	0.2	13.047	0.148	30233.418	0.010	1.148	0.980/0.998

6. Conclusions

Based on the derivation of the nonlinear creep constitutive model for asphalt mixtures and the experimental validation results, the following conclusions are obtained.

1. The strain yield point of viscoelastic plastic materials is used as the basis for dividing the creep process, solving the problem of strain characteristics at different stages of the creep process. For asphalt mixtures with viscoelastic plastic mechanical characteristics, their complete creep strain properties can be divided into different types including instantaneous elasticity, viscoelasticity, and viscoelastic-plasticity.
2. The coexistence mechanism of multiple strains in the subsequent yield phase of viscoelastic-plastic materials was proposed. This method can provide a new analytical

method for studying the coupling effect of viscoelastic viscoplastic mechanical behavior. Based on the theory of fractional calculus, a special fractional order viscoelastic element consisting of two fractional order viscoelastic elements in parallel has been established. It can more flexibly characterize the elastic, viscoelastic, and viscous mechanical behaviors of materials, providing a generalized model reference for the study of material mechanical behavior.

3. Through short-term creep tests of CA mortar indoors and fitting results with multi stress creep tests of asphalt mixtures composed of different materials in the literature, it is shown that under small stress, the creep process of asphalt mixtures exhibits more viscoelastic deformation. Under large stress, the creep process of asphalt mixtures has a complete three-phase property. When the strain value exceeds the yield strain, plastic deformation occurs.

4. The creep constitutive model of asphalt mixture established in this paper has excellent fitting accuracy for the creep process of CA mortar, asphalt mixture (applied to pavement engineering), and asphalt sand. The minimum fitting correlation values R^2 are 0.9976, 0.981, and 0.979, respectively. The fitting results also indicate that the model proposed in this paper has good adaptability for different types of asphalt matrix mixtures with creep properties.

5. This article only verified the established model through the creep behavior of asphalt matrix mixtures and whether it is suitable for the creep behavior of other materials still needs to be studied. At the same time, when analyzing the creep behavior of materials using the model established in this paper, it is necessary to obtain the yield strain value of the material under corresponding conditions.

Author Contributions: Conceptualization, J.Z. and K.X.; methodology, J.Z.; software, J.Z. and Y.Y.; validation, J.Z.; resources, W.Z. and Y.Y.; data curation, J.Z. and K.X.; writing—original draft preparation, J.Z.; writing—review and editing, W.Z. and K.X.; project administration, Y.Y.; funding acquisition, W.Z., K.X., Y.Y. and J.Z. All authors have read and agreed to the published version of the manuscript.

Funding: This research was supported by the National Natural Science Foundation of China, grant number 51978423 and 52008272; the Natural Science Foundation of Hebei Province, grant number E2022210046 and E2021210090; the Central Government Foundation for Guiding Local Science and Technology Development, grant number 226Z5401G; the Graduate Innovation Funding Project of Hebei Province, grant number CXZZBS2021116; the technology development project of Shuohuang Railway Development Co., Ltd., grant number GJNY-20-230.

Informed Consent Statement: Informed consent was obtained from all subjects involved in the study.

Data Availability Statement: Because the research data in this paper involves the interests and privacy of some enterprise projects, it is not convenient for publication at the moment.

Conflicts of Interest: The authors declare no conflict of interest.

References

1. Zhang, Y.; Sun, L. Assessing mechanical properties of hard asphalt mixtures with different design methods. *J. Mater. Civil Eng.* **2021**, *33*, 04021102. [CrossRef]
2. Fang, M.; Hu, T.; Fang, K.; Hu, Z.; Fang, K.; Hu, Z.; Zhang, X.; Zhang, J.; Xiao, J. Mechanical properties of coarse asphalt Mixture-Paved trackbed under high-speed moving train loads. *Constr. Build. Mater.* **2022**, *357*, 129389. [CrossRef]
3. Luan, Y.; Zhang, W.; Zhao, Y.; Pan, Z.; Niu, Z.; Zeng, K.; Chen, X.; Mohammad, L.N. Mechanical property evaluation for steel slag in asphalt mixture with different skeleton structures using modified marshall mix design methodology. *J. Mater. Civil Eng.* **2022**, *34*, 04021382. [CrossRef]
4. Judycki, J. Non-linear viscoelastic behaviour of conventional and modified asphaltic concrete under creep. *Mater. Struct.* **1992**, *25*, 95–101. [CrossRef]
5. Zhang, Q.; Gu, X.; Liang, J.; Yu, Z.; Dong, Q.; Jiang, J. Application of a stochastic damage model to predict the variability of creep behavior for asphalt mixtures. *Case Stud. Constr. Mat.* **2023**, *18*, e02078. [CrossRef]
6. Zhu, Y.; Guo, Z.; Wang, J. Creep Test and research on asphalt mixture at high temperature and heavy load. *J. Build. Mater.* **2008**, *11*, 545–549.

7. Celauro, C.; Fecarotti, C.; Pirrotta, A.; Collop, A.C. Experimental validation of a fractional model for creep/recovery testing of asphalt mixtures. *Constr. Build. Mater.* **2012**, *36*, 458–466. [CrossRef]
8. Zhang, D.; Feng, X. Research on viscoelastic properties of asphalt rubber base on creep test. *Highway* **2014**, *59*, 195–198.
9. Xiao, M.; Cheng, W.; Yang, L. Creep characteristics considering porosity of asphalt mixture and modified Burgers model. *Sci. Technol. Eng.* **2023**, *23*, 1698–1708.
10. Yin, Y.; Zhang, X. Study on constitutive relation of asphalt mixtures based on dynamic creep test. *J. Funct. Mater.* **2014**, *23*, 23020–23024.
11. Xu, J.; Zhu, Y. Finite element analysis on constitutive model of serial viscoelastic-viscoplastic asphalt mixture. *J. Chongqing Jiaotong Univ. Nat. Sci.* **2019**, *38*, 27–34+50.
12. Zhang, J.; Huang, X. Viscoelastoplastic-damage mechanics model of permanent deformation in asphalt mixture. *J. Southeast Univ. Nat. Sci. Ed.* **2010**, *40*, 185–189.
13. An, W.; Sheng, D.; Zhang, S.; Liu, B.; Cheng, J. Creep model of asphalt mixture under the influence of stress and temperature. *J. Funct. Mater.* **2021**, *52*, 11113–11119.
14. Zhou, Z.; Liu, X.; Luo, S.; Sha, X. Effect of water intrusion on performance of asphalt mixture. *J. Cent. South Univ. Sci. Technol.* **2016**, *47*, 1359–1367.
15. Li, P.; Zhang, Z.; Wang, B. High-temperature creep deformation behavior and its mechanism of asphalt mixture. *J. Build. Mater.* **2012**, *15*, 422–426.
16. Li, P.; Zhang, Z.; Wang, B. Analysis of visco-elastic response influencing factors of asphalt mixture. *J. Zhengzhou Univ. Eng. Sci.* **2010**, *31*, 96–100.
17. Zhao, Y.; Chen, J. Effect of stress state on viscoelastic properties of asphalt mixtures. *J. Build. Mater.* **2009**, *12*, 428–432.
18. Wang, P.; Xu, H.; Xie, K.; Zeng, X. Research on the dynamic compressive test of China railway track system (CRTS) I type CA mortar and its constitutive relationship. *J. Railw. Eng. Soc.* **2014**, *5*, 35–40+58.
19. Zhang, M.; Tian, Q.; Qu, M.; Qi, S.; Yao, T.; Xu, G.; Deng, D. Test study on stress-strain constitutive relations of cement asphalt mortar. *Mater. Rep.* **2022**, *36*, 47–51.
20. Zhang, S.; Fu, Q.; Zhou, X.; Long, G.; Xie, Y. Strain rate sensitivity and constitutive relation of mechanical properties of CA mortar. *J. South China Univ. Technol. Nat. Sci. Ed.* **2014**, *42*, 106–112.
21. Zhang, L.; Liu, Y.; Xue, L.; Yang, Q. A viscoelastic model and its fundamental properties based on thermodynamics with internal state variable. *Sci. Sin. Phys. Mech. Astron.* **2015**, *45*, 65–72.
22. Liu, G.; Gao, H.; Hu, Y. Thermodynamic requirements for the creep constructive equations of ageing visco-elastic materials. *Acta Mech. Solida Sin.* **2003**, *24*, 205–210. [CrossRef]
23. Zhu, H.; Sun, L. A viscoelastic–viscoplastic damage constitutive model for asphalt mixtures based on thermodynamics. *Int. J. Plast.* **2013**, *40*, 81–100. [CrossRef]
24. Sun, L.; Zhu, Y. A serial two-phase viscoelastic–viscoplastic constitutive model with thermodynamical consistency for characterizing time-dependent deformation behavior of asphalt concrete mixtures. *Constr. Build Mater.* **2013**, *40*, 584–595. [CrossRef]
25. Liu, H.; Zeiada, W.; Al-Khateeb, G.; Abdallah, S.; Samarai, M. A framework for linear viscoelastic characterization of asphalt mixtures. *Mater. Struct.* **2020**, *53*, 32. [CrossRef]
26. Li, P.; Jiang, X.; Guo, K.; Xue, Y.; Dong, H. Analysis of viscoelastic response and creep deformation mechanism of asphalt mixture. *Constr. Build. Mater.* **2018**, *171*, 22–32. [CrossRef]
27. Bhaskara, V.; Lakshmib, R.; Eyad, M.; Ramamania, R. Analysis of reclaimed asphalt blended binders using linear and nonlinear viscoelasticity frameworks. *Mater. Struct.* **2020**, *53*, 122.
28. Zhang, J.; Li, Z. Viscoelastic-plastic damage constitutive model of asphalt mixture under cyclic loading. *J. Northeast. Univ. Nat. Sci.* **2019**, *40*, 1496–1503.
29. Zhang, Q.; Gu, X.; Ding, J.; Hu, D. Creep damage model and damage evolution of asphalt mixtures. *J. Traffic Transp. Eng.* **2021**, *21*, 104–113.
30. An, W.; Sheng, D.; Zhang, S.; Liu, B. Nonlinear viscoelastic-plastic creep model for asphalt mixtures. *J. Mater. Sci. Eng.* **2022**, *40*, 1030–1033+1054.
31. Chen, S.; Xu, L.; Jia, S.; Wang, J. Characterization of the nonlinear viscoelastic constitutive model of asphalt mixture. *Case Stud. Constr. Mat.* **2023**, *18*, e01902. [CrossRef]
32. Liu, J.; Li, Q.; Li, X. A new creep model of asphalt mixture based on statistical damage theory. *J. Highw. Transp. Res. Dev.* **2014**, *31*, 13–18. [CrossRef]
33. Dong, M.; Lu, J.; Ling, T.; Sun, Z.; Li, L. Parametric model for asphalt mixtures considering temperature effect. *Eng. Mech.* **2016**, *33*, 180–185+193.
34. Tian, Q.; Deng, D.; Peng, J. Relationship between elastic modulus and composition parameters of cement-asphalt cementitious material. *China Railw. Sci.* **2016**, *37*, 1–7.
35. Gu, L.; Zhang, W.; Ma, T.; Qiu, X.; Xu, J. Numerical simulation of viscoelastic behavior of asphalt mixture using fractional constitutive model. *J. Eng. Mech.* **2021**, *147*, 04021027. [CrossRef]
36. Quan, W.; Zhao, K.; Ma, X.; Dong, Z. Fractional viscoelastic models for asphalt concrete: From parameter identification to pavement mechanics analysis. *J. Eng. Mech.* **2022**, *148*, 04022036. [CrossRef]

37. Liang, S.; Luo, R.; Luo, W. Fractional differential constitutive model for linear viscoelasticity of asphalt and asphalt mastic. *Constr. Build. Mater.* **2021**, *306*, 124886. [CrossRef]
38. Zeng, G.; Yang, X.; Bai, F.; Yin, A. Test researches on a visco-elastoplastic creep damage constitutive model of asphalt mastic. *Eng. Mech.* **2013**, *30*, 249–253.
39. Xie, Y.; Fu, Q.; Long, G.; Zheng, K.; Song, H. Creep properties of cement and asphalt mortar. *Constr. Build. Mater.* **2014**, *70*, 9–16. [CrossRef]
40. Xu, H.; Wang, P.; Xie, K.; Zeng, X. Test and model parameter analysis of cement and emulsified asphalt mortar with short-term creep property. *J. China Railw. Soc.* **2015**, *37*, 114–118.
41. Li, H.; Jiang, C.; Wu, S.; Xie, Y.; Yang, F.; Zheng, X.; Huang, W.; Xie, Y.; Deng, D.; Zhu, H. *Temporary Technical Conditions of Cement Emulsified Asphalt Mortar for CRTS Type I Slab Ballastless Track on Passenger Dedicated Line Railway*; China Railway Publishing House: Beijing, China, 2008.
42. Song, H.; Zeng, X.; Xie, Y. Creep characteristics of cement emulsified asphalt mortar under long-term load. *J. Build. Mater.* **2020**, *23*, 271–278.

Disclaimer/Publisher's Note: The statements, opinions and data contained in all publications are solely those of the individual author(s) and contributor(s) and not of MDPI and/or the editor(s). MDPI and/or the editor(s) disclaim responsibility for any injury to people or property resulting from any ideas, methods, instructions or products referred to in the content.

Article

Evaluation of the Adhesion between Aggregate and Asphalt Binder Based on Image Processing Techniques Considering Aggregate Characteristics

Min Li [1], Jian Wang [1], Zibao Guo [2], Jingchun Chen [1], Zedong Zhao [3] and Jiaolong Ren [1,*]

[1] School of Civil Engineering and Geomatics, Shandong University of Technology, Zibo 255000, China; 22507020009@stumail.sdut.edu.cn (M.L.); 21507020787@stumail.sdut.edu.cn (J.W.); 21507020774@stumail.sdut.edu.cn (J.C.)
[2] CCCC First Highway Northwest Engineering Co., Ltd., Xi'an 710110, China; cdcfllmdqs@163.com
[3] School of Transportation and Vehicle Engineering, Shandong University of Technology, Zibo 255000, China; 20402010140@stumail.sdut.edu.cn
* Correspondence: worjl@sdut.edu.cn

Abstract: Aggregate–asphalt adhesion plays an important role in the water stability of asphalt concrete. In various test standards of different countries, it is evaluated via the subjective judgment of testers using the boiling water test. The subjective judgment in the test method is detrimental to the accuracy of the adhesion evaluation. However, there is no quantitative evaluation method for the aggregate–asphalt adhesion in existing studies. Moreover, the effects of aggregate shape on adhesion are also not discussed and stipulated. Hence, an innovative method based on the Chinese boiling water test and image processing technique is put forward to quantificationally evaluate the aggregate–asphalt adhesion. Moreover, the effects of aggregate shapes on adhesion are also investigated via the proposed method from a view of aspect ratio and homogeneity. Results show that the peeling of the asphalt membrane on the aggregate surface is more serious as the complexity of the aggregate shape increases after the boiling water tests, while the effect degree gradually decreases. The effect of aspect ratio on the peeling status of asphalt membrane is lower than that of aggregate homogeneity.

Keywords: aggregate–asphalt adhesion; evaluation method; image processing techniques; aggregate characteristics; water boiling test

1. Introduction

Pavement diseases greatly influence the use quality, passenger comfort, and traffic safety of pavement structures. The inspection of raw materials before pavement construction can significantly control the occurrence of pavement diseases [1]. Potholes are one of the main diseases in asphalt pavement at present, resulting from a lack of water stability of asphalt concrete [2]. The water stability of asphalt concrete mainly depends on the adhesion between aggregates and asphalt binders [3]. In order to ensure the water stability of asphalt concrete, the adhesion between aggregates and asphalt binders is a required inspection in the standards of different countries. Tables 1–3 illustrate the adhesion evaluation method, namely the boiling water test, in the Chinese test standard [4] and the ASTM test standard [5], respectively.

As shown in Tables 1–3, it can be found that there is only standard practice of the boiling water test in the ASTM test standard, while there is no evaluation method for the adhesion. In the Chinese test standard, although an evaluation method for the adhesion level is provided according to the peeling area of the asphalt membrane on the surface of the aggregates, there is no detailed calculation method for the peeling area. In other words, the adhesion can only be evaluated via the subjective judgment of testers and observers

Citation: Li, M.; Wang, J.; Guo, Z.; Chen, J.; Zhao, Z.; Ren, J. Evaluation of the Adhesion between Aggregate and Asphalt Binder Based on Image Processing Techniques Considering Aggregate Characteristics. *Materials* 2023, 16, 5097. https://doi.org/10.3390/ma16145097

Academic Editor: Gilda Ferrotti

Received: 29 June 2023
Revised: 14 July 2023
Accepted: 18 July 2023
Published: 19 July 2023

Copyright: © 2023 by the authors. Licensee MDPI, Basel, Switzerland. This article is an open access article distributed under the terms and conditions of the Creative Commons Attribution (CC BY) license (https://creativecommons.org/licenses/by/4.0/).

for the peeling area. Obviously, the uncertain impact of subjective factors on the adhesion evaluation is inescapable. In fact, the explanation items of the Chinese test standard point out the reason that there is no evaluation method in the ASTM test standard is due to the effect of subjective factors. It is detrimental to the accuracy of the adhesion evaluation, especially for the criticality of different adhesion levels.

Table 1. Testing method of the adhesion in the Chinese test standard.

Procedure	Practice
Step 1	Heat clean aggregates (13.2–19 mm) to 105 °C.
Step 2	Immerse one aggregate into the asphalt binder (130–150 °C) for 45 s, and take out the aggregate to cool at room temperature for 15 min.
Step 3	Put the aggregate with asphalt membrane into micro-boiling water for 3 min.
Step 4	Evaluate the adhesion level according to the peeling degree (see Table 2) of the asphalt membrane on the surface of the aggregate.

Table 2. Evaluation method of the adhesion in the Chinese test standard.

Adhesion Status	Adhesion Grade
The asphalt membrane is intact.	5
The peeling area is less than 10% of the aggregate superficial area.	4
The peeling area reaches 10–30% of the aggregate superficial area.	3
The peeling area is more than 30% of the aggregate superficial area.	2
The asphalt membrane is completely moved, and the aggregates are bare.	1

Table 3. Testing method of the adhesion in the ASTM test standard.

Procedure	Practice
Step 1	Heat 500 mL distilled water to boiling (80–100 °C).
Step 2	Put 250 g of loose asphalt mixture into boiling water for 10 min.
Step 3	Take out the asphalt mixture and obtain the coating rate of asphalt on aggregates.

In this case, some new methods are put forward to evaluate the adhesion between aggregates and asphalt binders. Hefer et al. and Bhasin et al. [6,7] put forward a new method, namely the Wihelmy hanging piece method, to determine the adhesion via advance angles and retreat angles of asphalt binder. Hamzah et al. [8] adopted the direct tension test to establish a new method to evaluate the adhesion grade. Shen et al. [9] evaluated the adhesion via tension test, net adsorption test, and scanning electron microscope. Liu et al. [10] proposed an improved evaluation method for adhesion via measuring the thickness of asphalt membrane based on the contact angle test, scanning electron microscopy, and energy spectrum analysis. Ingrassia et al. [11] investigated the adhesion properties between aggregate substrates and the binder composed of reclaimed wood bio-oils and bio-adhesives after different aging degrees based on the asphalt bond strength (BBS) tests. D'Angelo et al. [12] adopted the same method to analyze the adhesion characteristics between plastomeric binder blends and aggregates. Liu et al. [13] judged the adhesion level based on the PosiTestAT-A adhesion test. Ji et al. [14] and Anastasiya et al. [15,16] evaluated the effect of bio-oil on asphalt adhesion using the sessile drop method from a view of surface free energy of asphalt, bio-oil, and aggregates. In addition, Anastasiya et al. [17,18] also studied the adhesive properties of the bio-oil/asphalt blend via a probe tack test and lap shear test. Although these methods could provide quantitative evaluation methodologies for adhesion inspection, the adopted test methods were not common for engineering applications. It was not beneficial to the engineering application.

Subsequently, with the development of computer science [17–20], image-processing techniques were tried to investigate the adhesion between aggregates and asphalt binders. Park et al. [21] and Nazirizad et al. [22] analyzed the effect of anti-stripping additives on the water stability of asphalt concrete based on the standard measurement method (water boiling water test) and image processing technique. They considered that the image processing techniques could obtain better efficacy than unaided viewing when evaluating the adhesion. However, although these studies proved that the image processing techniques had the potential to improve the evaluation method for the adhesion, they mainly focused on investigating the efficiency of different anti-stripping additives and did not essentially optimize the current evaluation methods. The issues of the evaluation method for the adhesion in current standards have not been solved.

In addition, the characteristics of different aggregates have significant differences, even for the aggregates with the same specification. It inevitably affects the adhesion evaluation. However, only Shen et al. [9] and Ji et al. [23] investigated the effect of microscopic structures and water-absorption characteristics of aggregates on the adhesion, respectively. Unfortunately, the effects of aggregate shape on adhesion have not been discussed and stipulated in existing studies and standards.

Hence, the objective of this study contains the following two points:
- Establish an evaluation method for the adhesion to reduce the impact of subjective factors considering the usability and popularization;
- Investigate the effect of aggregate shape on the adhesion to further improve the proposed evaluation method.

Based on this, an innovative method based on the Chinese boiling water test and image processing technique is put forward to quantificationally evaluate the adhesion between aggregate and asphalt binder. Moreover, the effects of aggregate shapes on adhesion are also investigated via the proposed method from a view of aspect ratio and homogeneity.

2. Materials

In this study, the #70 base asphalt binder and basalt aggregates are adopted to implement the adhesion experiments. Their technical parameters are listed in Tables 4 and 5, respectively.

Table 4. Technical parameters of asphalt binder.

Penetration at 25 °C (0.1 mm)	Softening Point (°C)	Ductility (cm)		Viscosity at 60 °C (Pa·s)	After the RTFOT		
		10 °C	15 °C		Residual Penetration Ratio at 25 °C (%)	Residual Ductility at 10 °C (cm)	Mass Loss (%)
67.4	47.4	37.2	>100	31.6	79.3	33.8	0.17

Table 5. Technical parameters of aggregates.

Crushed Stone Value (%)	Los Angeles Abrasion Value (%)	Ruggedness (%)	Flat-Elongated Particles Content (%)	<0.075 mm Particle Content (%)	Water Absorption (%)
14.4	16.2	6.2	6.1	0.5	0.81

3. Evaluation Method of the Adhesion Based on the Image Processing Technology

In this study, the evaluation method of the adhesion contains two parts: lab experiment and its result analysis. Considering the usability and popularization, the lab experiment is optimized and implemented based on the Chinese boiling water test. Moreover, according to the existing studies, an image processing technique is adopted to treat the experiment results to evaluate the adhesion.

3.1. Lab Experiments (The Chinese Water Boiling Test)

In the Chinese test standard, the boiling water test is adopted to judge the adhesion level between aggregates and asphalt binders via the peeling degree of asphalt membrane on the surface of one coarse aggregate in the boiling water, as shown in Figure 1.

Figure 1. A sketch of the Chinese boiling water test.

During the test, the factors affecting the test results are boiling time and boiling temperature. The boiling time has been set at 3 min in the Chinese test standard, while the boiling temperature is not specified. It only describes that the water should be under a micro-boiling state but without generating bubbles. However, the micro-boiling state is difficult to judge, which is detrimental to test standardization. Hence, the boiling water tests at different temperatures (85 °C, 90 °C, 92 °C, 94 °C, 96 °C, and 98 °C) are implemented in this study to determine the boiling temperature, as shown in Figure 2. The rate of heating used in this study is 10 °C/min.

Figure 2. Boiling state at different temperatures.

As shown in Figure 2, it can be found that there is no bubble and vapor at 85 °C. When the temperature reaches 90 °C and 92 °C, the vapor can be observed and gradually increases while there are still no bubbles. When the temperature reaches 94 °C, the bubbles begin to generate. Although dissociative asphalt membrane does not appear in the water, many bubbles have been generated on the surface of the aggregate. When the temperature exceeds 96 °C, lots of dissociative asphalt membranes can be observed, accompanied by a large number of bubbles and vapors, which is not permitted in the Chinese boiling water test. Hence, the recommended boiling temperature in this study is 92 ± 1 °C.

3.2. Adhesion Evaluation

In this study, the image processing technique based on the software "Image J 1.51j8" [24–26] is adopted to obtain the peeling area of the asphalt membrane on the surface of the aggregate. The process of the method is as follows.

- Put the aggregate after the boiling water test on a white slab, as shown in Figure 3. Considering the aggregate is underslung during the boiling water test, there is no peeling zone at the bottom surface of the aggregate owing to the fluidity of the asphalt membrane at high temperatures. Hence, the bottom surface should be in contact with the slab. It should be explained that the soft asphalt binder may not conform to the assumption that there is no peeling zone at the bottom of the aggregate owing to its high fluidity at high temperatures. Hence, one limitation of this study is that the proposed method may not be appropriate for the asphalt binder with a low softening point. However, in fact, soft asphalt is rarely adopted in pavement engineering in China because the softening point of the used asphalt must meet minimum standards.

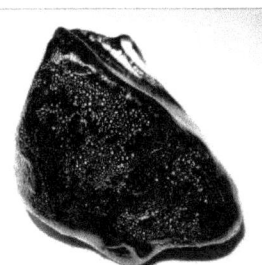

Figure 3. The aggregate after boiling water test on a white slab.

- Apply a stable light source to illuminate the aggregate. In order to reduce the shadow, it is better to apply the light source from three directions. If the condition is limited, the light source should be in a vertical direction.
- Obtain at least three images of the aggregate via camera equipment from different angles. The image should be larger than 96 dpi.
- Owing to the obtained images using traditional camera equipment, they are usually 24-bit depth and are difficult to be treated using the "Image J". Hence, these images should be transformed to 8-bit grayscale depth via "Image J" command "Type/8-bit", as shown in Figure 4. It should be explained that the aggregate is placed on a white paper with a rough texture. The rough texture on the paper can moderate the intensity of light reflection so as to reduce the glare as low as possible. In addition, the color of the area where asphalt peels from aggregate is different from that of the area where the glare occurs. The former is mostly gray and cyan, and the latter is white. They can be distinguished by adjusting the threshold of Image J.
- Eliminate the background of the 8-bit depth image via the "Image J" command "Process/Subtract Background/Rolling ball radius 20% pixels" (Figure 5). The threshold value is very important for the effect of eliminating background. Figure 6 plots the images after eliminating the background using different threshold values.

Figure 4. The transformation of the image from 24-bit depth to 8-bit depth.

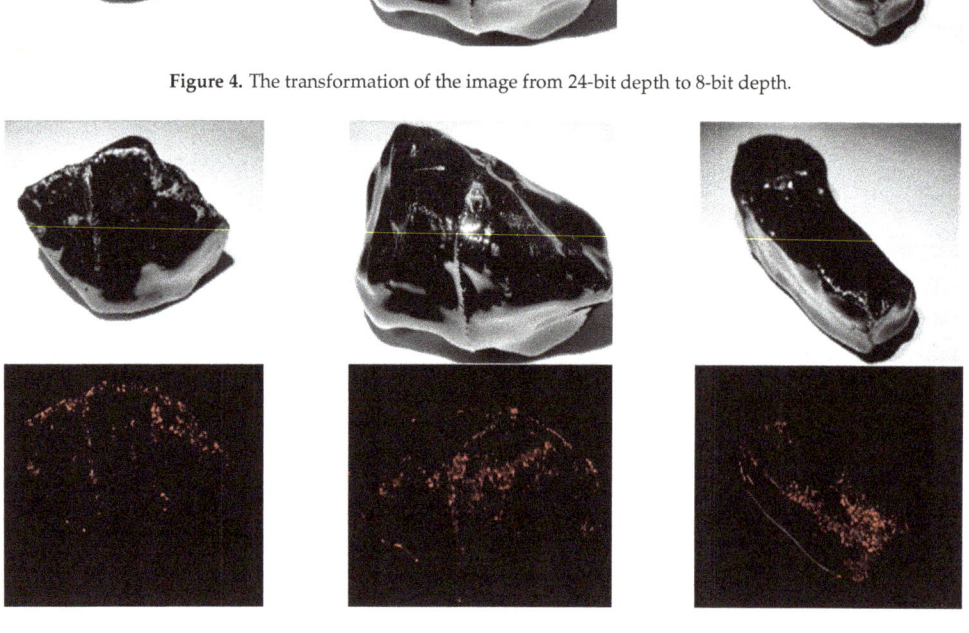

Figure 5. The images after eliminating the background.

In Figure 6, the red zone represents the peeling zone of the asphalt membrane. It can be found that the area of the red zone decreases as the threshold value increases [27–30]. When the threshold value exceeds 30%, the identifiability of the red zone will be reduced largely. However, it does not mean that the smaller the threshold value is, the better. The sensibility of the red zone increases as the threshold value decreases. When the threshold value is lower than 15%, many useless messages are contained in the red zone. Hence, by comparing the area of the red zone and the real peeling area of the asphalt membrane, the optimal threshold value is selected as 20%.

- The aggregate outline can be automatically identified by the "Image J", as shown in Figure 7.

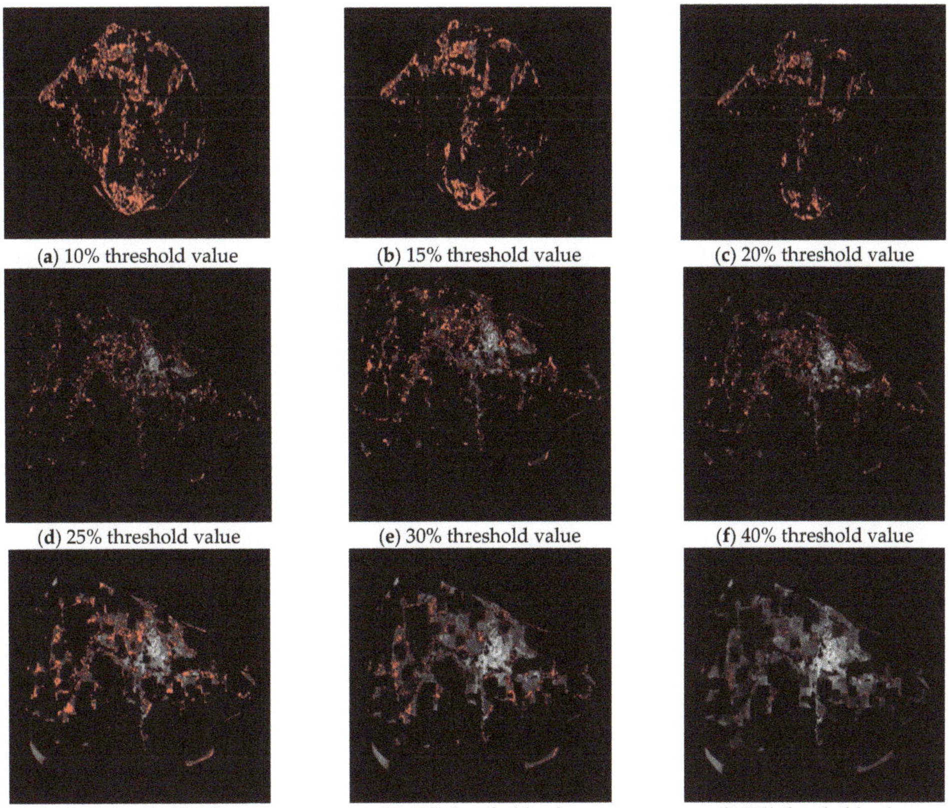

(a) 10% threshold value   **(b)** 15% threshold value   **(c)** 20% threshold value

(d) 25% threshold value   **(e)** 30% threshold value   **(f)** 40% threshold value

(g) 50% threshold value   **(h)** 60% threshold value   **(i)** 70% threshold value

Figure 6. The images after eliminating the background using different threshold values.

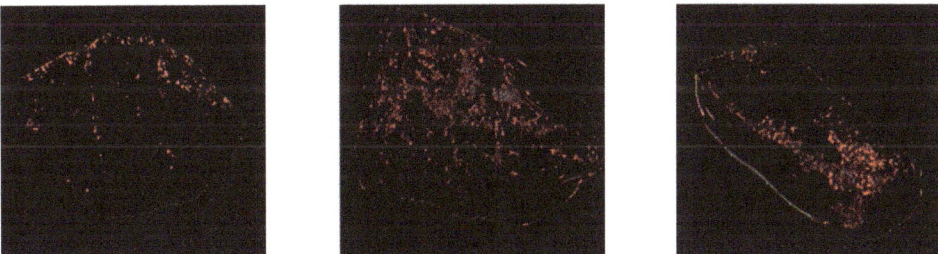

Figure 7. Aggregate outline in the image.

- Finally, the area of the red zone (peeling zone of the asphalt membrane) and the area of the aggregate can be obtained using the "Image J" command "Analyze/Measure" [31–33]. Moreover, the adhesion grade can be quantificationally evaluated according to the peeling area ratio of the aggregate and Table 2. The peeling area ratio RP can be calculated using Equation (1). It should be explained that the above process can be executed in batches.

$$R_P = \frac{A_P}{A_A} \times 100\% \tag{1}$$

where, A_P is the area of the red zone (peeling zone of the asphalt membrane), and A_A is the area of the aggregate.

The proposed method for the adhesion in this study can be illustrated in Figure 8.

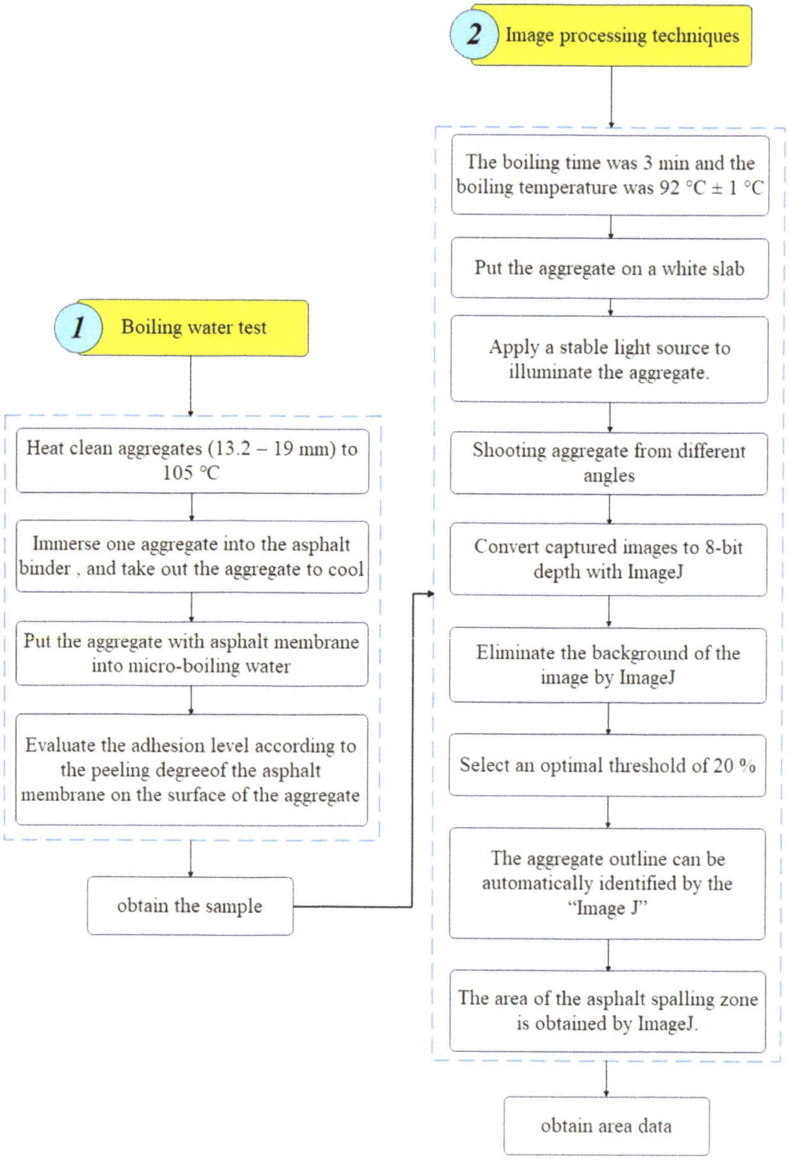

Figure 8. Flow chart of the proposed evaluation method for the adhesion.

4. Effect of Aggregate Characteristics on Adhesion Evaluation

The aggregate shape affects the aggregate–asphalt adhesion; e.g., the asphalt membrane at the sharp zone of the aggregate will be easier to peel during the boiling water test. It also brings some difficulty for the subjective evaluation of the aggregate–asphalt adhesion in traditional methods. Hence, in order to reveal the influence of aggregate shape on the aggregate–asphalt adhesion and explain the feasibility of the proposed method, the aggregates are divided into two categories according to the different homogeneities and aspect ratios. The adhesions of the two types of aggregates are evaluated using the proposed method in this study. The homogeneity and aspect ratio can be characterized using Equations (2) and (3), respectively.

$$H = \frac{\frac{\pi}{6}\left(\frac{l_1+l_2}{2}\right)^3}{V} = \frac{\pi\rho(l_1+l_2)^3}{48m} \tag{2}$$

$$A_t = \frac{l_1}{l_2} \tag{3}$$

where, H is a self-defined index and is no more less than 1, which represents the homogeneity of the target aggregate. It derives from the Zhang's study [34]. The lower the H is, the more quasi-spherical the aggregate is. A_t is the aspect ratio of the target aggregate and is also no more lessthan 1.0. The larger the A_t is, the more elongated and flaky the aggregate is. l_1, l_2, V, m, and ρ are the longest axis, shortest axis, volume, mass, and density of the target aggregate, respectively.

Some examples of the two types of aggregates are listed in Tables 6 and 7. Type A (Table 6) represents the aggregates that have similar A_t and different H. It can be adopted to reveal the effect of aggregate homogeneity on the aggregate–asphalt adhesion in this study. Type B (Table 7) represents the aggregates that have similar H and different A_t. It can be adopted to reveal the effect of aggregate angularity on the aggregate–asphalt adhesion. Tables 6 and 7 also present the appearances and general features of different aggregates.

Table 6. Type A aggregates.

No.	H	Appearance	Feature
A-I	1.0		Surface smooth; approximately spherical.
A-II	1.5		Not obvious angular; polyhedral.
A-III	2.0		Approximate cube.

Table 6. *Cont.*

No.	H	Appearance	Feature
A-IV	2.5		Tetrahedral; obvious angular.
A-V	3.0		Irregular shape.

Table 7. Type B aggregates.

No.	A_t	Appearance	Feature
B-I	1.5		Approximately cylindrical.
B-II	2.0		Obvious edges.

Table 7. *Cont.*

No.	A_t	Appearance	Feature
B-III	2.5		Rectangular shape.
B-IV	3.0		Irregular shape.
B-V	3.5		

4.1. Effect of Homogeneity

Table 8 lists the peeling area ratio and the adhesion grade of the type A aggregates (with different values of H) after boiling water tests. The peeling area ratio is obtained according to the proposed method (Section 3.2) in this study. The adhesion grade is obtained according to the traditional method (Table 2) of the Chinese experiment standard. In order to ensure the reliability of the adhesion grade, it is evaluated by a professional team (two professors and one senior experimentalist).

The effect of the values of H on the peeling area ratio is shown in Figure 9. The peeling area ratio is calculated as the average value, the same value as H.

As shown in Figure 9, it can be found that the peeling area ratio generally increases as the value of H increases, especially when the value of H is lower than 1.5. Then, a ten-percentage point increase for the value of H makes the peeling area ratio 8.1% increase. When the value of H is larger than 1.5, a ten-percentage point increase for the value of H only makes the peeling area ratio 1.9% increase. It shows that the increase in aggregate angularity can intensify the peeling of the asphalt membrane while the effect degree decreases as the value of H increases. When the value of H is lower, the aggregate surface is close to smooth, and the asphalt membrane on the aggregate surface is smooth. After boiling, the peeling of the asphalt membrane is not obvious. When the value of H is higher, the edges and corners of the aggregate are more significant, and the asphalt membrane on the aggregate surface is relatively thicker. After boiling, the asphalt membrane easily flows downward and gathers at the pit of the aggregate. As a result, the edges and corners of the aggregate are exposed during the boiling water test. This will influence the evaluation of aggregate–asphalt adhesion grade. Hence, in order to ensure the stability of adhesion evaluation, it is better to select the aggregate with the obvious angularity during the boiling water tests.

Table 8. Experiment results of type A aggregates.

No.	H	Peeling Area Ratio (%)	Adhesion Grade
A1	1.0	9.45	4
A2	1.0	11.25	3
A3	1.5	8.79	4
A4	1.5	8.93	4
A5	1.2	8.06	4
A6	1.2	9.07	3
A7	1.5	10.73	3
A8	1.5	12.77	3
A9	1.5	9.14	4
A10	1.5	9.38	4
A11	2.0	8.04	4
A12	2.0	12.87	3
A13	1.0	10.36	3
A14	1.0	10.60	3
A15	1.5	12.07	3
A16	1.5	12.23	3
A17	2.0	13.46	3
A18	2.0	12.94	4
A19	2.5	11.99	4
A20	2.5	14.36	3
A21	2.0	14.10	3
A22	2.0	15.89	3
A23	1.5	16.13	3
A24	1.5	18.78	3
A25	3.0	18.42	3
A26	3.0	15.58	3
A27	2.5	14.15	3
A28	2.5	18.45	3
A29	2.0	18.54	3
A30	2.0	19.50	3
A31	3.0	15.07	3
A32	3.0	17.27	3
A33	2.0	20.86	2
A34	2.0	18.30	3
A35	2.5	19.38	3
A36	2.5	18.29	3
A37	1.5	18.55	2
A38	1.5	19.28	3
A39	2.0	17.21	3
A40	2.0	18.33	3
A41	1.5	17.32	3
A42	1.5	18.53	3
A43	2.0	19.08	2
A44	2.0	15.23	3
A45	1.5	17.29	3
A46	1.5	17.96	3
A47	1.5	14.82	3
A48	1.5	15.66	3
A49	2.0	15.42	3
A50	2.0	19.68	3
A51	1.5	15.60	3
A52	1.5	15.10	3

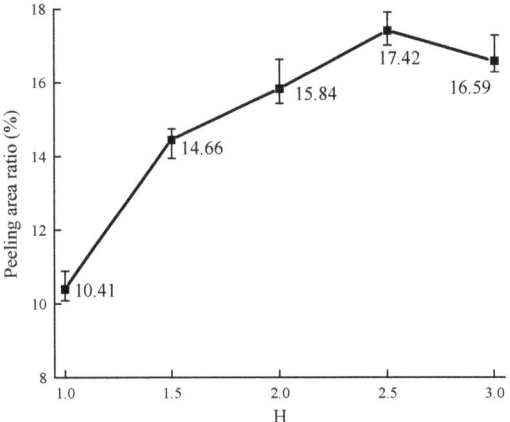

Figure 9. Effect of the value of H on the peeling area ratio.

In addition, the peeling area ratio slightly decreases when the value of H reaches 3.0. It may be due to the experiment errors resulting from a lack of the number of aggregate samples (the shape of aggregates is not common, and the aggregates are not easy to search when the value of H exceeds 3.0).

4.2. Effect of Aspect Ratio

Table 9 lists the peeling area ratio and the adhesion grade of the type B aggregates (with different values of A_t) after boiling water tests.

As shown in Figure 10, it can be found that the peeling area ratio generally increases as the value of A_t increases. The peeling of the asphalt membrane at both ends of the aggregate is more serious. However, the effect of the value of A_t on the peeling area ratio is lower than that of the value of H. A ten-percentage point increase for the value of A_t on average makes the peeling area ratio 3.4% increased, while on average, 4.5% increased for the value of H. It shows that the effect of aggregate flatness on the aggregate–asphalt adhesion is weaker than that of aggregate angularity. Similarly, with Figure 9, the peeling area ratio slightly decreases when the value of A_t reaches 4.0. It also may be due to the experiment errors resulting from a lack of the number of aggregate samples. The shape of aggregates is not common when the value of A_t exceeds 4.0.

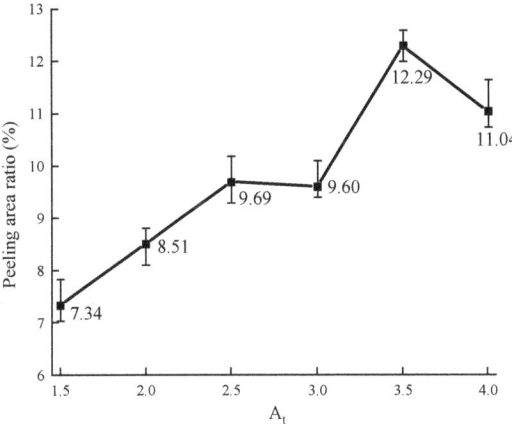

Figure 10. Effect of the value of A_t on the peeling area ratio.

Table 9. Experiment results of type B aggregates.

No.	A_t	Peeling Area Ratio (%)	Adhesion Grade
B1	2.5	8.38	3
B2	2.5	7.17	4
B3	3.0	9.94	3
B4	3.0	7.71	4
B5	3.0	9.05	3
B6	3.0	11.68	3
B7	4.0	10.16	4
B8	4.0	11.91	3
B9	2.5	12.44	3
B10	2.5	10.75	4
B11	2.0	12.07	3
B12	2.0	10.75	3
B13	3.5	11.47	4
B14	3.5	12.44	3
B15	1.5	11.98	4
B16	1.5	10.70	3
B17	2.0	13.80	3
B18	2.0	13.04	3
B18	3.5	11.85	4
B20	3.5	13.40	3

The effect of the values of A_t on the peeling area ratio is shown in Figure 10.

In addition, as shown in Tables 8 and 9 (No. A6, A18, A19, A37, A43, B1, B3, B5, B7, B10, B13, B15, and B18), some adhesion grades obtained by the professional team are different from the results of peeling area ratio obtained by the proposed method of this study. The proposed method is beneficial to evaluate the peeling status of asphalt membrane to accurately judge the adhesion grade.

5. Conclusions

In this study, an innovative method based on the Chinese boiling water test and image processing technique is put forward to quantificationally investigate the peeling status of the asphalt membrane. Moreover, two indexes are proposed to characterize the aggregate features (homogeneity and aspect ratio). The effects of aggregate features on the aggregate–asphalt adhesion are analyzed via the proposed method and conventional evaluation method. The proposed method is beneficial for evaluating the peeling status of asphalt membranes to accurately judge the adhesion grade. However, one disadvantage of the proposed method is the high requirement for light sources.

The peeling of the asphalt membrane on the aggregate surface is more serious as the complexity of the aggregate shape increases after the boiling water tests, while the effect degree gradually decreases. When the value of H is lower than 1.5, a ten-percentage point increase raises the peeling area ratio by 8.1%. When the value of H is larger than 1.5, it only raises the peeling area ratio by 1.9%. In order to ensure the stability of adhesion evaluation, it is better to select the aggregate with the obvious angularity during the boiling water tests. In addition, the effect of aspect ratio on the peeling status of asphalt membrane is lower than that of aggregate homogeneity. A ten-percentage point increase for the value of A_t makes the peeling area ratio 3.4% increased, while 4.5% increased for the value of H.

However, the adhesion can be influenced by the chemical interaction between aggregate and asphalt. The chemical interaction depends on asphalt type (e.g., base asphalt with different grades, modified asphalt, etc.) and aggregate lithology (e.g., limestone, basalt, granite, etc.). The effects of these factors on the adhesion will be systematically carried out in our future studies.

Author Contributions: Conceptualization, M.L. and J.R.; methodology, J.R.; validation, J.R.; formal analysis, J.R.; investigation, J.W., Z.G. and Z.Z.; resources, Z.Z.; data curation, M.L., J.W., Z.G. and J.C.; writing—original draft preparation, M.L., J.W., Z.G., J.C., Z.Z. and J.R.; writing—review and editing, J.R.; visualization, J.R.; supervision, J.R.; project administration, J.R.; funding acquisition, J.R. All authors have read and agreed to the published version of the manuscript.

Funding: This study is sponsored in part by the National Natural Science Foundation of China under grant 51808326, for which the authors are very grateful.

Institutional Review Board Statement: Not applicable.

Informed Consent Statement: Not applicable.

Data Availability Statement: The data presented in this study are available in the article.

Conflicts of Interest: The authors declare no conflict of interest.

References

1. Ren, J.; Zhang, X.; Peng, C.; Wang, Y.; Wang, Y.; Zhao, H.; Xu, X.; Xia, L.; Wang, C.; Li, G.; et al. Short-term aging characteristics and mechanisms of SBS-modified bio-asphalt binder considering time-dependent effect. *Constr. Build. Mater.* **2022**, *352*, 129048. [CrossRef]
2. Al-Kafaji, M.; Al-Busaltan, S.; Ewadh, H.A. Evaluating Water Damage in Acrylic Polymer-Modified Cold Bituminous Emulsion. *J. Mater. Civil. Eng.* **2021**, *12*, 04021337. [CrossRef]
3. Wu, H.; Li, P.; Nian, T.; Zhang, G.; He, T.; Wei, X. Evaluation of asphalt and asphalt mixtures' water stability method under multiple freeze-thaw cycles. *Constr. Build. Mater.* **2019**, *228*, 117089. [CrossRef]
4. JTG E20-2011; Standard Test Methods of Bitumen and Bituminous Mixtures for Highway Engineering. Ministry of Transport of the People's Republic of China: Beijing, China, 2011.
5. D3625; Standard Practice for Effect of Water on Bituminous-Coated Aggregate Using Boiling Water. American Society for Testing and Materials: Conshohocken, PA, USA, 2005.
6. Hefer, A.W.; Bhasin, A.; Little, D.N. Bitumen Surface Energy Characterization Using a Contact Angle Approach. *J. Mater. Civ. Eng.* **2006**, *18*, 759–767. [CrossRef]
7. Bhasin, A. Development of Methods to Quantify Bitumen-Aggregate Adhesion and Loss of Adhesion due to Water. Ph.D. Thesis, Texas A&M University, College Station, TX, USA, 2006.
8. Hamzah, M.O.; Kakar, M.R.; Quadri, S.A.; Valentin, J. Quantification of moisture sensitivity of warm mix asphalt using image analysis technique. *J. Clean. Prod.* **2014**, *68*, 200–208. [CrossRef]
9. Shen, A.; Zhai, C.; Guo, Y.; Yang, X. Mechanism of adhesion property between steel slag aggregate and rubber asphalt. *J. Adhes. Sci. Technol.* **2018**, *32*, 2727–2740. [CrossRef]
10. Liu, Z.; Huang, X.; Sha, A.; Wang, H.; Chen, J.; Li, C. Improvement of Asphalt-Aggregate Adhesion Using Plant Ash Byproduct. *Materials* **2019**, *12*, 605. [CrossRef]
11. Ingrassia, L.P.; Cardone, F.; Canestrari, F.; Lu, X. Experimental investigation on the bond strength between sustainable road bio-binders and aggregate substrates. *Mater. Struct.* **2019**, *52*, 80. [CrossRef]
12. D'angelo, S.; Ferrotti, G.; Cardone, F.; Canestrari, F. Asphalt Binder Modification with Plastomeric Compounds Containing Recycled Plastics and Graphene. *Materials* **2022**, *15*, 516. [CrossRef]
13. Liu, W.; Li, H.; Zhu, H.; Xu, P. The Interfacial Adhesion Performance and Mechanism of a Modified Asphalt-Steel Slag Aggregate. *Materials* **2020**, *13*, 1180. [CrossRef]
14. Ji, J.; Yao, H.; Liu, L.; Suo, Z.; Zhai, P.; Yang, X.; You, Z. Adhesion Evaluation of Asphalt-Aggregate Interface Using Surface Free Energy Method. *Appl. Sci.* **2017**, *7*, 156. [CrossRef]
15. Yadykova, A.Y.; Ilyin, S.O. Rheological and adhesive properties of nanocomposite bitumen binders based on hydrophilic or hydrophobic silica and modified with bio-oil. *Constr. Build. Mater.* **2022**, *342*, 127946. [CrossRef]
16. Yadykova, A.Y.; Ilyin, S.O. Bitumen improvement with bio-oil and natural or organomodified montmorillonite: Structure, rheology, and adhesion of composite asphalt binders. *Constr. Build. Mater.* **2022**, *364*, 129919. [CrossRef]
17. Zhao, H. A reduced order model based on machine learning for numerical analysis: An application to geomechanics. *Eng. Appl. Artif. Intell.* **2021**, *100*, 104194. [CrossRef]
18. Ren, J.; Zhao, H.; Zhang, L.; Zhao, Z.; Xu, Y.; Cheng, Y.; Wang, M.; Chen, J.; Wang, J. Design optimization of cement grouting material based on adaptive boosting algorithm and simplicial homology global optimization. *J. Build. Eng.* **2022**, *49*, 104049. [CrossRef]
19. Ren, J.; Zhang, L.; Zhao, H.; Zhao, Z.; Wang, S. Determination of the fatigue equation for the cement-stabilized cold recycled mixtures with road construction waste materials based on data-driven. *Int. J. Fatigue* **2022**, *158*, 106765. [CrossRef]
20. Zhao, Z.; Wang, S.; Ren, J.; Wang, Y.; Wang, C. Fatigue characteristics and prediction of cement-stabilized cold recycled mixture with road-milling materials considering recycled aggregate composition. *Constr. Build. Mater.* **2021**, *301*, 124122. [CrossRef]

21. Park, D.-W.; Seo, W.-J.; Kim, J.; Vo, H.V. Evaluation of moisture susceptibility of asphalt mixture using liquid anti-stripping agents. *Constr. Build. Mater.* **2017**, *144*, 399–405. [CrossRef]
22. Nazirizad, M.; Kavussi, A.; Abdi, A. Evaluation of the effects of anti-stripping agents on the performance of asphalt mixtures. *Constr. Build. Mater.* **2015**, *84*, 348–353. [CrossRef]
23. Ji, J.; Dong, Y.; Zhang, R.; Suo, Z.; Guo, C.; Yang, X.; You, Z. Effect of Water Absorption and Loss Characteristics of Fine Aggregates on Aggregate-Asphalt Adhesion. *KSCE J. Civ. Eng.* **2021**, *25*, 2020–2035. [CrossRef]
24. Van de Linde, S. Single-molecule localization microscopy analysis with ImageJ. *J. Phys. D-Appl. Phys.* **2019**, *52*, 203002. [CrossRef]
25. Gao, L. QSIM: Quantitative structured illumination microscopy image processing in ImageJ. *Biomed. Eng. Online* **2015**, *14*, 4. [CrossRef] [PubMed]
26. Tajima, R.; Kato, Y. Comparison of threshold algorithms for automatic image processing of rice roots using freeware ImageJ. *Field Crop. Res.* **2011**, *121*, 460–463. [CrossRef]
27. Alam, J.; Shaheen, A.; Anwar, M.S. Accessing select properties of the electron with ImageJ: An open-source image-processing paradigm. *Eur. J. Phys.* **2014**, *35*, 15011. [CrossRef]
28. Fontenete, S.; Carvalho, D.; Lourenço, A.; Guimarães, N.; Madureira, P.; Figueiredo, C.; Azevedo, N.F. FISHji: New ImageJ macros for the quantification of fluorescence in epifluorescence images. *Biochem. Eng. J.* **2016**, *112*, 61–69. [CrossRef]
29. Schroeder, A.B.; Dobson, E.T.A.; Rueden, C.T.; Tomancak, P.; Jug, F.; Eliceiri, K.W. The ImageJ ecosystem: Open-source software for image visualization, processing, and analysis. *Protein Sci.* **2021**, *30*, 234–249. [CrossRef]
30. Rueden, C.T.; Schindelin, J.; Hiner, M.C.; Dezonia, B.E.; Walter, A.E.; Arena, E.T.; Eliceiri, K.W. ImageJ2: ImageJ for the next generation of scientific image data. *BMC Bioinform.* **2017**, *18*, 529. [CrossRef]
31. Aragón-Sánchez, J.; Quintana-Marrero, Y.; Aragón-Hernández, C.; Hernández-Herero, M.J. ImageJ: A Free, Easy, and Reliable Method to Measure Leg Ulcers Using Digital Pictures. *Int. J. Low. Extremity Wounds* **2017**, *16*, 269–273. [CrossRef]
32. Lam, J.; Katti, P.; Biete, M.; Mungai, M.; AshShareef, S.; Neikirk, K.; Lopez, E.G.; Vue, Z.; Christensen, T.A.; Beasley, H.K.; et al. A Universal Approach to Analyzing Transmission Electron Microscopy with ImageJ. *Cells* **2021**, *10*, 2177. [CrossRef]
33. Dos Santos, S.M.; Klinkhardt, U.; Schneppenheim, R.; Harder, S. Using imageJ for the quantitative analysis of flow-based adhesion assays in real-time under physiologic flow conditions. *Platelets* **2010**, *21*, 60–66. [CrossRef]
34. Zhang, D.; Huang, X.; Zhao, Y. Numerical Study on the Effect of Coarse-Aggregate Morphology on Shear Performance. *J. Test. Evaluation* **2015**, *43*, 20130067. [CrossRef]

Disclaimer/Publisher's Note: The statements, opinions and data contained in all publications are solely those of the individual author(s) and contributor(s) and not of MDPI and/or the editor(s). MDPI and/or the editor(s) disclaim responsibility for any injury to people or property resulting from any ideas, methods, instructions or products referred to in the content.

Article

Influence of Morphological Characteristics of Coarse Aggregates on Skid Resistance of Asphalt Pavement

Yuanshuai Dong [1,2,3], Zihao Wang [4], Wanyan Ren [4,5,*], Tianhao Jiang [6], Yun Hou [1,2,3] and Yanhong Zhang [1,2,3]

1. China Highway Engineering Consulting Group Co., Ltd., No. 17 Changyungong Road, Haidian District, Beijing 100089, China
2. Research and Development Center on Highway Pavement Maintenance Technology, CCCC, No. 116 Zizhuyuan Road, Haidian District, Beijing 100097, China
3. Research and Development Center of Transport Industry of Technologies, Materials and Equipments of Highway Construction and Maintenance, No. 116 Zizhuyuan Road, Haidian District, Beijing 100097, China
4. School of Civil and Transportation Engineering, Beijing University of Civil Engineering and Architecture, No. 1 Zhanlanguan Road, Xicheng District, Beijing 100044, China
5. Collaborative Innovation Center of Energy Conservation & Emission Reduction and Sustainable Urban-Rural Development in Beijing, No. 1 Zhanlanguan Road, Xicheng District, Beijing 100044, China
6. China Civil Engineering Construction Corporation, No. 4 Beifengwo Road, Haidian District, Beijing 100038, China
* Correspondence: renwanyan@bucea.edu.cn

Abstract: This research aims to improve the durability of skid resistance of asphalt pavement from the perspective of coarse aggregates based on on-site investigation. Firstly, the skid resistance of six representative actual roads was tested during two years by employing the Dynamic Friction Tester and the attenuation characteristics of skid resistance of different types of asphalt pavements were analyzed. Secondly, core samples were drilled onsite and coarse aggregates were extracted from the surface layer of the core samples. The morphological parameters of coarse aggregates were collected by a "backlighting photography" system and three-dimensional profilometer, and the variation rules of angularity and micro-texture of coarse aggregates were investigated. Finally, the correlation between the morphological characteristics of coarse aggregates and the pavement skid resistance was established based on the grey correlation entropy. The research results show that with the increase in service time, the attenuation rate of skid resistance of asphalt pavement gradually slows down; the angularity of coarse aggregates gradually decreases, and the micro-texture on the wearing surface gradually wears away. The grey correlation entropy between all the micro-texture indexes of coarse aggregates and dynamic friction coefficient, as well as between the roundness and skid resistance is more than 0.7, whereas the correlation between other evaluation indicators and the dynamic friction coefficient is poor, indicating that compared with the angularity of coarse aggregates, the micro-texture affects the skid resistance of actual asphalt pavement more greatly. In engineering applications, the use of coarse gradation, coarse aggregates with high roughness or high anti-wear performance can slow down the attenuation of pavement skid resistance, so that the pavement can maintain superior long-term anti-skidding performance.

Keywords: actual pavement; skid resistance; coarse aggregate; morphological characteristics; attenuation characteristics; grey correlation entropy analysis

Citation: Dong, Y.; Wang, Z.; Ren, W.; Jiang, T.; Hou, Y.; Zhang, Y. Influence of Morphological Characteristics of Coarse Aggregates on Skid Resistance of Asphalt Pavement. *Materials* 2023, 16, 4926. https://doi.org/10.3390/ma16144926

Academic Editor: Gilda Ferrotti

Received: 8 June 2023
Revised: 5 July 2023
Accepted: 7 July 2023
Published: 10 July 2023

Copyright: © 2023 by the authors. Licensee MDPI, Basel, Switzerland. This article is an open access article distributed under the terms and conditions of the Creative Commons Attribution (CC BY) license (https://creativecommons.org/licenses/by/4.0/).

1. Introduction

Traffic safety issues have always been a common challenge faced by countries around the world, and insufficient skid resistance of pavement is one of the main factors leading to frequent road traffic accidents [1–4]. In essence, excellent pavement skid resistance is ensured by good and stable macro- and micro-textures on its surface layer. The surface texture with a wavelength range of 0.5–50 mm is referred to as macro-texture, while the wavelength range of 0–0.5 mm is referred to as micro-texture. They affect the skid resistance of the vehicle when driving at high and low speeds, respectively, and sufficient friction cannot be generated by the contact between the tire and pavement in the case of too scarce pavement surface texture [5,6]. For this reason, in recent years, many scholars at home and abroad have carried out extensive research on the surface texture and anti-skidding performance of asphalt pavement. For instance, the skid resistance of pavement, especially in wet road conditions on rainy days, is closely related to macro-texture which provides a drainage system for water on the pavement surface, thus preventing a buildup of water between the tire and the pavement and resultant hydroplaning [7–9]. In the meantime, researchers believe that the type, apparent morphology and mineral composition of aggregates in the surface layer of pavement play an important role in the formation of surface texture [5,10–13]. Based on the skid resistance measurement results in the laboratory, researchers have pointed out that the anti-skidding performance of asphalt pavement depended on the effective friction and texture depth of the pavement [14–17]. Besides, using steel slag and other materials with high wear resistance as the aggregates to prepare asphalt mixtures or construct asphalt pavement can also improve the skid resistance [18–20]. By conducting laboratory accelerated loading tests on asphalt mixture specimens prepared with different aggregates, the results revealed that aggregates such as basalt and granite could provide excellent long-term pavement skid resistance for pavement [21–23]. In addition, different levels of abrasion tests on coarse aggregates have been carried out in the laboratory, and the correlation between their surface texture and the skid resistance has been analyzed [24,25]. The research results showed that the surface texture features of aggregates have a significant impact on long-term pavement skid resistance [26]. Moreover, the mineral composition, wear resistance and surface texture of various aggregates have been investigated, and the results demonstrate that the mineral hardness of aggregates has a great impact on their wear resistance [11,27–31]. Furthermore, by conducting accelerated loading tests on asphalt mixtures prepared with different aggregates, the results revealed that improving the wear resistance of aggregates could help reduce the attenuation rate of the skid resistance of asphalt pavement.

In summary, although many scholars at home and abroad have conducted in-depth research on the skid resistance of asphalt pavement, most of them are carried out in the laboratory, and there is a lack of studies based on the skid resistance of actual pavements, especially for morphological features of coarse aggregates. In this case, the environment, as a key factor affecting the skid resistance deterioration, has not been fully considered. At the same time, the wear effect of vehicle loads on pavement surfaces is different from that of laboratory-accelerated loading equipment. During the construction and operation stages of asphalt pavement, the skid resistance varies significantly among different actual asphalt pavements due to differences in pavement material properties and construction conditions, as well as factors such as changes in the operating environment. Therefore, this study aims to explore the attenuation law of skid resistance of different types of asphalt pavements through long-term tracking of skid resistance of actual roads under real service conditions, and to analyze the correlation between morphological features of coarse aggregates in the surface layer and skid resistance of actual pavements, thus contributing to the evaluation of aggregate morphological features and improvement in skid resistance.

2. Research Methods

2.1. Test Roads

The test roads selected are all located in Beijing, including Shidan (SD) Road, Shuinan (SN) Road, Luanchi (LC) Road, Yangyan (YY) Road, Xiyuan (XY) Road and Shunping (SP) Side Road. The basic information of the test roads is shown in Table 1, and the mixture composition information is shown in Table 2. Among the six roads, Shidan road is a newly constructed road while all the other five roads are put into service after major repair or upgrading projects and they have been milled and repaved with surface layers. Limestone is used in most of the roads in Beijing, so only one road with basalt was selected.

Table 1. Basic information of test roads.

Road Name	SD	SN	LC	YY	XY	SP
Road type	New road	Major repair	Major repair	Upgrading	Major repair	Upgrading
Road class	Trunk	Secondary	Secondary	Trunk	Secondary	Trunk
Sample length/km	13.00	2.00	9.70	1.65	5.00	6.32
Number of lanes	6	4	2	4	2	8
Opening time	August 2019	Octobr 2019	August 2019	August 2019	Octobr 2019	August 2019
Annual average daily traffic volume/vehicle	27,389	8500	495	6350	5630	11,663

Table 2. Type and design information of asphalt mixture of test road sections.

Road Name	SD	SN	LC	YY	XY	SP
Mixture type in the surface layer	UTWC-10	AC-16C	WAC-16C	AC-13	RAC-13	RAC-13C
Asphalt type	Modified	70#	90#	70#	70#	70#
Coarse aggregate type	Basalt	Limestone	Limestone	Limestone	Limestone	Limestone
Nominal maximum aggregate size/mm	9.5	16	16	13.2	13.2	13.2
Coarse aggregate ratio/%	70	54.5	55.2	55.8	49.7	47.3
Coarse aggregate crushing value/%	13.2	15.1	13.3	16.3	18.6	17.5
Coarse aggregate Los Angeles wear value/%	14.3	18.5	14.1	14.1	18.2	16.4
Asphalt-aggregate ratio/%	4.8	4.6	4.5	4.8	4.8	4.8
Voids/%	13.8	4.8	5.4	4.5	4.6	4.3

Notes: Similar abbreviations use the same naming method. RAC-13 denotes Recycled Asphalt Concrete with a nominal maximum aggregate size of 13.2 mm; AC denotes Asphalt Concrete; WAC denotes Warm Asphalt Concrete; UTWC denotes Ultra-Thin Wearing Course.

2.2. Measurement Time and Positions

The selected six roads were basically completed in August 2019. Their pavement skid resistance was tracked during the two years from opening to traffic to September 2021. During the two years, the pavement skid resistance was tested four times, November 2019, September 2020, December 2020 and September 2021, respectively, and set as stages I, II, III and IV. According to the Chinese Specification JTG 3450-2019 [32], the measurement and sampling positions were determined according to the uniform method in T0902-2019, and the measurement and core-drilling positions for the subsequent test stage the latter three measurement stages were kept to approach the initial measuring position in stage I.

2.3. Measurement Methods

2.3.1. Measurement of Pavement Skid Resistance

The pavement skid resistance was measured by the Dynamic Friction Tester (DFT) (as shown in Figure 1) according to T0968-2008 in the Chinese Specification JTG 3450-2019 [32]. The Dynamic Friction (DF) can be measured by the DFT to characterize the skid resistance of vehicles at various speeds within the range of 0~80 km/h in steps of 1 km/h. For actual roads, the pavement skid resistance when vehicles are running at a high speed is usually characterized by DF_{60}, namely the pavement DF coefficient at 60 km/h. Hence, DF_{60} was employed as the representative value of pavement skid resistance.

2.3.2. Acquisition Method of Coarse Aggregates

Using a pavement core-drilling machine, as shown in Figure 2, in-situ core-drilling was carried out on the measuring positions where the on-site pavement skid resistance was conducted at different measurement times. The coarse aggregates in core samples were then extracted according to T 0735-2011 in Chinese Specification JTG E20-2011 [33]. The extraction process of coarse aggregates is shown in Figure 2.

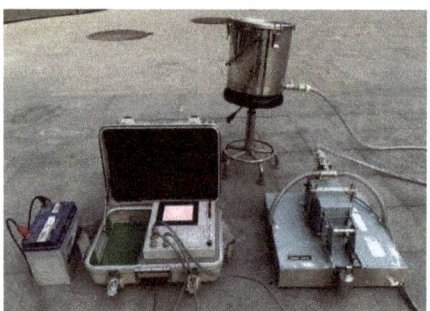

Figure 1. Dynamic Friction Tester.

Figure 2. Extraction process of coarse aggregates in the core samples from the asphalt pavement. (**a**) In-situ core-drilling; (**b**) Drilled core sample; (**c**) Coarse aggregates extracted in core sample.

Due to the fact that the tires only directly contact with the surface layer of the pavement, the aggregates in other layers are almost unaffected by the wheel wear. Therefore, the exposed coarse aggregates on the surface layer of core samples were taken as test samples that would be subjected to wear caused by vehicle loads, while the coarse aggregates 3~5 cm below the surface layer were collected as the control groups, that is, the aggregates were collected in the original state without being subjected to the vehicle load wear after the pavement was completed and opened to traffic. Coarse aggregates in the control group were obtained from core samples drilled in November 2019.

2.3.3. Measurement of Coarse Aggregate Angularity

(a) Digital image acquisition

The images of coarse aggregates were acquired using a digital camera with 24 million pixels and a self-made "backlighting photography" system. The purpose of adopting the "backlighting photography" is to eliminate the influence of external light sources and obtain high-contrast coarse aggregate images through an internal single light source, so that the profile of coarse aggregates can be accurately identified. The "backlighting photography" image acquisition system is shown in Figure 3.

(b) Digital image processing

Image-Pro Plus 6.0 (IPP 6.0) software was utilized to process the collected coarse aggregate images. The processing flow was divided into image gray processing, image denoising and image binarization (as shown in Figure 4). Besides, the area, perimeter, major axis length and minor axis length of each aggregate were obtained through the statistical system of IPP. The image processing process is shown in Figure 4.

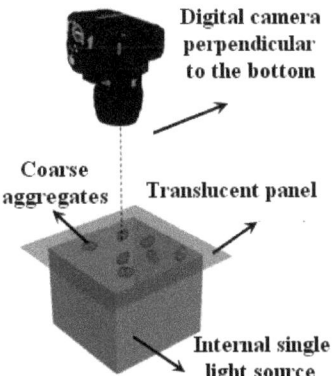

Figure 3. "Backlighting photography" image acquisition system.

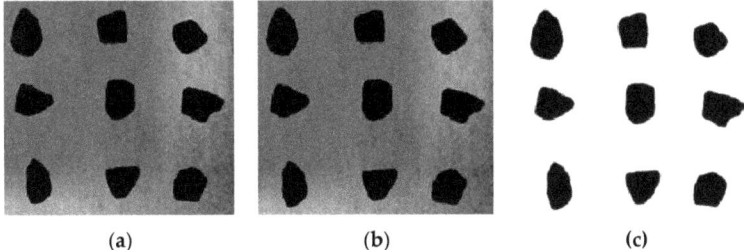

Figure 4. Image processing steps. (**a**) Gray processing; (**b**) Denoising; (**c**) Binarization.

(c) Angularity evaluation indicators

Relevant research has revealed that convexity and axiality can directly characterize the angularity of aggregates, while roundness can indirectly express the angularity of aggregates by characterizing the degree to which the profile curve of the aggregate surface is close to a circle [34]. Therefore, roundness, convexity and axiality were selected as angularity evaluation indicators based on the calculated parameters using IPP 6.0 software [35]. They were calculated according to Equations (1)–(3), respectively.

$$R = \frac{4\pi A}{P_A^2} \quad (1)$$

where R denotes roundness; A denotes the area of aggregate images; P_A is the perimeter of aggregate images and π denotes the Pi.

$$P = \frac{P_{convex}}{P_{ellipse}} \quad (2)$$

where P denotes convexity; P_{convex} denotes the perimeter of the protrusion and $P_{ellipse}$ denotes the perimeter of the equivalent ellipse.

$$AS = \frac{L_{maj}}{L_{min}} \quad (3)$$

where AS denotes the axiality; L_{maj} denotes the length of the major axis of the equivalent ellipse and L_{min} denotes the length of the minor axis of the equivalent ellipse.

2.3.4. Measurement of Coarse Aggregate Micro-Texture

(a) Measurement device

The Contour GT-X white light interferometer (as shown in Figure 5) was employed to measure the three-dimensional morphology of coarse aggregates. The interferometer can measure the optical path difference according to the interference principle, based on which relevant physical parameters can be obtained. It is characterized by the fast acquisition of information and high accuracy in morphological information measurement.

Figure 5. Three-dimensional topography profiler.

(b) Micro-texture measurement method

Figure 6 is a schematic diagram of the coarse aggregate wear position. The position subjected to wear caused by wheel loads is different due to the differences in the position and shape of aggregates in asphalt mixtures. By observing the surface texture of coarse aggregates after burning and washing, it was found that the worn surface of coarse aggregates subjected to wheel loads for a long time was smooth and dull, which could be clearly distinguished by the naked eye. Therefore, such kind of positions were selected as measurement surfaces, and combined with the measurement requirements of the three-dimensional topography profiler, coarse aggregates with a large wear area and flat wear position were selected from different sizes of aggregates obtained from drilled samples of each test road to measure the aggregate micro-texture.

Figure 6. Schematic diagram of coarse aggregate wear position.

(c) Micro-texture evaluation indicators

Research has shown that the maximum profile peak height (R_p), the maximum profile valley depth (R_v) and the total height of the profile (R_t) reflect the height and depth of the profile, while the arithmetic mean deviation (R_a) and the root mean square deviation of the profile (R_q) reflect the dispersion degree of variations in the profile. These five evaluation indicators can effectively characterize the micro-texture of coarse aggregate surfaces [36–39]. Hence, the five indicators, including the two indicators calculated by Equations (4) and (5) and the three indicators shown in Figure 7, were chosen as the micro-texture evaluation indicators of coarse aggregates. In Figure 7, R_p is the maximum profile peak height within a sampling length; R_v is the maximum profile valley depth within a sampling length and R_t is the total height of the profile, namely the sum of R_p and R_v within an evaluation length.

$$R_a = \frac{1}{l}\int_0^1 |Z(x)|dx \tag{4}$$

where R_a is the arithmetic mean deviation; l is the sampling length and $Z(x)$ is the corresponding ordinate value at the x coordinate.

$$R_q = \sqrt{\frac{1}{l}\int_0^1 Z^2(x)dx} \tag{5}$$

where R_q is the root mean square deviation of the profile.

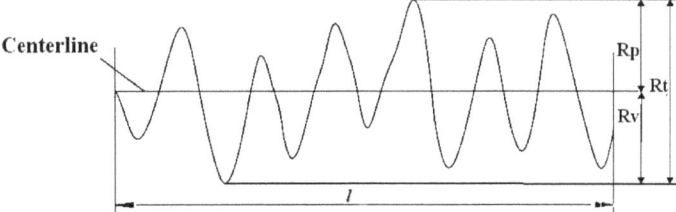

Figure 7. Schematic diagram of the calculation principle of R_p, R_v and R_t.

2.4. Grey Correlation Entropy Analysis Method

The grey correlation entropy analysis is an improved algorithm based on the grey correlation system theory, which can quantitatively describe and compare the impact degree and contribution of various factors in the system through the correlation degree. In this research, the grey correlation entropy analysis method was employed to analyze the impact of various morphological characteristics of coarse aggregates on the anti-skidding performance of the pavement. The specific calculation steps are as follows [40]:

- Determination of the reference sequence and comparison sequence

The reference sequence X_0 is a data sequence that can reflect the characteristics of the target system, while the comparison sequence X_i is a data sequence composed of factors that affect the system's behavior. In this research, the skid resistance was regarded as a reference sequence and the results of various indicators of morphological characteristics of coarse aggregates were considered as comparison sequence.

- Dimensionless processing of raw data

To improve the accuracy of the analysis and simplify the calculations, the averaging method was employed to achieve dimensionless transformation of various data sequences.

- Calculation of grey correlation coefficient

The calculation of grey correlation coefficient between each reference sequence and comparison sequence were calculated based on their absolute value, the maximum and minimum value among the absolute values of all the reference sequence and comparison sequence.

- Calculation of distribution density of grey entropy correlation coefficient

The distribution density of the grey entropy correlation coefficient was the ratio of a grey correlation coefficient to the sum of the grey correlation coefficient in the same sequence.

- Calculation of grey entropy correlation degree

The grey entropy correlation degree was calculated based on the distribution density of grey entropy correlation coefficient calculated above.

3. Results and Discussion

3.1. Attenuation Law of Pavement Skid Resistance

Figure 8 displays the changes of DF_{60} of six roads with the service time. For each measurement position, at least six repeated tests were conducted. In order to analyze the impact of different types of asphalt mixtures on pavement skid resistance, the six roads were divided into UTWC, AC-16 and AC-13 pavement according to their mixture types.

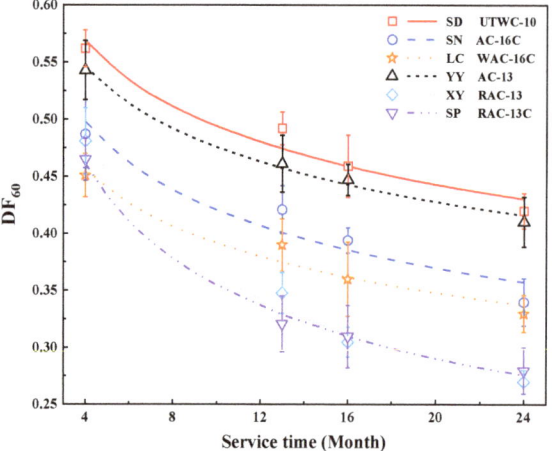

Figure 8. The changes of DF_{60} with service time.

As can be seen from Figure 8, compared with the initial pavement skid resistance, the decline rate of skid resistance of each test road after two years of service varied between 24% and 43%. This is due to differences in pavement materials, mixture design, construction techniques and traffic volume of each test road.

Figure 8 also shows that the pavement skid resistance of SD Road maintained a high level throughout the whole service time. This is mainly because the coarse aggregates used in SD Road are basalt, which is different from that (limestone) used in other test roads. Compared with limestone, basalt has higher roughness and wear resistance [10,23], enabling SD Road to still exhibit excellent pavement skid resistance after two years of service, even under the condition of withstanding several times the traffic volume of other test roads. Meanwhile, due to the fact that the initial skid resistance of the newly constructed asphalt pavement after operation is mainly affected by the asphalt film on the pavement surface, the modified asphalt used in SD Road has higher adhesion than the matrix asphalt used in other test roads, thus enhancing the anti-erosion and anti-stripping performance of the asphalt mixture, which has an important impact on slowing down the initial attenuation of pavement skid resistance. In addition, as shown in Figure 9, a high porosity was found in the surface layer of SD Road after two years of service, signifying a good macro-texture depth of SD Road surface [41]. This is because discontinuous gradation is adopted in UTWC pavement, in which coarse aggregates with a particle size above 4.75 mm account for 70%, and its coarse gradation effectively enhances the biting force and internal friction between aggregate particles [8], which contributes to the highest level of pavement skid resistance of SD Road after two years of service to some extent.

Figure 9. Core samples of ShiDan Road at different service period. (**a**) November 2019; (**b**) September 2021.

As for the attenuation rate of pavement skid resistance of AC-16 pavement, the attenuation rate of pavement skid resistance of SN Road was 21.5% higher than that of LC Road in the whole test period. This is mainly attributed to the fact that LC Road is a secondary highway connecting nearby towns and villages, and its annual average daily traffic is only 5.8% of that of SN Road. In this case, the LC Road suffers less wear from vehicle loads and still has a rich surface texture.

In terms of AC-13 pavement, YY Road exhibited excellent pavement skid resistance, as shown in Figure 8. By the fourth on-site skid resistance test, the attenuation rates of pavement skid resistance of XY Road and SP Side Road were 1.51 and 1.79 times higher than that of YY Road, respectively. This is mainly because RAC-13 is used as the top layer of asphalt pavement on XY Road and SP Side Road. When the asphalt film on the pavement gradually falls off, the aggregates are exposed and directly contact the vehicle, and sufficient friction between tires and pavements cannot be offered by the recycled aggregates with poor angularity and micro-texture, which leads to the continuous and rapid attenuation of pavement skid resistance. Moreover, it was found through investigation that the distance between the construction site of YY Road and the mixing plant is short, with a linear distance of only 2 km, which greatly reduces the temperature loss of hot-mixed asphalt mixture during transportation, thus effectively avoiding the problems of paving and compaction quality caused by construction at low temperature [42] and ensuring the superior and stable texture depth and surface texture of YY Road. This is also the key factor to keep the skid resistance of YY Road at a high level within two years of service.

3.2. Variation in Angularity of Coarse Aggregates

In this study, angularity tests were carried out on coarse aggregates with particle sizes of 4.75~9.5 mm, 9.5~13.2 mm and 13.2~16 mm. It failed to extract coarse aggregates with required particle sizes from some test roads due to the difference in gradation composition of asphalt mixtures, so only coarse aggregates from SD Road, SN Road, YY Road and XY Road were subjected to angularity tests, with the calculation results of the angularity evaluation indicators of the four test roads shown in Figure 10. In Figure 10, similar abbreviations are named in the same way. For example, "13.2 mm AS" represents the axiality of coarse aggregates with a particle size of 13.2~16 mm; "9.5 mm AS" stands for the axiality of coarse aggregates with a particle size of 9.5~13.2 mm and "4.75 mm AS" represents the axiality of coarse aggregates with a particle size of 4.75~9.5 mm.

As shown in Figure 10, among angularity indicators of coarse aggregates, the roundness (R) declined overall with the increase in service time. It suggests that the two-dimensional profile of coarse aggregates tends to be round after wearing, with gradually decreased angularity, which results in the decrease in micro-cutting between tires and pavement and the decline of pavement skid resistance.

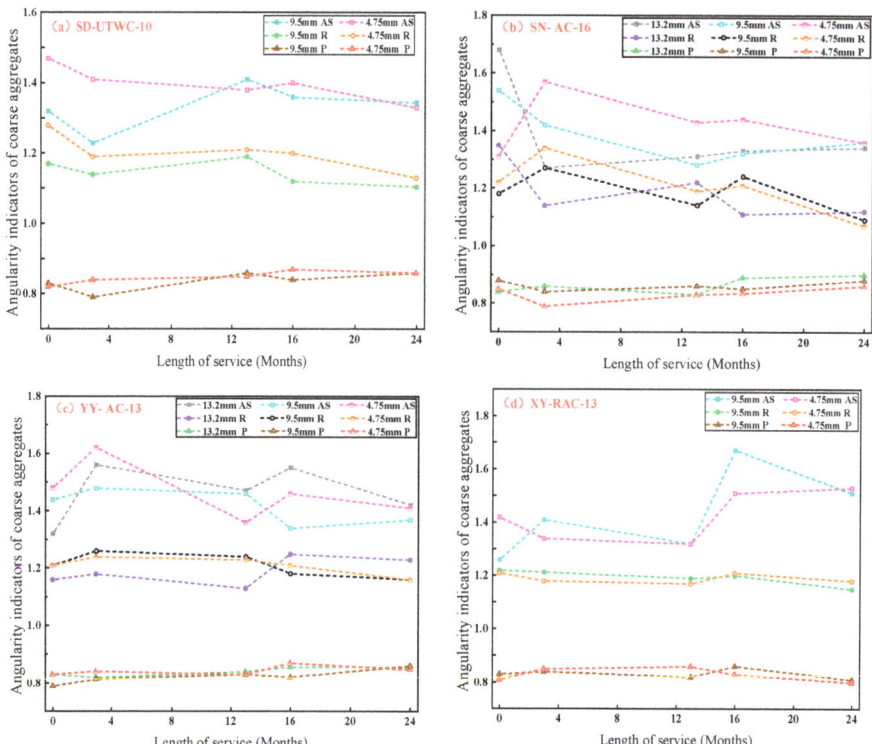

Figure 10. The changes of angularity of coarse aggregates with service time for different roads.

Figure 10 also shows that there were no obvious change law in convexity (P) and axiality (AS). It is attributed to the fact that the worn surface of aggregates exposed on the pavement is relatively small during the actual service process. The real wear position of coarse aggregates may not be collected by the " backlighting photography" system due to the effect of the placement position of coarse aggregates. As a result, the convexity and axiality represented the macro-morphology of coarse aggregates in essence. Therefore, the convexity and axiality of coarse aggregates should not be used separately as evaluation indicators characterizing the attenuation law of coarse aggregate angularity.

3.3. Variation in Micro-Texture of Coarse Aggregates

Micro-texture measurement was performed on coarse aggregates with particle sizes of 4.75~9.5 mm, 9.5~13.2 mm and 13.2~16 mm for each test road. The micro-texture of coarse aggregates obtained from four roads, SD, SN, YY and XY were measured. Figure 11 shows the calculation results of micro-texture evaluation indicators of the four test roads. In Figure 11, similar abbreviations are named in the same way. For example, "13.2 mm R_t" represents the total height of the profile of coarse aggregates with a particle size of 13.2~16 mm; "9.5 mm R_t" represents the total height of the profile of coarse aggregates with a particle size of 9.5–13.2 mm; and "4.75 mm R_t" represents the total height of the profile of coarse aggregates with a particle size of 4.75~9.5 mm.

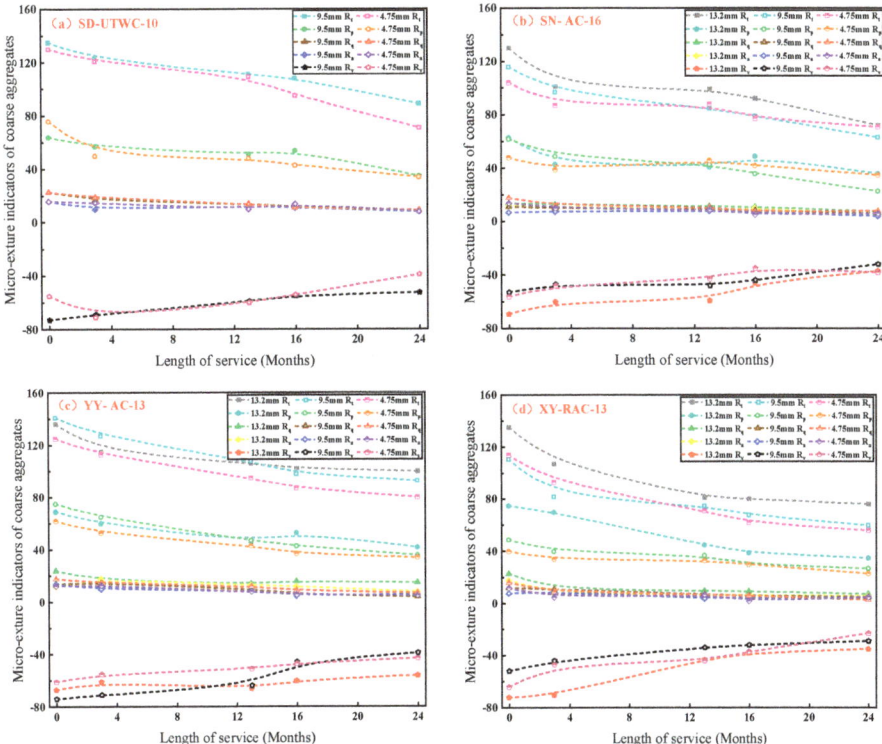

Figure 11. The changes of micro-texture of coarse aggregates with service time for different roads.

As can be seen from Figure 11, during the whole service period, all micro-texture indicators showed a trend of rapid change in the early stage and gradual change in the later stage, which was specifically reflected as the decrease in the maximum profile peak height (R_p) and the total height of the profile (R_t) and the increase in the maximum profile valley depth (R_v). This indicates that with the increase in service time, the worn surface of coarse aggregates tends to be smooth since the asperities on the worn surface of coarse aggregates are gradually worn, and some depressions are filled.

As shown in Figure 11, SD Road with basalt as coarse aggregates still exhibited excellent micro-texture after two years of service, even under the condition of withstanding several times the traffic volume of other test roads. Meanwhile, as shown in Figure 11c,d, compared with YY Road with the same mixture type of AC-13, the attenuation rate of each micro-texture indicator of XY Road was 2~3 times that of YY Road after ten months of service, indicating that when the asphalt film on the pavement falls off, the recycled aggregates with poor angularity and micro-texture are exposed and directly contact with vehicles, resulting in the rapid attenuation of pavement skid resistance. It illustrates that the pavement skid resistance is greatly affected by the type and wear condition of aggregates. By selecting high-quality aggregates with high wear resistance, the initial pavement skid resistance can be improved and its attenuation rate can be reduced.

In addition, the micro-texture indicators of coarse aggregates with different aggregate sizes in the initial state were compared and the results demonstrated that the difference in micro-texture indicators of test coarse aggregates with different particle sizes were all within 10%, as shown in Figure 11b,c, signifying poor correlation between the micro-texture and particle size of coarse aggregates at this stage. However, with the increase in service time, a lower micro-texture attenuation rate was detected in the coarse aggregates with large particle size, indicating that the pavement surface can still maintain excellent skid resistance under long-term wheel loads by increasing the nominal maximum aggregate size or the proportion of coarse aggregates of the mixture.

3.4. Correlation between Pavement Skid Resistance and Morphological Characteristics of Coarse Aggregates

The variation tendency of angularity and micro-texture of coarse aggregates is related to the attenuation law of pavement skid resistance. In order to identify the morphological indicators of coarse aggregates which can more accurately characterize pavement skid resistance, the grey correlation entropy analysis method was adopted to compare the influence of each morphological indicator of coarse aggregates on pavement skid resistance. The calculation results of grey correlation entropy between the morphological indicators of coarse aggregates and pavement skid resistance DF_{60} are shown in Table 3.

Table 3. Grey correlation entropy between morphological characteristics of coarse aggregates and skid resistance.

Morphological Indicators of Coarse Aggregates	R_a	R_p	R_q	R_t	R_v	P	AS	R
Correlation degree	0.715	0.705	0.728	0.747	0.752	0.668	0.671	0.731

Table 3 shows that the morphological indicator R_v had the greatest effect on the pavement skid resistance DF_{60}, followed by R_t, R, R_q, R_a, R_p, AS and P in turn. The correlation degree between the five micro-texture indicators (including R_v, R_t, R_q, R_a and R_p) and the pavement skid resistance was greater than 0.7, indicating that the micro-texture of coarse aggregates is a crucial factor affecting pavement skid resistance. Among them, the variation range of coarse aggregates micro-texture has the most significant effect on pavement skid resistance, followed by the dispersion degree of variations in the micro-texture. However, among the evaluation indicators of coarse aggregate angularity, only roundness had a correlation over 0.7 with pavement skid resistance, while other angularity evaluation indicators displayed poor correlations with pavement skid resistance.

After asphalt film on the pavement surface disappears due to wear, the vehicle can only contact the worn surface of aggregates on the pavement surface, and pavement skid resistance is ensured by the friction between the tire and pavement provided by aggregate angularity and the micro-texture on the worn surface, as well as rolling friction provided by the pavement macro-texture. With the increasing number of wheel loads, the asperities on the worn surface of coarse aggregates on the pavement surface were gradually worn away, whereas some depressions were filled, manifested as the reduced total height of the profile (R_t) and maximum profile peak height (R_p) and increased maximum profile valley depth (R_v), which represented the variation range of the micro-texture. The variation was directly related to the friction between the tire and pavement, leading to the decline in pavement skid resistance. In addition, the root mean square deviation of the profile (R_q) and the arithmetic mean deviation (R_a) representing the dispersion degree of micro-texture variations were also strongly related to pavement skid resistance, implying that there were some similarities in the wear characteristics of micro-texture of coarse aggregates on the pavement surface. Therefore, compared with the angularity of coarse aggregates, the micro-texture has a greater impact on the skid resistance of actual asphalt pavement. Using aggregates with high roughness can effectively improve pavement skid resistance. However, it is undeniable that this will also increase the rolling resistance of the pavement surface, which will lead to an increase in fuel consumption [43]. Therefore, in the pavement design stage, the morphological characteristics of aggregates should be considered to ensure driving safety while balancing environmental protection and economy.

In terms of the evaluation indicators of coarse aggregate angularity, only the roundness rose under vehicle loads compared with that of aggregates in the initial state, i.e., the exposed aggregates on the pavement surface became more rounded and smooth, whereas the macro-morphological characteristics, namely convexity and axiality, displayed no significant variations. Therefore, the indicators of coarse aggregate angularity are generally not suitable for characterizing the skid resistance of actual asphalt pavement.

4. Conclusions

In this research, the skid resistance of six actual roads was observed during the two years to analyze the skid resistance deterioration characteristics. In the meantime, core samples were drilled onsite to investigate the variation of aggregate morphological characteristics of coarse aggregates, including the angularity and micro-texture. Finally, the correlation between the morphological characteristics of coarse aggregates and the pavement skid resistance was established. Based on the test results and analysis, the following conclusions could be drawn.

With the increase in service time, the attenuation of pavement skid resistance gradually slows down, and finally is maintained near a certain value for a long time. Compared with the initial value of pavement skid resistance, the attenuation rate of pavement skid resistance of each test road after two years of service varies between 24% and 43%. The skid resistance of UTWC pavement is better than that of AC pavement, and the skid resistance of new pavement is better than that of recycled pavement. The attenuation rate of pavement skid resistance can be slowed down by increasing the proportion of coarse aggregates, using coarse aggregates with a large particle size, or using basalt as coarse aggregates.

With the increase in service time, the coarse aggregate angularity indicator roundness (R) decreases overall, with a decline of 15 percent, while convexity (P) and axiality (AS) have no obvious variation law. In the actual service process of pavement, the two-dimensional profile of coarse aggregates tends to be round, with angularity gradually decreasing. Therefore, the convexity and axiality of coarse aggregates should not be used as evaluation indicators characterizing the attenuation law of aggregate angularity alone.

With the increasing service time, the asperities on the surface of coarse aggregates gradually wear away, and the surface of coarse aggregates tends to be smooth, which is manifested by the decrease in the maximum profile peak height (R_p) and the total height of the profile (R_t) and the increase in the maximum profile valley depth (R_v). The wear rate of micro-texture can be effectively slowed down by selecting coarse aggregates with high roughness and high wear resistance so that pavement skid resistance can keep a high level for a long time.

The grey correlation entropy between all the micro-texture indicators and angularity indicator roundness of coarse aggregates and skid resistance is more than 0.7, while the other angularity indicators have poor correlations with pavement skid resistance. It signifies that among the evaluation indicators of morphological characteristics of coarse aggregates, the micro-texture affects pavement skid resistance more greatly, while angularity indicators of coarse aggregates are generally not suitable for characterizing the skid resistance of actual asphalt pavement.

5. Future Research

Skid resistance attenuation of different types of asphalt pavements through long-term tracking on actual roads was explored and its correlation with morphological characteristics of coarse aggregates in the surface layer was analyzed. In future research, more roads with different materials compositions will be included in the investigation. We will explore the similarities and differences between the attenuation characteristics of skid resistance and the surface texture of asphalt pavement by combining the on-site investigation and laboratory simulation, which contribute to the improvement of skid resistance and determination of pavement maintenance time.

Author Contributions: Conceptualization, W.R.; methodology, Z.W., W.R. and T.J.; formal analysis, Z.W, W.R. and T.J.; investigation, Z.W. and T.J.; resources, Y.D., Y.H. and Y.Z.; data curation Z.W., W.R. and T.J.; writing—original draft preparation, Z.W. and W.R.; writing—review and editing, Z.W. and W.R.; visualization, Z.W. and T.J.; supervision, Y.D., Y.H. and Y.Z.; project administration, W.R. and Y.D.; funding acquisition, Y.D., Y.H. and Y.Z. All authors have read and agreed to the published version of the manuscript.

Funding: This research was funded by the National Natural Science Foundation of China [grant numbers 52108391, 52078025]; the Pyramid Talent Training Project of Beijing University of Civil Engineering and Architecture [grant number JDYC20220810]; the Program for Chang-jiang Scholars and Innovative Research Team in Universities [grant number IRT-17R06]; the Beijing Postdoctoral Research Foundation [grant number 2020-zz-096]; the China Highway Engineering Consulting Corporation [grant number YFZX-2019-06].

Institutional Review Board Statement: Not applicable.

Informed Consent Statement: Not applicable.

Data Availability Statement: All data, models, and codes generated or used in this study are included in the submitted manuscript.

Conflicts of Interest: The authors declare no conflict of interest.

References

1. Kumar, A.; Tang, T.; Gupta, A.; Anupam, K. A state-of-the-art review of measurement and modelling of skid resistance: The perspective of developing nation. *Case Stud. Constr. Mater.* **2023**, *18*, e02126. [CrossRef]
2. Hussein, N.; Hassan, R.; Fahey, M.T. Effect of pavement condition and geometrics at signalised intersections on casualty crashes. *J. Saf. Res.* **2021**, *76*, 276–288. [CrossRef] [PubMed]
3. Najafi, S.; Flintsch, G.W.; Medina, A. Linking roadway crashes and tire–pavement friction: A case study. *Int. J. Pavement Eng.* **2017**, *18*, 119–127. [CrossRef]
4. Huang, X.; Zheng, B. Research status and progress for skid resistance performance of asphalt pavements. *China J. of Highway Transport* **2019**, *32*, 32–49.
5. Li, Q.J.; Zhan, Y.; Yang, G.; Wang, K.C. Pavement skid resistance as a function of pavement surface and aggregate texture properties. *Int. J. Pavement Eng.* **2020**, *21*, 1159–1169. [CrossRef]
6. Edjeou, W.; Cerezo, V.; Do, M.-T.; Zahouani, H.; Ropert, C.; Augris, P. Multiscale analyse of the relation between skid resistance and pavements surfaces texture evolution with polishing. *Road Mater. Pavement Des.* **2023**, 1–25. [CrossRef]
7. Liu, X.Y.; Cao, Q.Q.; Wang, H.; Chen, J.Y.; Huang, X.M. Evaluation of vehicle braking performance on wet pavement surface using an integrated tire-vehicle modeling approach. *Transp. Res. Rec.* **2019**, *2673*, 295–307. [CrossRef]
8. Chen, S.; Liu, X.; Luo, H.; Yu, J.; Chen, F.; Zhang, Y.; Ma, T.; Huang, X. A state-of-the-art review of asphalt pavement surface texture and its measurement techniques. *J. Road Eng.* **2022**, *2*, 156–180. [CrossRef]
9. Ji, J.; Ren, W.; Jiang, T.; Dong, Y.; Hou, Y.; Li, H. Establishment and Analysis of the Relationship Model between Macro-Texture and Skid Resistance Performance of Asphalt Pavement. *Coatings* **2022**, *12*, 1464. [CrossRef]
10. Li, P.; Yi, K.; Yu, H.; Xiong, J.; Xu, R. Effect of aggregate properties on long-term skid resistance of asphalt mixture. *J. Mater. Civ. Eng.* **2021**, *33*, 04020413. [CrossRef]
11. Zong, Y.; Li, S.; Zhang, J.; Zhai, J.; Li, C.; Ji, K.; Feng, B.; Zhao, H.; Guan, B.; Xiong, R. Effect of aggregate type and polishing level on the long-term skid resistance of thin friction course. *Constr. Build. Mater.* **2021**, *282*. [CrossRef]
12. El-Ashwah, A.S.; Broaddus, K.; Abdelrahman, M. Predicting the friction coefficient of high-friction surface treatment application aggregates using the aggregates' characteristics. *J. Mater. Civ. Eng.* **2023**, *35*, 04023089. [CrossRef]
13. Sun, P.; Zhang, K.; Han, S.; Xiao, Y. Aggregate geometrical features and their influence on the surface properties of asphalt pavement. *Materials* **2022**, *15*, 3222. [CrossRef] [PubMed]
14. Yu, M.; Xiao, B.; You, Z.; Wu, G.; Li, X.; Ding, Y. Dynamic friction coefficient between tire and compacted asphalt mixtures using tire-pavement dynamic friction analyzer. *Constr. Build. Mater.* **2020**, *258*, 119492. [CrossRef]
15. Mezgeen, R.; Franziska, S.; Silvia, I.; Lucas, A.; Boumediene, N.; Malal, K.; Christophe, C. Progress and Monitoring Opportunities of Skid Resistance in Road Transport: A Critical Review and Road Sensors. *Remote Sensing* **2021**, *13*.
16. Kane, M.; Lim, M.; Do, M.T.; Edmonsond, V. A new predictive skid resistance model (PSRM) for pavement evolution due to texture polishing by traffic. *Constr. Build. Mater.* **2022**, *342*, 128052. [CrossRef]
17. Ren, W.; Han, S.; He, Z.; Li, J.; Wu, S. Development and testing of a multivariable accelerated abrasion machine to characterize the polishing wear of pavement by tires. *Surf. Topogr.: Metrol. Prop.* **2019**, *7*, 035006. [CrossRef]
18. Zhao, Z.; Wang, Z.; Wu, S.; Xie, J.; Yang, C.; Li, N.; Cui, P. Road performance, VOCs emission and economic benefit evaluation of asphalt mixture by incorporating steel slag and SBS/CR composite modified asphalt. *Case Stud. Constr. Mater.* **2023**, *18*, e01929. [CrossRef]

19. Haritonovs, V.; Tihonovs, J. Use of unconventional aggregates in hot mix asphalt concrete. *Balt. J. Road Bridge Eng.* **2014**, *9*, 276–282. [CrossRef]
20. Cui, P.; Wu, S.; Xiao, Y.; Yang, C.; Wang, F. Enhancement mechanism of skid resistance in preventive maintenance of asphalt pavement by steel slag based on micro-surfacing. *Constr. Build. Mater.* **2020**, *239*, 117870. [CrossRef]
21. Li, S.; Xiong, R.; Dong, X.; Sheng, Y.; Guan, B.; Zong, Y.; Xie, C.; Zhai, J.; Li, C. Effect of chemical composition of calcined bauxite aggregates on mechanical and physical properties for high friction surface course. *Constr. Build. Mater.* **2021**, *302*. [CrossRef]
22. Liu, H.; Wang, Z.; Yang, C.; Chen, S.; Yu, H.; Huang, T.; Li, X.; You, Z. Effect of coarse aggregate characteristics on skid resistance deterioration of the ultrathin wearing course. *J. Mater. Civ. Eng.* **2021**, *33*, 04021051. [CrossRef]
23. Qi, L.; Liu, Y.; Liu, Z.; Zhang, C.; Chen, Z.; Lv, J.; Wan, H. Skid resistance attenuation of asphalt pavement based on multifactor accelerated wear Test. *Coatings* **2023**, *13*, 717. [CrossRef]
24. Ergin, B.; Gökalp, İ.; Uz, V.E. Effect of aggregate microtexture losses on skid resistance: Laboratory-based assessment on chip seals. *J. Mater. Civ. Eng.* **2020**, *32*, 04020040. [CrossRef]
25. Edjeou, W.; Cerezo, V.; Zahouani, H.; Do, M.-T. Contribution of multiscale analysis to the understanding of friction evolution of aggregates surfaces. *Surf. Topogr.: Metrol. Prop.* **2023**, *11*, 014006. [CrossRef]
26. Wang, D.; Zhang, Z.; Kollmann, J.; Oeser, M. Development of aggregate micro-texture during polishing and correlation with skid resistance. *Int. J. Pavement Eng.* **2020**, *21*, 629–641. [CrossRef]
27. He, Y.; Xing, C.; Hong, B.; Tan, Q.; Wang, D.; Oeser, M. Influence of polishing value of coarse and fine aggregate on long-term skid resistance of asphalt pavement. *China J. of Highway Transport* **2022**, *35*, 215–223.
28. Zhu, S.Y.; Ji, X.P.; Yuan, H.Z.; Li, H.L.; Xu, X.Q. Long-term skid resistance and prediction model of asphalt pavement by accelerated pavement testing. *Constr. Build. Mater.* **2023**, *375*, 131004. [CrossRef]
29. Tan, L.; Wu, M.; Deng, D.; Deng, H.; Lu, X. Experimental study on skid-resistance and wear-resistance performance attenuation rule of coarse aggregate. *J. Highway Transp. Res. Dev.* **2020**, *37*, 20–24, 42.
30. Roy, N.; Sarkar, S.; Kuna, K.K.; Ghosh, S.K. Effect of coarse aggregate mineralogy on micro-texture deterioration and polished stone value. *Constr. Build. Mater.* **2021**, *296*, 123716. [CrossRef]
31. Wang, C.; Wang, H.; Oeser, M.; Mohd Hasan, M.R. Investigation on the morphological and mineralogical properties of coarse aggregates under VSI crushing operation. *Int. J. Pavement Eng.* **2021**, *22*, 1611–1624. [CrossRef]
32. MTPRC. *Field Test Methods of Highway Subgrade and Pavement*; JTG 3450-2019: 2019; China Communications Press Co., Ltd.: Beijing, China, 2019.
33. MTPRC. *Standard Test Methods of Bitumen and Bituminous Mixture for Highway Engineering*; JTG E20-2011: 2011; China Communications Press Co., Ltd.: Beijing, China, 2011.
34. Yang, H.; Mei, Y.; Lu, X. Research on the shape features of coarse aggregate based on image and sensor. *J. Qingdao Univ. Technol.* **2021**, *41*, 55–60.
35. Yang, H.; Mei, Y.; Feng, G.; Dou, P.; Zheng, Y.; Zhao, H. Intelligent numerical computation method and quantitative study on morphological characteristics of multi-scale aggregate. *J. Phys. Conf. Ser.* **2021**, *2083*, 042056.
36. De Oliveira, R.; Albuquerque, D.; Cruz, T.; Yamaji, F.; Leite, F. Measurement of the nanoscale roughness by atomic force microscopy: Basic principles and applications. In *Atomic Force Microscopy-Imaging, Measuring and Manipulating Surfaces at the Atomic Scale*; IntechOpen: London, UK, 2012; Volume 3.
37. Menezes, P.L.; Kishore; Kailas, S.V. Influence of roughness parameters and surface texture on friction during sliding of pure lead over 080 M40 steel. *Int. J. Adv. Manuf. Technol.* **2009**, *43*, 731–743. [CrossRef]
38. Pawlus, P.; Reizer, R.; Wieczorowski, M. Characterization of the shape of height distribution of two-process profile. *Measurement* **2020**, *153*, 107387. [CrossRef]
39. Martins, A.M.; Rodrigues, P.C.; Abrão, A.M. Influence of machining parameters and deep rolling on the fatigue life of AISI 4140 steel. *Int. J. Adv. Manuf. Technol.* **2022**, *121*, 6153–6167. [CrossRef]
40. Ren, W. Study on the Abrasion Characteristic of Surface Texture and Its Effect on Noise for Asphalt Pavements. Ph.D. dissertation, Chang'an University, Xi'an, China, 2019.
41. Cui, W.; Wu, K.; Cai, X.; Tang, H.; Huang, W. Optimizing gradation design for ultra-thin wearing course asphalt. *Materials* **2020**, *13*, 189. [CrossRef]
42. Nevalainen, N.; Pellinen, T. The use of a thermal camera for quality assurance of asphalt pavement construction. *Int. J. Pavement Eng.* **2016**, *17*, 626–636. [CrossRef]
43. Ziyadi, M.; Ozer, H.; Shakiba, M.; Al-Qadi, I.L. Vehicle excess fuel consumption due to pavement deflection. *Road Mater. Pavement Des.* **2023**, *24*, 609–630. [CrossRef]

Disclaimer/Publisher's Note: The statements, opinions and data contained in all publications are solely those of the individual author(s) and contributor(s) and not of MDPI and/or the editor(s). MDPI and/or the editor(s) disclaim responsibility for any injury to people or property resulting from any ideas, methods, instructions or products referred to in the content.

Article

Mesoscopic Mechanical Properties of Aggregate Structure in Asphalt Mixtures and Gradation Optimization

Jingchun Chen [1], Jian Wang [1], Min Li [1], Zedong Zhao [2] and Jiaolong Ren [1,*]

[1] School of Civil Engineering and Geomatics, Shandong University of Technology, Zibo 255000, China
[2] School of Transportation and Vehicle Engineering, Shandong University of Technology, Zibo 255000, China
* Correspondence: worjl@sdut.edu.cn

Citation: Chen, J.; Wang, J.; Li, M.; Zhao, Z.; Ren, J. Mesoscopic Mechanical Properties of Aggregate Structure in Asphalt Mixtures and Gradation Optimization. *Materials* **2023**, *16*, 4709. https://doi.org/10.3390/ma16134709

Academic Editor: Giovanni Polacco

Received: 10 June 2023
Revised: 25 June 2023
Accepted: 26 June 2023
Published: 29 June 2023

Copyright: © 2023 by the authors. Licensee MDPI, Basel, Switzerland. This article is an open access article distributed under the terms and conditions of the Creative Commons Attribution (CC BY) license (https://creativecommons.org/licenses/by/4.0/).

Abstract: Particle media are widely used in engineering and greatly influence the performance of engineering materials. Asphalt mixtures are multi-phase composite materials, of which coarse aggregates account for more than 60%. These coarse aggregates form a stable structure to transfer and disperse traffic loads. Therefore, knowing how to adjust the structural composition of coarse aggregates to optimize their performance is the key to optimize the performance of asphalt mixtures. In this study, the effects of different roughness and different sizes on the interlocking force and contact force of coarse aggregates were investigated through means of simulation (DEM), and then the formation-evolution mechanism of the coarse aggregate structure and the role of different sizes of aggregates in the coarse aggregate structure were analyzed. Subsequently, the optimal ratio of coarse aggregates was explored through indoor tests, and finally, the gradation of asphalt mixture based on the optimization of fine structure was formed and verified through indoor tests. The results showed that the major model can effectively reveal the role of different types of aggregates in the fine structure and the relationship between the strength of contact forces between them and clarify that the strength of the fine structure increases with the increase in aggregate roughness. Hence, the coarse aggregate structure can be regarded as a contact force transmission system composed of some strong and sub-strong contact forces. Their formation-evolution mechanism can be regarded as a process of the formation of strong and sub-strong contact forces and the transformation from sub-strong contact force to strong contact force. Moreover, the dynamic stability of the optimized graded asphalt mixture was increased by 30%, and the fracture toughness was increased by 26%.

Keywords: coarse aggregate structure; DEM; mechanical properties; gradation optimization; interlocking force; contact force

1. Introduction

As the most widely used road pavement materials, the asphalt mixture plays an important role in infrastructure construction. In the asphalt mixture as an asphalt–void–aggregate composition of multi-phase complex materials (particulate materials), its coarse aggregates account for more than 60%, and these coarse aggregates can form a stable structure to transfer and disperse the traffic load. Therefore, there is a significant correlation between coarse aggregate and the performance of the asphalt mixture. Shen et al. [1,2] specifically pointed out that the coarse aggregate structure is the most critical factor affecting the performance of asphalt mixtures, which was also proved by Pouranian [3] and Lira et al. [4] In addition, existing studies have shown that the key factor affecting the aggregate is not the number of contact points but the number and distribution of strong contact forces. Hence, it is reasonable to believe that the aggregate structure can be optimized more effectively from the perspective of mechanical properties.

Shashidhar et al. [5] applied an image approach to analyzing the contact force transmission paths of the aggregate structure. Sun et al. [6] studied the effect of coarse aggregate meso-structure on the characteristics of the asphalt mixture through the technology of CT and DCT. However, the complexity and uncertainty of the asphalt mixture materials could not be fundamentally revealed only through conventional macroscopic tests. Therefore, more and more studies [7,8] analyze the asphalt mixtures from a mesoscopic perspective and try to establish effective relationships between mesoscopic properties and macroscopic behavior. Ren et al. [9,10] revealed the effect of different void ratios and void sizes on the fracture behavior of the asphalt mixture. Liu et al. [11] discussed the effect of particle size and specimen scale on the crack resistance behavior of asphalt mixtures. Wang et al. [12] constructed a model by DEM to analyze the deformation mechanism of the asphalt mixture during the loading test from the fine view level. Unfortunately, existing studies have explored the variation of properties more through material properties and proportions but not from the perspective of aggregate structure.

Currently, only a few studies have analyzed the structural mechanical properties of aggregates. Cal et al. [13] established a model to quantify the effect of gradation and estimated the shear modulus of asphalt materials. Chen et al. [14] revealed the properties of asphalt mixtures with different sizes of aggregate particles and different gradations based on DEM simulation. Pouranian et al. [15] established a new framework to study packing behavior for the asphalt mixture. All these studies help to deepen our understanding of the structural mechanical properties of aggregates. However, due to the complexity of the aggregate contact structure, the analysis of the mechanical properties of the aggregate structure is not a simple task. Some shortcomings of the existing studies are as follows: In road projects, aggregates are divided into many different sizes according to their particle size, and the ratio of different sizes of aggregates greatly affects the mechanical properties of road materials. However, the existing studies have neglected to investigate the effect on aggregate gradation. In addition, the formation-evolution mechanism of the aggregate structure has not been revealed.

Hence, this study analyzed the macro-skeleton interlocking forces and meso-aggregate contact force of the coarse aggregate structure based on the DEM simulation and clarified the formation–evolution mechanism and the function of different types of aggregates on the coarse aggregate structure. On this basis, the coarse aggregate structure is optimized based on the principle of optimal aggregate skeleton strength, and a better asphalt mixture gradation is proposed, which passed the indoor test verification.

2. Materials

2.1. Asphalt

In this paper, AH-70 asphalt is adopted, and the basic performance of AH-70 asphalt is tested according to the specification [16]. Its indexes meet the requirements shown in Table 1.

Table 1. Technical properties of asphalt.

	Test Project	Results	Technical Specifications
	Needle penetration (25 °C, 100 g, 5 s)/0.1 mm	65.2	60~80
	Softening point (Universal method)/°C	51.9	≥45
	Ductility(5 cm/min, 10 °C)/cm	21.4	≥15
	Needle penetration index	−0.818	−1.5~+1.0
	60 °C power viscosity/(/Pa·s)	411	≥180
	Density (25 °C)/(g/cm^3)	1.01	-
	Wax content (%)	1.55	<2.2
After RTFOT	Mass loss (%)	0.33	−1%~1
	Needle penetration ratio (%)	62.1	≥61
	Ductility (5 cm/min, 10 °C)/cm	6.71	≥6

2.2. Aggregate

The physical and mechanical properties of the aggregate affect the road characteristics of the asphalt mixture. Coarse aggregate is mainly used to form the skeleton structure, fine aggregate is mainly used to fill the skeleton structure of coarse aggregate, mineral powder and asphalt to form asphalt slurry filled in the gaps between coarse and fine aggregate and the aggregate bonded together, so that the asphalt mixture to produce the ability to resist the traffic load. The aggregate is obtained from limestone in Shandong Province, China. The test results [17] of the basic characteristics of the aggregates are shown in Table 2.

Table 2. Technical properties of aggregates.

Properties	Index		Requirements	Test Value	Test Method
Coarse aggregate	Robustness (%)		≤12	7.7	T 0314
	Stone crushing value (%)		≤26	12.1	T 0316
	Los Angeles abrasion loss (%)		≤28	14.4	T 0317
	Water absorption rate (%)		≤2.0	1.2	T 0304
Fine aggregate	Angularity		≥30	45	T 0345
	sand equivalent (%)		≥60	79	T 0334
	Robustness (%)		≥12	15.4	T 0340
Mineral powder	Hydrophilic coefficient		<1	0.51	T 0353
	Moisture content (%)		≤1	0.44	T 0103
	Particle size (%)	<0.6 mm	100	100	T 0351
		<0.15 mm	90~100	94.3	
		<0.075 mm	75~100	86.7	

3. Numerical Model

3.1. Numerical Model

Due to the particularity of DEM numerical calculation, the complexity of model operation depends on the number of particles and contact points. The DEM simulation of unbonded particles is usually random, and these particles need to reach the equilibrium state of contact force under gravity before the calculation begins. However, the existing numerical models [18–20] involve a large number of particles, so it takes a lot of time to find the equilibrium state of contact force. Therefore, a new simplified modeling method is adopted to study the mechanical properties of aggregate structures by referring to the experience of existing studies [21,22]. The specific process is as follows:

The two-dimensional mapping area (S_i) of coarse aggregate is calculated according to the mass, apparent density and size of different coarse aggregates, as shown in Equation (1). Then, the built-in command "ball" of PFC^{2D} is used to generate circular particles in the simulated cylinder and make them meet the particle size requirements of a coarse aggregate.

$$S_i = \frac{4m_i}{\pi \rho_i W} \quad (1)$$

The initial model is generated by combing different aggregates with particle sizes including 13.2~19 mm, 9.5~13.2 mm, 4.75~9.5 mm and 2.36~4.75 mm, as shown in Figure 1.

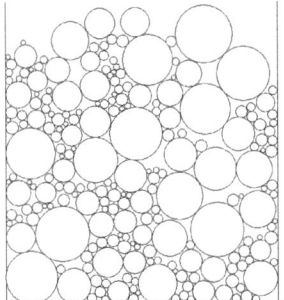

Figure 1. The initial model of aggregate structure.

Secondly, the built-in command "b_id" of PFC^{2D} is used to automatically search the particles numbered t in the i-th type coarse aggregate and read their center coordinates, radius and so on. Moreover, according to the surface non-uniformity coefficient (a) of the coarse aggregate, four sub-particles are generated by the built-in command "ball" of PFC^{2D}. The structure of irregular particles is shown in Figure 2.

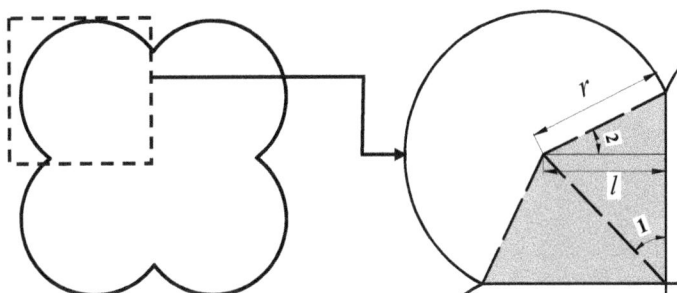

Figure 2. The structure of irregular particles.

As shown in Figure 2, the area of irregular aggregate (S) can be represented by Equation (2). Then, S is set qual to the initial circular particle area (S_o), as shown in Equation (3). On this basis, the radius of the particle is obtained by Equation (4).

$$S = 4r_{it}^2[a\sqrt{1-a^2} + a^2/\tan(\pi/4) + 3\pi/4 - \arccos a] \tag{2}$$

$$S = S_o = \pi R_{it}^2 \tag{3}$$

$$r_{it} = \sqrt{\frac{\pi R_{it}^2}{4[a\sqrt{1-a^2} + a^2/\tan(\pi/4) + 3\pi/4 - \arccos a]}} \tag{4}$$

Subsequently, the center coordinates of the four small particles are calculated according to Equations (5)–(8).

$$x_{1,it} = X_{it} - ar_{it} \tag{5}$$

$$x_{2,it} = X_{it} + ar_{it} \tag{6}$$

$$y_{1,it} = Y_{it} - ar_{it} \tag{7}$$

$$y_{2,it} = Y_{it} + ar_{it} \qquad (8)$$

A series of particles of different a values are shown in Figure 3. The greater roughness of the aggregate is seen obviously with the larger a value.

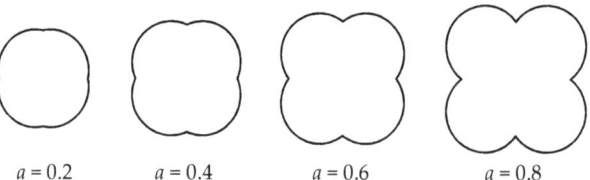

$a = 0.2$ $a = 0.4$ $a = 0.6$ $a = 0.8$

Figure 3. The irregular aggregate of different a values.

In the end, the irregular particles are composed of four sub-particles by using "clump" in the PFC^{2D}. Then, initial particles of number t are removed by using "delete" in the PFC^{2D} and replaced by the corresponding irregular particles. When the aggregate structure reaches the static equilibrium state under the action of gravity, the numerical model of the irregular particle structure can be obtained, as shown in Figure 4.

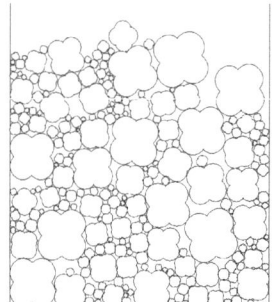

Figure 4. The numerical model of irregular particle structure.

In addition, the simplified Hertz–Mindlin constitutive is adopted because it can better describe the nonlinear characteristics of aggregate structures, as shown in Equations (9)–(11).

$$K_n = \frac{2}{3} \frac{G\sqrt{2U\overline{R}}}{1 - \varepsilon} \qquad (9)$$

$$K_s = \frac{2\left[3G^2(1-\varepsilon)\overline{R}\right]^{1/3}}{2-\varepsilon} |f_n|^{1/3} \qquad (10)$$

$$f_s = \mu|f_n| \qquad (11)$$

3.2. Numerical Test and Verification

3.2.1. Numerical Tests

According to Section 2.1, numerical sample is randomly generated in a square area, as shown in Figure 5.

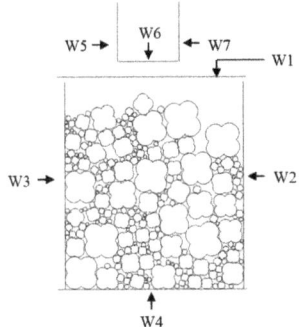

Figure 5. Numerical sample.

In order to simulate the confining pressure effect in the test, the speed of walls (W1, W2, W3 and W4) are controlled by the servo mechanism. When the aggregate structure is compacted to reach the target porosity, the servo mechanism of the W1 and W4 stops running, while that of the W2 and W3 keeps running during the whole numerical test. The servo mechanism can be listed as follows:

$$v^{t+1} - v^t = \frac{\Delta \sigma \alpha l}{k_n N} \tag{12}$$

The simulated indenter is penetrated into the numerical model of aggregate structure at a constant speed, and the contact force and displacement of the indenter are recorded during the whole simulated penetration process. The numerical simulation results are shown in Figure 6. The experiment is a reference specification [23]. The loading speed of both virtual and indoor tests in this study is 1 mm/min for a better comparison.

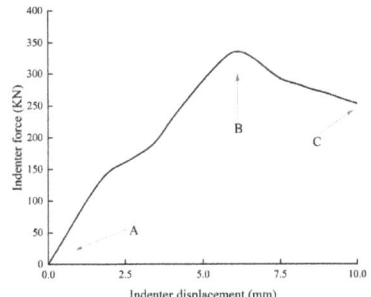

(**a**) Relationship between force and displacement of the indenter

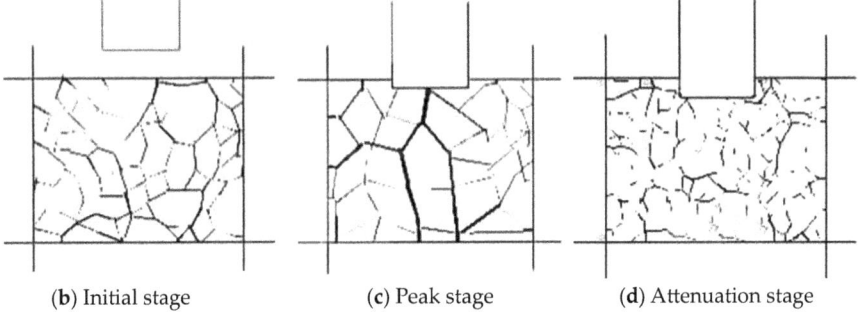

(**b**) Initial stage (**c**) Peak stage (**d**) Attenuation stage

Figure 6. Numerical simulation results.

(1) The relationship between force and displacement of the indenter

Figure 6a shows the relationship between force and displacement of the indenter, where the peak of indenter force is the interlocking force of aggregate structure. It can be found obviously that there are two stages: A-B stage is a stable increase, while B-C stage is the opposite. The evolution process of aggregate contact force in different stages is revealed further in the Figure 6b–d. The black network is the mesoscopic contact force between the particles, and the darker the color, the greater the value of the contact force.

(2) Initial stage

It can be seen that the particle contact force is uniformly distributed in Figure 6b. When the simulated indenter began to penetrate into the numerical model of aggregate structure, the particle contact force gradually increased to resist the penetration of the indenter, which corresponds to the curve state of A-B.

(3) Peak stage

The particle contact force is not uniformly distributed under the action of load and has an obvious phenomenon of concentration of contact force, as shown in Figure 6c. The particle contact force network composed of a series of robust particle contact forces in the red circle can be regarded as a representation of aggregate structure. In other words, the aggregate structure can be considered as a contact force transmission structure composed of a series of strong particles. With the increase in head displacement, the contact force will reach the peak point (B), which indicates that the aggregate structure reached the ultimate bearing capacity.

(4) Attenuation stage

Subsequently, when the contact force of the indenter exceeds the peak value, the aggregate structure begins to collapse, which corresponds to the curve state of B-C. Correspondingly, in Figure 6d, there is no significant contact force of coarse particles.

3.2.2. Results Verification

Based on the conventional triaxial test system, this paper reproduces the above numerical simulation loading process by setting an improved loading head to measure the interlocking forces of the aggregate structure. The schematic diagram of the test device is shown in Figure 7.

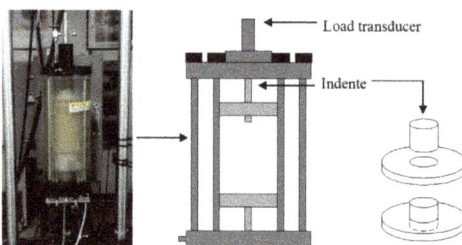

Figure 7. Test device.

The interlocking forces of the aggregate structure are evaluated by using a series of laboratory and numerical tests. As shown in Equations (9)–(11), three DEM contact parameters (shear modulus, Poisson's ratio and friction coefficient) need to be assigned to the numerical model of aggregate structure. The shear modulus and Poisson's ratio can be measured by corresponding tests. However, the friction coefficient is obtained through the indoor test inversion after the change in a value and follows the principle of the highest simulation accuracy. The comparison of simulation results and test results of aggregate structures with different a values is shown in Figure 8.

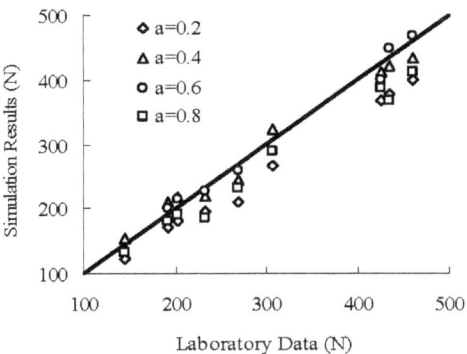

Figure 8. Comparison of simulation results and test results.

As shown in Figure 8, the simulation results (a = 0.6) have the best accuracy, with an average error of only 4.9%. Hence, the parameters corresponding to a = 0.6 (shear modulus of 21.5 GPa, Poisson's ratio of 0.3 and friction coefficient of 0.5) are used in this section for the later numerical simulations. In addition, the same parameters are used for other a value to facilitate the comparison between the simulation results.

However, the above simulations are also flawed in that 2D aggregate structural bodies are referenced instead of 3D structural bodies, which are known to affect the contact force paths. Moreover, 3D model has the problems of long simulation time and high complexity. Many existing studies [24–26] of the aggregate structural bodies have shown that 2D models can achieve better results with savings in simulation time.

3.3. Simulation Plan

In this section, the effect of different aggregates and a values on the mechanical properties of the aggregate structures are discussed. There are five kinds of aggregates, including 13.2~19 mm (D1), 9.5~13.2 mm (D2), 4.75~9.5 mm (D3), 2.36~4.75 mm (D4) and 1.18~2.36 mm (D5). In addition, the a value also includes five kinds, which are 0, 0.2, 0.4, 0.6 and 0.8. The simulation plan is as follows:

Step 1: Denote the numerical model of D1 mixing as P1 and record the corresponding aggregate interlocking force and contact force.

Step 2: Obtain the aggregate interlocking force and contact force of the aggregate structure mixed with different proportions of P1 and D2; determine the optimal ratio of P1 and D2 by following the principle of maximum aggregate interlocking force and denote the above ratio of P1 and D2 as P2.

Step 3: Denote the optimal ratio of P2 and D3 as P3 by following the above principle and record the corresponding force.

Step 4: By analogy, complete the design of P4 and P5 and record the corresponding results.

4. Results and Discussion

4.1. Aggregate Interlocking Forces

The relationship between the aggregate interlocking force (f) of P1~P5 and the particle size (d_{min}) of the minimum-size aggregate in the aggregate structure is shown in Figure 9. Moreover, Figure 10 shows the relationship between values f and a.

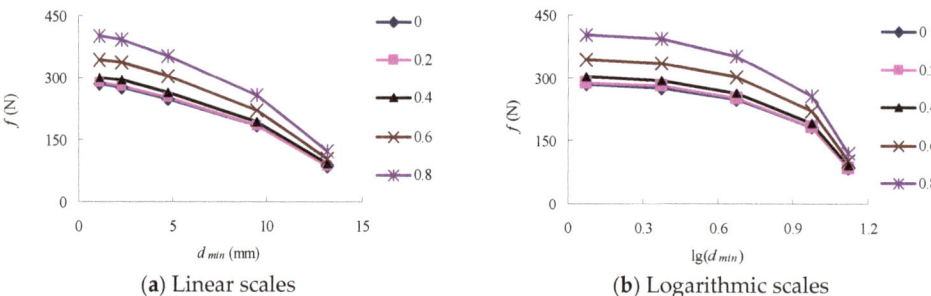

(a) Linear scales (b) Logarithmic scales

Figure 9. The relationship between f and d_{min}.

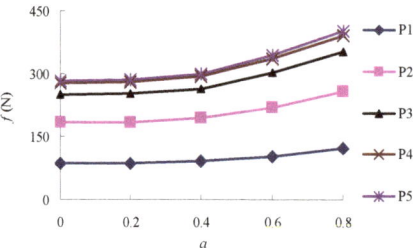

Figure 10. The relationship between f and a.

As shown in Figure 9, the aggregate interlocking force increases linearly with the incorporation of D1~D3, and the average of the extruding force of P3 is 2.6 times that of P1. However, with the addition of D4 and D5, the increasing trend of the aggregate interlocking force decreases gradually. The average extruding force of P5 is only 1.33 times and 1.04 times of that of P3 and P4. This indicates that D1~D3 is the main part of the aggregate structure, D4 is the secondary structure of the aggregate structure and plays a role in supporting the main structure, while D5 has little influence on the aggregate interlocking force.

Hence, although the aggregate interlocking force contributed by D4 is lower than that contributed by D1 + D2 + D3, it still accounts for 24.8% of the total aggregate interlocking force. On the other hand, with the addition of D5, the extruding force of the aggregate structure almost no longer increases. The aggregate extruding force contributed by D5 is only 3.6% of the total extruding force, indicating that D5 has little effect on the formation of aggregate skeleton. The above results can provide a basis for dividing coarse and fine aggregates: 2.36 mm is a suitable dividing point. When the aggregate particle size is greater than 2.36 mm, it mainly plays the role of building the main body and secondary skeleton. When the aggregate particle size is less than 2.36 mm, its contribution to skeleton formation is limited.

As shown in Figure 10, the aggregate interlocking force increases with the increase in a value, which indicates that the roughness of the aggregate can increase the strength of the aggregate structure. In addition, the contribution of D1~D3 to the strength increase of the aggregate structure increases with the increase in aggregate roughness, but the change in D4 and D5 is not significant.

4.2. Aggregate Contact Forces

The aggregate contact forces play an important role in the strength of granular media. In this section, the distribution and number of aggregate contact forces in the numerical model of aggregate structure are studied when the aggregate structure reaches the ultimate load strength. The aggregate contact force is divided into three levels [27]: weak (0, <{F}), slightly strong ({F}, double {F}), strong (double {F}, +∞), where {F} is the average value of the aggregate contact force.

Table 3 shows the number and distribution of aggregate contact forces for P3, P4 and P5 ($a = 0.6$). Table 4 shows the number and distribution of aggregate contact forces using P5 with different a values. The distribution probability of aggregate contact force [$P(f)$] is given by Equation (13).

$$P(f) = \frac{N_i}{N_t} \times 100\% \tag{13}$$

Table 3. The aggregate contact force of different aggregate structures ($a = 0.6$).

Packing	Index	Grade 1: (Double {F}, +∞)	Grade 2: ({F}, Double {F})	Grade 3: (0, <{F})
P3	$P(f)/\%$	17.97	36.94	45.09
	N_i	108	222	271
P4	$P(f)/\%$	14.04	33.24	52.72
	N_i	155	367	582
P5	$P(f)/\%$	9.93	24.17	65.87
	N_i	162	393	1071

Table 4. The aggregate contact force of P5 with different a values.

a	Index	Grade 1: (Double {F}, +∞)	Grade 2: ({F}, Double {F})	Grade 3: (0, <{F})
0	$P(f)/\%$	7.13	21.57	71.30
	N_i	119	360	1190
0.2	$P(f)/\%$	7.73	22.69	69.59
	N_i	127	373	1144
0.4	$P(f)/\%$	8.62	23.51	67.87
	N_i	139	379	1094
0.6	$P(f)/\%$	9.93	24.17	65.87
	N_i	162	393	1071
0.8	$P(f)/\%$	11.65	24.49	63.85
	N_i	196	412	1074

As shown in Table 3, the strong and slightly strong contact forces of the aggregate structure increase continuously with the addition of D4, while the growth slows down. However, the distribution probability of strong and slightly strong contact forces decrease with the sharply increase in the number of weak contact force. This indicates that D4 converts part of the slightly strong contact force composed of D1~D3 into strong contact force and provides additional slightly strong contact force. Moreover, this also explains why the strength of the aggregate structure increases with the increase in D4, but the strengthening trend begins to decline.

With the addition of D5, the number of weak contact forces in the aggregate structure increases sharply, while the number of strong and slightly strong contact forces increases very slowly. This leads to a decrease in the distribution probability of strong and slightly strong contact forces, while the distribution probability of weak contact force increases obviously. Furthermore, it can be seen from the above research that P5 and P4 have little difference in aggregate interlocking forces. This indicates that the strong and slightly strong contact forces are mainly provided by D1~D4. The weak contact force has little effect on the strength enhancement of aggregate structure. The D5 type mainly provides weak contact force for the aggregate structure but makes little contribution to strong and slightly strong contact forces. This explains why the interlocking force of the aggregate structure does not increase obviously with the addition of D5. Therefore, the formation and evolution mechanism of aggregate structure can be regarded as a process of the formation of strong and slightly strong contact forces and the transformation from slightly strong contact force to strong contact force.

Table 4 shows the total number of contact force increases with the increase in a value, which indicates that the particle roughness can increase the degree of interlocking of the aggregate. In addition, $P(f)$ and N_i of the strong contact force and N_i of the slightly strong contact force increase with the increase in a value, but $P(f)$ of the slightly strong and weak contact force decreases with the increase in a value. This shows that the increase in the aggregate roughness promotes the transformation from slightly strong and weak contact forces to strong contact force and slightly strong contact force, thus increasing the interlocking forces of the aggregate structure.

4.3. Aggregate Structure Composition

Figures 11 and 12 show that the optimal mass percentage relationship of D1~D5 in P5 and the relationship between a value and optimal mass percentage, respectively.

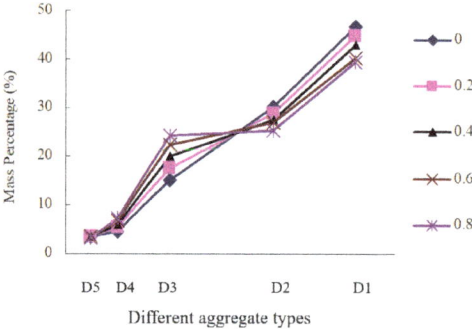

Figure 11. The optimal mass percentage of different types of aggregate.

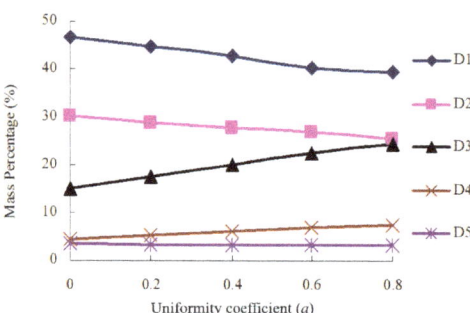

Figure 12. The relationship between a value and optimal mass percentage of aggregate.

As shown in Figures 11 and 12, the optional mass percentage of D1, D2, D3, D4 and D5 are 39.48~46.72%, 25.46~30.15%, 15.08~24.38%, 4.58~7.41% and 3.27~3.48% respectively. It can be found that D1 + D2 + D3 accounts for 89.32~91.94% of the total mass of aggregate structure, and there is no significant difference in the total mass percentage of D1 + D2 + D3 under different coarse particles. Moreover, the optimal content of D1 and D2 decreases with the increase in particle roughness, while the optimal content of D3 is the opposite. This is due to the fact that the main aggregate structure composed of it is relatively stable, while it is little affected by the other types of the aggregate.

5. The Effect of Ratio of Coarse–Fine Aggregate (CF)

The ratio of CF aggregate (including mineral powder) is determined to be 90:10, 85:15, 80:20, 75:25, 70:30, 65:35 and 60:40. It should be noted that since real aggregate is generally rarely round or round-like, the a values of 0.4, 0.6 and 0.8 are used in the calculation of coarse aggregate gradation. Hence, the calculation table of coarse aggregate is shown in Table 5, and the proposed gradation is shown in Table 6.

Table 5. Coarse aggregate calculation.

a	D1:D2:D3:D4	Coarse Aggregate Ratio	Sieve Passing Rate (%)				
			19 mm	13.2 mm	9.5 mm	4.75 mm	2.36 mm
0.4	42.30:27.64:20.00:6.08		100	60.4~73.6	34.5~56.3	15.7~43.8	10~40
0.6	40.40:26.93:22.44:6.82	60~90	100	62.4~74.9	37.3~58.2	16.4~44.2	10~40
0.8	39.48:25.46:24.38:7.41		100	63.3~75.5	39.6~59.7	16.9~44.6	10~40

Table 6. Initial proposed gradation.

CF Aggregate Ratio	Sieve Passing Rate (%)									
	19 mm	13.2 mm	9.5 mm	4.75 mm	2.36 mm	1.18 mm	0.6 mm	0.3 mm	0.15 mm	0.075 mm
90:10	100	62.0	37.1	16.3	10	7.30	5.58	4.43	3.68	3.24
85:15	100	64.1	40.6	21.0	15	10.96	8.38	6.65	5.53	4.86
80:20	100	66.2	44.1	25.6	20	14.61	11.17	8.87	7.37	6.48
75:25	100	68.3	47.6	30.3	25	18.26	13.96	11.08	9.21	8.11
70:30	100	70.5	51.1	34.9	30	21.91	16.75	13.30	11.05	9.73
65:35	100	72.6	54.6	40.0	35	25.56	19.55	15.52	12.89	11.35
60:40	100	74.7	58.1	44.2	40	29.22	22.34	17.73	14.74	12.97

The performance of the asphalt mixture with different CF ratios is shown in Table 7.

Table 7. Test results.

CF Ratio	Dynamic Stability	Fracture Toughness
90:10	848	0.94
85:15	1190	1.14
80:20	1336	1.26
75:25	1446	1.41
70:30	1125	1.09
65:35	834	0.79
60:40	768	0.88

According to Table 7, it can be concluded that the asphalt mixture has good road performance when the CF ratio is 80:20 and 75:25. Therefore, the above proportions can be determined as the final ratios of CF.

6. Asphalt Mixture Gradation Optimization and Test Verification

The grading range determined based on the previous test results can be seen in Table 8. The standard gradation (SG) and recommended gradation (RG) are used to compare the performance of the asphalt mixture.

Table 8. Recommended gradation range.

Gradation	Sieve Passing Rate (%)									
	19 mm	13.2 mm	9.5 mm	4.75 mm	2.36 mm	1.18 mm	0.6 mm	0.3 mm	0.15 mm	0.075 mm
RG	100	64~70	42~49	25~31	20~25	14~18	11~14	9~12	7~10	6~9
SG	100	70~92	60~80	34~62	20~48	13~36	9~26	7~18	5~14	4~8

Based on Table 8, it can be seen that the recommended gradation with the "two more and one less" characteristic, which means more coarse aggregate and mineral powder content and less fine aggregate content. This indicates that the structure of coarse aggregate and mineral powder significantly affect the performance of the asphalt mixture and the mortar, respectively, while the impact of fine aggregate on both is not significant. At the same time, compared with the standard gradation, the recommended gradation has a narrower gradation range, which makes it possible to control the aggregate ratio more strictly in the gradation design.

In order to compare the performance indicators of SG and RG, the results of the rutting test(dynamic stability) and SCB test(fracture toughness) are used to detect the high and low temperature performance of the asphalt mixture, as shown in Table 9.

Table 9. Road performance of asphalt mixture.

Technical Properties	RG	SG	Performance Ratio
Dynamic stability (KN)	3187	2451	1.30
Fracture toughness (Mpa·m$^{0.5}$)	1.122	0.893	1.26

As shown in Table 9, it can be seen that compared with the SG, in the asphalt mixture using the RG, the dynamic stability increased by 30%, and the fracture toughness increased by 26%, which indicates a better quality of use and proves the superiority of the proposed grade in this study.

Although some useful conclusions and findings can be obtained from the numerical simulation study at the two-dimensional scale, the performance pattern of the asphalt mixture will definitely be affected by the three-dimensional scale, so the subsequent study will strive to expand from two-dimensional scale to three-dimensional scale.

7. Conclusions

In this study, the mechanical properties of aggregate structures, pavement properties of asphalt mortar and the effect of CF ratio on material properties are studied by means of numerical simulation and laboratory test, respectively. A new gradation based on meso-structure optimization is formed and then verified by laboratory tests. The main conclusions are as follows:

- The coarse aggregate structure can be regarded as a contact force transmission system composed of some strong and sub-strong contact forces. Moreover, the formation–evolution mechanism of coarse aggregate structure can be regarded as a process of the formation of strong and sub-strong contact forces and the transformation from sub-strong contact force to strong contact force.
- The main body of coarse aggregate consists of 4.75~19 mm aggregate, which is composed of a relatively stable structural system and is less affected by other sizes of coarse aggregate. The 2.36~4.75 mm aggregate acts as the secondary structure of the aggregate structure body, which plays the role of supporting the main structure and, at the same time, induces the secondary strong contact force provided by 4.75~19 mm aggregate to transform it into strong contact force and provide additional secondary strong contact force.
- The weak contact force is mainly provided by the aggregates with the particle size less than 2.36 mm, indicating that this aggregate contributes slightly to the formation of coarse aggregate structural bodies. From the point of view of the contribution to the formation of the aggregate structure body, 2.36 mm can be used as the dividing point of coarse and fine aggregates.
- The strength of the aggregate structure increases with the increase in the roughness of the aggregate, which is due to the transformation of sub-strong and weak contact forces to strong and sub-strong contact forces.

- In the asphalt mixture with the recommended gradation, dynamic stability increases by 30%, and fracture toughness increases by 26%, when compared to the mixtures with the standard gradation.
- The effects of other aggregate (recycled aggregate) properties and different temperatures on the aggregates are considered in our subsequent studies.

Author Contributions: Conceptualization, J.C. and J.R.; Methodology, J.C.; Validation, J.C.; Formal analysis, J.W. and J.R.; Investigation, J.C., M.L. and J.R.; Data curation, M.L. and Z.Z.; Writing—original draft, J.C., J.W., Z.Z. and J.R.; Writing—review & editing, J.C., Z.Z. and J.R.; Visualization, J.R. All authors have read and agreed to the published version of the manuscript.

Funding: This study is sponsored in part by the National Natural Science Foundation of China under grants 51808326, to which the authors are very grateful.

Data Availability Statement: The data presented in this study are available in the article.

Conflicts of Interest: The authors declare no conflict of interest.

Abbreviations

S_i (i:1,2,3)	two-dimensional mapping area of the ith type coarse aggregate
m_i	coarse aggregate mass of the ith type
W	particle formation area width
ρ_i	apparent density of the ith type coarse aggregate
G	shear modulus
U	amount of overlap between particle units
f_n	contact force in each direction
E	Poisson's ratio
v^{t+1}/v^t	target wall speed at time step $t+1$ and t
$\Delta\sigma$	difference between the current pressure and the target pressure
A	relaxation factor
L	the length of target wall
k_n	average stiffness
N	total number of contacts between the target wall and particles
N_i	aggregate contact force number of grade "i"
N_t	total number of aggregate contact forces
R	average radius of contact particles
R_{it}/r_{it}	radius of particle and small particle at the ith type coarse aggregate
X_{it}/Y_{it}	X and Y coordinates of particle at the ith type coarse aggregate
a	uniformity coefficient

References

1. Shen, S.; Yu, H. Analysis of aggregate gradation and packing for easy estimation of hot-mix-asphalt voids in mineral aggregate. *J. Mater. Civ. Eng.* **2011**, *23*, 664–672. [CrossRef]
2. Shen, S.; Yu, H. Characterize packing of aggregate particles for paving materials: Particle size impact. *Constr. Build. Mater.* **2011**, *25*, 1362–1368. [CrossRef]
3. Pouranian, M.; Haddock, J. Effect of aggregate gradation on asphalt mixture compaction parameters. *J. Mater. Civ. Eng.* **2020**, *32*, 04020244. [CrossRef]
4. Lira, B.; Ekblad, J.; Lundström, R. Evaluation of asphalt rutting based on mixture aggregate gradation. *Road Mater. Pavement Des.* **2021**, *22*, 1160–1177. [CrossRef]
5. Shashidhar, N.; Gopalakrishnan, K. Evaluating the aggregate structure in hot-mix asphalt using three-dimensional computer modeling and particle packing simulation. *Can. J. Civ. Eng.* **2006**, *33*, 945–954. [CrossRef]
6. Sun, Z.; Qi, H.; Li, S.; Tan, Y.; Yue, Z.; Lv, H. Estimating the effect of coarse aggregate meso-structure on the thermal contraction of asphalt mixture by a hierarchical prediction approach. *Constr. Build. Mater.* **2022**, *342 Pt A*, 128048. [CrossRef]
7. You, Z.; Buttlar, W. Discrete element modeling to predict the modulus of asphalt concrete mixtures. *J. Mater. Civ. Eng.* **2004**, *16*, 140–146. [CrossRef]
8. You, Z.; Adhikari, S.; Dai, Q. Three-dimensional discrete element models for asphalt mixtures. *J. Eng. Mech.* **2008**, *134*, 1053–1063. [CrossRef]

9. Ren, J.; Sun, L. Characterizing air void effect on fracture of asphalt concrete at low-temperature using discrete element method. *Eng. Fract. Mech.* **2016**, *170*, 23–43. [CrossRef]
10. Sun, L.; Ren, J.; Zhang, S. Fracture characteristics of asphalt concrete in mixed-loading mode at low-temperature based on discrete-element method. *J. Mater. Civ. Eng.* **2018**, *30*, 04018321. [CrossRef]
11. Liu, W.; Gao, Y.; Huang, X. Effects of aggregate size and specimen scale on asphalt mixture cracking using a micromechanical simulation approach. *J. Wuhan Univ. Technol.-Mater. Sci. Ed.* **2017**, *32*, 1503–1510. [CrossRef]
12. Wang, H.; Zhou, Z.; Huang, W.; Dong, X. Investigation of asphalt mixture permanent deformation based on three-dimensional discrete element method. *Constr. Build. Mater.* **2021**, *272*, 121808. [CrossRef]
13. Cai, X.; Chen, L.; Zhang, R.; Huang, W.; Ye, X. Estimation of shear modulus of asphalt mixture based on the shear strength of the aggregate interface. *Constr. Build. Mater.* **2020**, *248*, 118695. [CrossRef]
14. Chen, J.; Li, H.; Wang, L.; Wu, J.; Huang, X. Micromechanical characteristics of aggregate particles in asphalt mixtures. *Constr. Build. Mater.* **2015**, *91*, 80–85. [CrossRef]
15. Pouranian, M.R.; Haddock, J.E. A new framework for understanding aggregate structure in asphalt mixtures. *Int. J. Pavement Eng.* **2019**, *22*, 1090–1106. [CrossRef]
16. *JTC E-20-2011*; Highway Engineering Asphalt and Asphalt Mixture Test Procedures. Ministry of Transport of the People's Public of China: Beijing, China, 2011.
17. *JTC B42-2005*; Highway Engineering Aggregate Test Procedure. Ministry of Transport of the People's Public of China: Beijing, China, 2005.
18. Xue, B.; Pei, J.; Zhou, B.; Zhang, J.; Li, R.; Guo, F. Using random heterogeneous dem model to simulate the SCB fracture behavior of asphalt concrete. *Constr. Build. Mater.* **2020**, *236*, 117580. [CrossRef]
19. Yang, X.; You, Z.; Jin, C.; Diab, A.; Hasan, M. Aggregate morphology and internal structure for asphalt concrete: Prestep of computer-generated microstructural models. *Int. J. Geomech.* **2018**, *18*, 06018024. [CrossRef]
20. Wei, H.; Li, J.; Wang, F.; Zheng, J.; Tao, Y.; Zhang, Y. Numerical investigation on fracture evolution of asphalt mixture compared with acoustic emission. *Int. J. Pavement Eng.* **2021**, *23*, 3481–3491. [CrossRef]
21. Azéma, E.; Estrada, N.; Radjai, F. Nonlinear Effects of particle shape angularity in sheared granular media. *Phys. Rev. E* **2012**, *86*, 1115. [CrossRef]
22. Visseq, V.; Martin, A.; Iceta, D.; Azema, E.; Dureisseix, D.; Alart, P. Dense granular dynamics analysis by a domain decomposition approach. *Comput. Mech.* **2012**, *49*, 709–723. [CrossRef]
23. *JTC 3430-2020*; Test Methods of Soils for Highway Engineering. Ministry of Transport of the People's Public of China: Beijing, China, 2020.
24. Le, J.-L.; Hendrickson, R.; Marasteanu, M.O.; Turos, M. Use of fine aggregate matrix for computational modeling of low temperature fracture of asphalt concrete. *Mater. Struct.* **2018**, *51*, 152.1–152.13. [CrossRef]
25. Ma, T.; Zhang, Y.; Wang, H.; Huang, X.; Zhao, Y. Influences by air voids on the low-temperature cracking property of dense-graded asphalt concrete based on micromechanical modeling. *Adv. Mater. Sci. Eng.* **2016**, *2016*, 6942696. [CrossRef]
26. Zhang, H.; Liu, H.; You, W. Microstructural behavior of the low-temperature cracking and self-healing of asphalt mixtures based on the discrete element method. *Mater. Struct.* **2022**, *55*, 18. [CrossRef]
27. Sun, Q.; Jing, F.; Wang, G.; Zhang, G. Force chains in a uniaxially compressed static granular matter in 2D. *Acta Phys. Sin.* **2010**, *59*, 30–37.

Disclaimer/Publisher's Note: The statements, opinions and data contained in all publications are solely those of the individual author(s) and contributor(s) and not of MDPI and/or the editor(s). MDPI and/or the editor(s) disclaim responsibility for any injury to people or property resulting from any ideas, methods, instructions or products referred to in the content.

MDPI AG
Grosspeteranlage 5
4052 Basel
Switzerland
Tel.: +41 61 683 77 34

Materials Editorial Office
E-mail: materials@mdpi.com
www.mdpi.com/journal/materials

Disclaimer/Publisher's Note: The statements, opinions and data contained in all publications are solely those of the individual author(s) and contributor(s) and not of MDPI and/or the editor(s). MDPI and/or the editor(s) disclaim responsibility for any injury to people or property resulting from any ideas, methods, instructions or products referred to in the content.

www.ingramcontent.com/pod-product-compliance
Lightning Source LLC
LaVergne TN
LVHW070446100526
838202LV00014B/1681